Integrated Photonics and Optoelectronics

Integrated Photonics and Optoelectronics

Guest Editors

He Yang
Xinyang Su
Yizhong Huang

MDPI
Basel • Beijing • Wuhan • Barcelona • Belgrade • Novi Sad • Cluj • Manchester

Guest Editors

He Yang
Beihang Univesity
Beijing
China

Xinyang Su
Beijing Jiaotong University
Beijing
China

Yizhong Huang
The University of Chicago
Chicago, IL
United States

Editorial Office
MDPI AG
Grosspeteranlage 5
4052 Basel, Switzerland

This is a reprint of the Special Issue, published open access by the journal *Micromachines* (ISSN 2072-666X), freely accessible at: https://www.mdpi.com/journal/micromachines/special_issues/5UL51L8721.

For citation purposes, cite each article independently as indicated on the article page online and as indicated below:

Lastname, A.A.; Lastname, B.B. Article Title. *Journal Name* **Year**, *Volume Number*, Page Range.

ISBN 978-3-7258-3139-5 (Hbk)
ISBN 978-3-7258-3140-1 (PDF)
https://doi.org/10.3390/books978-3-7258-3140-1

Contents

About the Editors

He Yang

Prof. He Yang received his doctoral degree from Aalto University in Finland in 2018, and then continued his research work at the National Micro- and Nanofabrication Technology Center on the topic of Advanced On-Chip Photonic and Optoelectronic Devices. He has over 50 publications in top high-level journals, including *Science Advances*, *ACS Photonics*, and *Laser & Photonics Reviews*, among others. Now, he works as an associate professor in the School of Instrumentation and Optoelectronic Engineering at Beihang University, China. Prof. Yang's group is dedicated to the study of on-chip integrated linear and nonlinear photonic devices and their design, micro- and nanofabrication, the integration of various photonic devices, and the exploration of their applications in the filed of telecommunication, sensing, optical computing, etc.

Xinyang Su

Xinyang Su received his PhD degree from Beijing Jiaotong University. From Sept. 2016 to Feb. 2019, he was an international visiting PhD student at the University of Waterloo supervised by the 2018 Nobel Prize laureate Donna Strickland. He joined Beijing Jiaotong University as an assistant professor in 2020. His research focuses on fiber lasers, nonlinear optics, photonic materials and their applications.

Yizhong Huang

Yizhong Huang is a scientist in the field of photonics and quantum optics. His research interests mainly include the following: nanophotonics; nonlinear optics; ultrafast optics; lasers; hybrid quantum systems interfacing different quantum degrees of freedom; novel qubit platforms; and quantum networks.

micromachines

MDPI

Editorial

Editorial for the Special Issue on Integrated Photonics and Optoelectronics

He Yang [1,*] and Xinyang Su [2,*]

1 School of Instrumentation and Optoelectronic Engineering, Beihang University, Beijing 100191, China
2 School of Physical Science and Engineering, Beijing Jiaotong University, Beijing 100044, China
* Correspondence: yanghe@buaa.edu.cn (H.Y.); xysu@bjtu.edu.cn (X.S.)

Integrated photonic and optoelectronic technologies have become powerful tools in the drive to develop devices that are much smaller and more highly integrated, with lower power consumption and higher functionality. The trend in research and development (R&D) towards the miniaturization of optical and optoelectrical devices and systems [1–5], similarly to recent interest in microelectronics [6,7], has led to a continuous research interest in various integrated photonic and optoelectronic structures and devices [8–11]. Thus, various easy-to-integrate nanomaterials [12–17] with excellent optical and electrical properties, as well as high-precision nanofabrication techniques, have been developed in the last two decades, which have witnessed tremendous progress in integrated photonic and optoelectronic devices.

In this Special Issue (SI) on Integrated Photonics and Optoelectronics, we include 15 papers focusing on the latest research advances both in physical theory and device applications. Through the publication of these results, we hope to present the latest research developments in current challenges and future opportunities for the advancement of integrated photonic and optoelectronic devices, in order to promote the application of these devices in our daily life. Specifically, in this SI, we discuss three current topics in the field of light–matter interactions, including the fundamental theory, such as surface plasma, metasurfaces, and the active manipulation of the optical fields. Twelve publications are included that discuss the latest research on integrated photonic and optoelectronic devices, such as organic light-emitting devices, edge-emitting diode lasers, liquid crystal–quantum dot hybrids, a miniature Fourier transform spectrometer, rod fiber picosecond amplifiers, a thin-film integrated optical intraocular pressure sensor, polarization beam splitters (PBSs), hollow-core anti-resonant fibers, and Mach–Zehnder interferometer (MZI)-based matrix computation.

In particular, Liu et al. [18] numerically demonstrated and investigated a compact and high-power THz radiation source based on the excitation of the surface plasmon polaritons (SSP) mode on a roofed metallic grating by an electron beam. Wang et al. [19] increased the light extraction efficiency of organic light-emitting devices using electrochemically corroded patterned substrates. Shen et al. [20] demonstrated a double-strip array-based metasurface that supports the sharp quasi-bound states in the continuum (quasi-BICs) in terahertz regions. Han et al. [21] introduced the progress of edge-emitting diode lasers, based on a coupled-waveguide concept. Xi et al. [22] investigated an all-optical controlled THz modulator based on a Bi_2Te_3/Si heterostructure under 532 nm and 405 nm continuous-wave (CW) laser illumination. Schmelz et al. [23] presented and evaluated a possible fabrication process based on colloidal polystyrene (PS) nanosphere lithography for the fabrication of such anti-resonant (AR) structures on arbitrarily shaped fused silica substrates. Sun et al. [24] studied several of the most traditional hollow-core anti-resonant fiber (HC-ARF) structures, with the aim of achieving low confinement loss, single-mode performance, and a high insensitivity to bending in the 2 μm band. Bezrukov et al. [25] reported on the developing approaches to modulate the optical behavior of the microfluidic devices,

Citation: Yang, H.; Su, X. Editorial for the Special Issue on Integrated Photonics and Optoelectronics. *Micromachines* **2024**, *15*, 1090. https://doi.org/10.3390/mi15091090

Received: 23 July 2024
Accepted: 28 August 2024
Published: 29 August 2024

which is conducted by infusing smart hybrids of liquid crystal and quantum dots into microchannel confinement. Zhang et al. [26] proposed a miniature Fourier transform spectrometer using a thin-film lithium niobate electro-optical modulator, instead of the conventional modulator made by titanium diffusion in lithium niobate. Liu et al. [27] enhanced the coupling efficiency of the picosecond rod-type fiber amplifier by carefully setting the optimal mode–field matching. Xu et al. [28] designed a thin-film integrated optical intraocular pressure sensor based on the interferometry principle, which could read out the intraocular pressure value from interference patterns and monitor the value changes in real time simultaneously. Kotlyar et al. [29] discussed the superposition of several parallel identical Laguerre–Gaussian beams with single rings. Sun et al. [30] investigated the spin thermal radiation in a twisted bilayer α-MoO3 metasurface. Mei et al. [31] proposed and realized a PBS based on surface plasmonic resonance in a designed photonic crystal fiber (PCF). Hou et al. [32] identified the main hardware error sources of MZI-based matrix computation, summarized the available hardware error correction methods from the perspective of the entire MZI meshes and a single MZI device, and proposed a new architecture that will largely improve the precision of MZI-based matrix computations without increasing the size of the MZI's mesh.

Funding: We thank the Fundamental Research Funds for the Central Universities (2021RC206, 2022YJS128, 501QBJH2023117011, 501XYGG2024117011) and the High-end Foreign Expert Introduction Program (G2022104010L, G2023104003L) for their financial support.

Acknowledgments: We would like to take this opportunity to thank all the authors for submitting their papers to this Special Issue, all the reviewers for dedicating their time and helping to improve the quality of the submitted papers, and our assistant editor Billy Yi for his kind invitation and support for the production of this Special Issue.

Conflicts of Interest: The authors declare no conflicts of interest.

References

1. Pérez, D.; Gasulla, I.; Das Mahapatra, P.; Capmany, J. Principles, fundamentals, and applications of programmable integrated photonics. *Adv. Opt. Photonics* **2020**, *12*, 709. [CrossRef]
2. Wang, X.; Cui, Y.; Li, T.; Lei, M.; Li, J.; Wei, Z. Recent Advances in the Functional 2D Photonic and Optoelectronic Devices. *Adv. Opt. Mater.* **2018**, *7*, 1801274. [CrossRef]
3. Liang, G.; Yu, X.; Hu, X.; Qiang, B.; Wang, C.; Wang, Q.J. Mid-infrared photonics and optoelectronics in 2D materials. *Mater. Today* **2021**, *51*, 294–316. [CrossRef]
4. Sinatkas, G.; Christopoulos, T.; Tsilipakos, O.; Kriezis, E.E. Electro-optic modulation in integrated photonics. *J. Appl. Phys.* **2021**, *130*, 010901. [CrossRef]
5. You, J.; Luo, Y.; Yang, J.; Zhang, J.; Yin, K.; Wei, K.; Zheng, X.; Jiang, T. Hybrid/Integrated Silicon Photonics Based on 2D Materials in Optical Communication Nanosystems. *Laser Photonics Rev.* **2020**, *14*, 2000239. [CrossRef]
6. Yang, L.; Hu, H.; Scholz, A.; Feist, F.; Cadilha Marques, G.; Kraus, S.; Bojanowski, N.M.; Blasco, E.; Barner-Kowollik, C.; Aghassi-Hagmann, J.; et al. Laser printed microelectronics. *Nat. Commun.* **2023**, *14*, 1103. [CrossRef] [PubMed]
7. Dieny, B.; Prejbeanu, I.L.; Garello, K.; Gambardella, P.; Freitas, P.; Lehndorff, R.; Raberg, W.; Ebels, U.; Demokritov, S.O.; Akerman, J.; et al. Opportunities and challenges for spintronics in the microelectronics industry. *Nat. Electron.* **2020**, *3*, 446–459. [CrossRef]
8. Liu, C.; Yan, X.; Song, X.; Ding, S.; Zhang, D.W.; Zhou, P. A semi-floating gate memory based on van der Waals heterostructures for quasi-non-volatile applications. *Nat. Nanotechnol.* **2018**, *13*, 404–410. [CrossRef]
9. Zhang, Z.; Chen, P.; Duan, X.; Zang, K.; Luo, J.; Duan, X. Robust epitaxial growth of two-dimensional heterostructures, multiheterostructures, and superlattices. *Science* **2017**, *357*, 788–792. [CrossRef]
10. Sena-Torralba, A.; Alvarez-Diduk, R.; Parolo, C.; Piper, A.; Merkoci, A. Toward Next Generation Lateral Flow Assays: Integration of Nanomaterials. *Chem. Rev.* **2022**, *122*, 14881–14910. [CrossRef]
11. Li, W.; Wang, C.; Lu, X. Integrated transition metal and compounds with carbon nanomaterials for electrochemical water splitting. *J. Mater. Chem. A* **2021**, *9*, 3786–3827. [CrossRef]
12. Elbanna, A.; Jiang, H.; Fu, Q.; Zhu, J.F.; Liu, Y.; Zhao, M.; Liu, D.; Lai, S.; Chua, X.W.; Pan, J.; et al. 2D Material Infrared Photonics and Plasmonics. *ACS Nano* **2023**, *17*, 4134–4179. [CrossRef] [PubMed]
13. Cheng, Z.; Cao, R.; Wei, K.; Yao, Y.; Liu, X.; Kang, J.; Dong, J.; Shi, Z.; Zhang, H.; Zhang, X. 2D Materials Enabled Next-Generation Integrated Optoelectronics: From Fabrication to Applications. *Adv. Sci.* **2021**, *8*, e2003834. [CrossRef] [PubMed]
14. Singh, A.; Jo, S.S.; Li, Y.; Wu, C.; Li, M.; Jaramillo, R. Refractive Uses of Layered and Two-Dimensional Materials for Integrated Photonics. *ACS Photonics* **2020**, *7*, 3270–3285. [CrossRef]

15. Huang, Y.; Sutter, E.; Shi, N.; Zheng, J.; Yang, T.; Englund, D.; Gao, H.; Sutter, P. Reliable exfoliation of large-area high-quality flakes of graphene and other two-dimensional materials. *ACS Nano* **2015**, *9*, 10612. [CrossRef]
16. Lee, H.Y.; Kim, S. Nanowires for 2D material-based photonic and optoelectronic devices. *Nanophotonics* **2022**, *11*, 2571–2582. [CrossRef]
17. Yan, C.; Gan, L.; Zhou, X.; Guo, J.; Huang, W.; Huang, J.; Jin, B.; Xiong, J.; Zhai, T.; Li, Y. Space-Confined Chemical Vapor Deposition Synthesis of Ultrathin HfS2 Flakes for Optoelectronic Application. *Adv. Funct. Mater.* **2017**, *27*, 1702918. [CrossRef]
18. Liu, Y.; Zhang, X.; Wang, Y.; Cai, H.; Sun, J.; Zhu, Y.; Li, L. Excitation of Terahertz Spoof Surface Plasmons on a Roofed Metallic Grating by an Electron Beam. *Micromachines* **2024**, *15*, 293. [CrossRef]
19. Wang, Y.; Li, Z.; Bai, Y.; Wang, Y. Increasing the Light Extraction Efficiency of Organic Light-Emitting Devices by Electrochemically Corroded Patterned Substrates. *Micromachines* **2024**, *15*, 67. [CrossRef]
20. Shen, Y.; Wang, J.; Sheng, H.; Li, X.; Yang, J.; Liu, H.; Liu, D. Double-Strip Array-Based Metasurfaces with BICs for Terahertz Thin Membrane Detection. *Micromachines* **2024**, *15*, 43. [CrossRef]
21. Han, L.; Wang, Z.; Gordeev, N.Y.; Maximov, M.V.; Tang, X.; Beckman, A.A.; Kornyshov, G.O.; Payusov, A.S.; Shernyakov, Y.M.; Zhukov, A.E.; et al. Progress of Edge-Emitting Diode Lasers Based on Coupled-Waveguide Concept. *Micromachines* **2023**, *14*, 1271. [CrossRef]
22. Xi, Y.; Zhou, Y.; Cao, X.; Wang, J.; Lei, Z.; Lu, C.; Wu, D.; Shi, M.; Huang, Y.; Xu, X. Broadband All-Optical THz Modulator Based on Bi_2Te_3/Si Heterostructure Driven by UV-Visible Light. *Micromachines* **2023**, *14*, 1237. [CrossRef] [PubMed]
23. Schmelz, D.; Jia, G.; Käsebier, T.; Plentz, J.; Zeitner, U.D. Antireflection Structures for VIS and NIR on Arbitrarily Shaped Fused Silica Substrates with Colloidal Polystyrene Nanosphere Lithography. *Micromachines* **2023**, *14*, 1204. [CrossRef]
24. Sun, T.; Su, X.; Meng, F.; Wang, Z.; Song, J.; Zhang, C.; Xu, T.; Zhang, Y.; Zhang, H.; Cui, M.; et al. Design of 2 μm Low-Loss Hollow-Core Anti-Resonant Fibers. *Micromachines* **2023**, *14*, 1198. [CrossRef]
25. Bezrukov, A.; Galyametdinov, Y. Dynamic Flow Control over Optical Properties of Liquid Crystal–Quantum Dot Hybrids in Microfluidic Devices. *Micromachines* **2023**, *14*, 990. [CrossRef] [PubMed]
26. Zhang, L.; Gou, G.; Chen, J.; Li, W.; Ma, W.; Li, R.; An, J.; Wang, Y.; Liu, Y.; Yan, W.; et al. Miniature Fourier Transform Spectrometer Based on Thin-Film Lithium Niobate. *Micromachines* **2023**, *14*, 458. [CrossRef]
27. Liu, D.; Mao, X.; Bi, G.; Li, T.; Zang, D.; Sun, N. Efficiency Enhancing Technique for Rod Fiber Picosecond Amplifiers with Optimal Mode Field Matching. *Micromachines* **2023**, *14*, 450. [CrossRef]
28. Xu, X.; Liu, Z.; Wang, L.; Huang, Y.; Yang, H. High—Accuracy Film—Integrated Optical Sensor for Real—Time Intraocular Pressure Monitoring. *Micromachines* **2023**, *14*, 353. [CrossRef]
29. Kotlyar, V.V.; Kovalev, A.A.; Kozlova, E.S.; Savelyeva, A.A. Tailoring the Topological Charge of a Superposition of Identical Parallel Laguerre–Gaussian Beams. *Micromachines* **2022**, *13*, 2227. [CrossRef] [PubMed]
30. Sun, Y.; Zhang, D.; Wu, B.; Liu, H.; Yang, B.; Wu, X. Metasurfaces Assisted Twisted α-MoO_3 for Spinning Thermal Radiation. *Micromachines* **2022**, *13*, 1757. [CrossRef]
31. Mei, C.; Wu, Y.; Yuan, J.; Qiu, S.; Zhou, X. Design of Compact and Broadband Polarization Beam Splitters Based on Surface Plasmonic Resonance in Photonic Crystal Fibers. *Micromachines* **2022**, *13*, 1663. [CrossRef] [PubMed]
32. Hou, H.; Xu, P.; Zhou, Z.; Su, H. Hardware Error Correction for MZI-Based Matrix Computation. *Micromachines* **2023**, *14*, 955. [CrossRef] [PubMed]

micromachines

Article

Excitation of Terahertz Spoof Surface Plasmons on a Roofed Metallic Grating by an Electron Beam

Yongqiang Liu *, Xutao Zhang, Yan Wang, He Cai, Jinhai Sun , Yong Zhu and Liangsheng Li

National Key Laboratory of Scattering and Radiation, Beijing 100854, China; 13240194055@163.com (X.Z.); wangyan2307@126.com (Y.W.); 18510165603@126.com (H.C.); jinhaisun@126.com (J.S.); 13810065549@163.com (Y.Z.); liliangshengbititp@163.com (L.L.)
* Correspondence: liuyq1990@126.com

Abstract: In this paper, both fundamental SSP modes on a roofed metallic grating and its effective excitation of the bounded SSP mode by an injected electron beam on the structure are numerically examined and investigated in the THz regime. Apart from the bounded SSP mode on the metallic grating with open space, the introduced roofed metallic grating can generate a closed waveguide mode that occupies the dispersion region outside the light line. The closed waveguide mode shifts gradually to a higher frequency band with a decreased gap size, while the bounded SSP mode line becomes lower. The effective excitation of the bounded SSP mode on this roofed metallic grating is also implemented and studied by using a particle-in-cell simulation studio. The output SSP power spectrums with various gap sizes by the same electron beam on this roofed metallic grating are obtained and analyzed. The simulation results reveal that the generated SSP spectra show a slight red shift with a decreased gap size. This work on the excitation of the SSP mode using an electron beam can benefit the development of high-power compact THz radiation sources by utilizing the strong near-field confinement of SSPs on metallic gratings.

Keywords: spoof surface plasmons (SSPs); roofed metallic grating; leaky wave SSP mode; terahertz radiation source; electron beam; high-power excitation

Citation: Liu, Y.; Zhang, X.; Wang, Y.; Cai, H.; Sun, J.; Zhu, Y.; Li, L. Excitation of Terahertz Spoof Surface Plasmons on a Roofed Metallic Grating by an Electron Beam. *Micromachines* **2024**, *15*, 293. https://doi.org/10.3390/mi15030293

Academic Editors: He Yang, Yizhong Huang and Xinyang Su

Received: 30 December 2023
Revised: 5 February 2024
Accepted: 7 February 2024
Published: 21 February 2024

1. Introduction

Spoof surface plasmon polaritons (SSPs) are a special kind of surface waves that propagate or localize on periodical textured metallic surfaces or related electromagnetic interfaces [1–3]. Due to their many excellent electromagnetic properties such as sub-wavelength wave guiding, enhanced light–matter interaction and versatile dispersion manipulation, etc., "spoof" plasmonics have attracted tremendous attention and become a frontier in both microwave and terahertz (THz) regimes [4,5]. SSP modes on various patterned metallic shapes have been investigated, including metal gratings [6], wedge structurse [7], cylindrical corrugated disks [8], "T"-shaped grooves [9], trapezoidal grooves [10], etc. Their fundamental guiding schemes such as strong near-field locations and low-loss propagations have been extensively pursued and demonstrated. Moreover, these novel waveguides and meta-structures have also been proposed for some functional devices and systems such as plasmonic lenses [11–15], on-chip filters and splitters [16–18], polarization converters [19,20], SSP antennas [21], OAM generators [22], etc.

Among the various SSP-based devices and applications, their excitation on corrugated metasurfaces using various approaches is an important aspect and has also been largely studied in recent years. In order to realize an efficient dispersion momentum match to excite the SSP mode, various phase gradient reflective or transmissive metasurfaces have been proposed and demonstrated in the microwave band [23–26]. In these works, dispersion engineering of metasurfaces is key to excite the SSP mode with high-efficiency and versatile manipulations. In addition, the excited SSP mode can also be amplified by inserting some active circuit chips or other components [27], which can also be termed as "active plasmonic metamaterials". Apart from the excitation or amplification mechanisms of SSP modes mentioned, high-power SSP emitters on meta-structure arrays using active photon or electron sources have also been largely demonstrated [28–33]. The working principles of the active excitation method rely on an effective dispersion match between the injected energy sources (photons or electrons) and SSP mode; thus, the interaction or amplification of the SSP mode can provide a new venue to obtain high-power radiation sources [29].

As an important SSP waveguide and interaction circuit, a metallic grating formed by single rectangular groove arrays has been proposed to develop free-electron-driven high-power radiation or amplifier sources [34–37]. It is shown that the SSP mode can be effectively excited and gradually amplified by an injected electron beam both on uniform and gradient meta-structures. These studies focus mainly on open space metallic grating structures (the inset in Figure 1a) and the interaction between an electron beam and the SSP mode is also simple. Some recent studies indicate that SSP properties and dispersion diagrams can be largely modified or tuned by introducing a closed metal plate in the vicinity of the metallic grating, namely on the roofed metallic grating (the inset in Figure 1b) [38–47]. Thus, the variation in SSP modes on this kind of roofed metallic grating with different gap sizes should be considered carefully. Additionally, the introduced metal plate may also influence the various plasmonic devices, including high-power radiation sources, based on the interaction between the SSP mode and injected electron beam [34–37,48,49], which needs to be studied in detail.

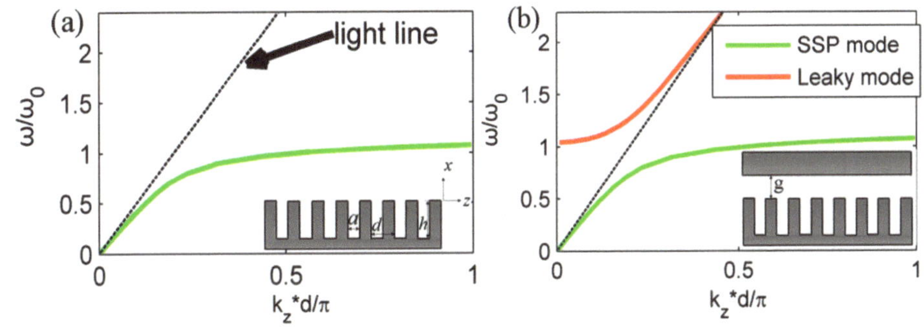

Figure 1. (**a**) SSP dispersion line on the conventional metallic grating with open space boundary. The inset is a detailed parameter of the rectangular metallic grating. The black line is the light line. (**b**) SSP dispersion lines on the considered roofed metallic grating for this study. The green line is the bounded SSP mode and red line is the leaky wave mode. The inset is a structure illustration with a metal plate in the vicinity of corrugated metallic surfaces with a gap size of g in between.

According to the above-mentioned insufficient studies in the existing literature, we here propose the excitation of the THz SSP mode on a roofed metallic grating using an injected electron beam. The basic SSP dispersion diagrams on the open or roofed metallic grating are investigated and compared first. In addition, the influence of the gap size between the bare metallic grating and metal plate on the SSP mode is specifically studied. Also, the excitation of the bounded SSP mode by an injected electron beam on the roofed metallic grating is also implemented and modeled by using particle-in-cell simulations. The output power spectrums of the SSP mode with different gap sizes are also given and analyzed. The presented studies can provide a new path to develop compact THz radiation sources induced by an injected electron beam based on the roofed SSP waveguide.

2. Theoretical Model and Dispersion Theory

The proposed SSP wave excitation by an injected electron beam is based on a rectangular groove periodical metallic grating with an open space or closed boundary at the top, as schematically plotted in the inset of Figure 1a,b, respectively. The metallic grating has the same groove depth of h, groove width of a and period of d, respectively, as shown in the inset of Figure 1. A metal plate is introduced to this plasmonic waveguide and the distance of the metal plate from the grating surface is marked by g in the inset of Figure 1b [38–44]. We first consider SSP mode dispersion on the x-z plane, which is assumed to be the transverse magnetic mode (TM) along z and metal is set as the ideal conductor for the study [39]. The whole structure is divided into two different regions: under and above the grating surface, respectively. The initial point of the x axis is on the grating surface. The dispersion characteristics of the SSP mode for the open space or roofed metallic grating structure are studied using an effective medium method [40] or rigorous field expansion method [39,42], as mentioned previously. We here employ a simplified mode matching method to obtain its dispersion property on the metallic grating. For the open space metallic grating or closed metallic grating in Figure 1, the fields under the metallic surface are the same and can be expressed as homogeneous, based on the fact that the lattice period is usually shorter than the wavelength in the THz band [34–45]:

$$E_z^u = A \sin k(x+h) e^{-j\beta_0 md} \tag{1}$$

Also, its magnetic field can be obtained based on its relation to the TM mode:

$$H_y^u = A \frac{k}{j\omega\mu} \cos k(x+h) e^{-j\beta_0 md} \tag{2}$$

where A is the unknown index and md (m = 0, 1, 2, 3. . .) is the axial distance, $k = \omega/c$ is the wave vector in free space, μ is vacuum permeability, β_0 is the propagation constant of the fundamental mode, ω is the angular frequency and j is the imaginary unit. For the fields above the metallic grating, its formulism is different because of the different open space or roofed metallic plate boundary condition. In the bare metallic grating of Figure 1a, its fields can be expressed as:

$$E_z^a = \sum_{n=-\infty}^{\infty} B_n e^{k_n x} e^{-j\beta_n z} \tag{3}$$

$$H_y^a = \sum_{n=-\infty}^{\infty} \frac{-j\omega\varepsilon_0}{k_n} B_n e^{k_n x} e^{-j\beta_n z} \tag{4}$$

where B_n is the unknown index and the periodical harmonic mode along the propagation direction is considered as $\beta_n = \beta_0 + 2n\pi/d$, while ${k^2}_n = {\beta^2}_n - k^2$ (n = 0, ±1, ±2, ±3. . .) is the propagation constant along the x axis. The axial field E_z above the grating surface decays exponentially along the x axis because of its infinite open space boundary condition.

If we consider the roofed metallic grating in Figure 1b, its fields above the metallic grating change into the following forms:

$$E_z^a = \sum_{n=-\infty}^{\infty} B_n \mathrm{sinh} k_n(g-x) e^{-j\beta_n z} \tag{5}$$

$$H_y^a = \sum_{n=-\infty}^{\infty} \frac{-j\omega\varepsilon_0}{k_n} B_n \cosh k_n(g-x) e^{-j\beta_n z} \tag{6}$$

It can be noted that the fields above the gating are closely related to the distance of the metal plate of g. Based on the rigorous tangential electric field continuity of E_z and magnetic flow conservation of H_y for one period in region I and II [36], the SSP dispersion expression can be obtained after tedious calculations, which eliminate the above unknown index (A and B_n). Based on the field expressions of Equations (1)–(4), the SSP dispersion relation to the open space grating in Figure 1a is as follows:

$$\sum_{n=-\infty}^{\infty} \frac{\sin c^2(\frac{\beta_n a}{2})}{k_n} = \frac{d}{a}\frac{1}{k} \cot(kh) \tag{7}$$

Also, the SSP dispersion expression on the roofed metallic grating shown in Figure 1b can also be calculated based on Equations (1), (2), (5) and (6) as follows:

$$\sum_{n=-\infty}^{\infty} \frac{\sin c^2(\frac{\beta_n a}{2})}{k_n \tanh(k_n g)} = \frac{d}{a}\frac{1}{k}\cot(kh) \tag{8}$$

The function sinc$(\beta_n a/2)$ is $\sin(\beta_n a/2)/(\beta_n a/2)$ in the dispersion expression and the other symbols are the same as those in the above-mentioned field expressions. Our derivation processes are simplified compared to previous studies for similar roofed metallic gratings and include the effective medium method or rigorous field expansion method, which consider the SSP harmonic mode [38–42]. Furthermore, it can be noted that the SSP mode in Expression (8) on the roofed metallic grating can be simplified to Equation (7) if the gap size of g is infinite [39].

According to the above analytic SSP dispersion expressions in Equations (7) and (8), the dispersion lines of SSP modes within the first Brillouin zone are plotted and presented in Figure 1a,b, respectively. For the bare metallic grating with open space, there is only one well-known bounded surface mode of the green line under the light line. For the roofed metallic grating in Figure 1b, there is a closed waveguide mode of the red line, which is outside the light line, excluding the well-known bounded SSP mode on the structure. This new leaky mode has also been talked about in a previous work under the name of a cavity mode [38,39] and can open new pathways to design novel plasmonic devices, including leaky wave antenna and fast-wave vacuum electronic devices. In the relevant calculations, the metallic grating parameters are as follows: a/d = 0.5, h/d = 2.2, g/d = 2.8, d = 30 μm, λ = 300 μm and $\omega_0 = 2\pi \times 1$ THz. Previous studies also indicate that the SSP mode on this roofed metallic grating can realize ultra-high Q resonant tunneling with enhanced emissivity, which may be of great importance to some SSP-based functional devices and systems such as compact and integrated optical sources [40].

The metallic grating parameters such as groove depth and width can influence SSP dispersion, which has been largely studied in some previous works but mostly limited to bare metallic gratings without cladding, as shown in Figure 1a [39–45]. Here, we investigate the influence of gap size on the SSP dispersion change as expressed in Equation (8). SSP dispersion lines of both the bounded mode and leaky mode with different gap sizes of g are plotted in Figure 2a. It can be seen that the cutoff frequency of the leaky SSP mode with $k_z = 0$ gradually shifts to a higher frequency band with a decreased gap size. On the other hand, the propagation constant of the bounded SSP mode becomes larger as the gap size decreases. In addition, the asymptotic frequency of the bounded SSP mode is almost unchanged. This distinct SSP dispersion mode variation with different structural parameters can find different applications that are dependent on the bounded or leaky wave characteristic of the SSP mode. Furthermore, the electric field distributions of the bounded SSP mode on the roofed metallic grating with different gap sizes near asymptotic frequency are also modeled using the finite integration method and are presented in Figure 2b–d with g = 15, 30 and 84 μm, respectively. A small gap size between the grating and metal plate can pose a great influence on the propagation of the SSP mode and thus needs to be studied specifically for some SSP-based functional devices and systems. Next, we will investigate the efficient SSP excitation on the roofed metallic grating with a moderately small gap size using an injected electron beam and also give various SSP output power spectrums with different structural or electron beam parameters.

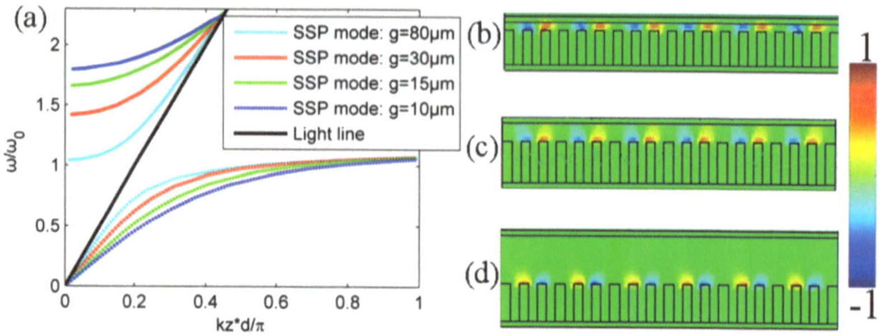

Figure 2. (**a**) SSP dispersion lines with different gap sizes of g in the inset of Figure 1b. (**b–d**) The normalized electric field distributions of the bounded SSP mode on the roofed metallic grating with different gap sizes of g = 15, 30 and 80 μm, respectively.

3. THz SSP Excitation on the Roofed Metallic Grating by Electron Beam

Based on above studies on SSP dispersion on the roofed metallic grating, a high-power THz radiation source based on the effective excitation of the bounded SSP mode by an injected electron beam is modeled and studied via particle-in-cell simulations based on the finite-difference-time-domain (FDTD) algorithm [35–37,50–56]. The software studio is also used for some traditional vacuum electronic devices such as backward-wave oscillators (BWO) [51], travelling-wave tubes (TWT) [52], klystrons [53,54], and so on. The working mechanism is similar to some previous works on open space metallic gratings, which is based on the matched dispersion relation between the SSP mode and electron beam [36]. Its schematic diagrams or the proposed functional device of THz SSP excitation by an electron beam on the structure are plotted in Figure 3a. The electron beam is emitted from the left side of the roofed metallic grating with a distance from the grating surface that is marked by H. The total interaction length of the roofed metallic grating structure is L. As shown in Figure 2, the gap size g = 30 μm; thus, its dispersion property of the SSP mode is determined. The corresponding dispersion lines of the SSP mode and injected electron beam are plotted in Figure 3b. It can be seen that the electrons are synchronous with the forward wave of the SSP mode. In the simulations, a 2D model is used and the boundary effect along y is ignored. Metal is assumed to be a perfect electric conductor and the propagation losses of the SSP mode are ignored [34–37]. The boundary condition around the interaction system is an absorbing boundary. The dispersion of the injected electron beam is given by $v_b = \beta * c$, $\beta = (1-(1+\gamma)^{-2})^{0.5}$, $\gamma = U/U_r(U_r = 5.11 \times 10^5$ V is electron energy), where U is the beam voltage and c is the light velocity. In order to realize a good dispersion match between the SSP mode and electron beam, the velocity of the electron beam v_b can be tuned according to its beam voltage and so the operation frequency is also set. Here, the beam voltage is first set as 18.40 kV and the operation frequency is near 1 THz. Then, its output SSP power spectrums with different beam voltages will be studied and presented. Previous studies show that the output SSP is very sensitive to the distance of the electron beam above the metallic grating of H on the open space metallic grating [36]. Here, we also examine the influence of this beam height on the generated SSP power at the end of the structure. Figure 4 gives the simulated SSP output power distributions along a beam height range from 3.4 to 4.0 μm for different interaction lengths of 2.7, 4.0 and 7.6 mm. The obtained SSP power is calculated by integrating the flux along the x direction from the bottom to the top of the roofed metallic grating. A magnetic field with 1 T along the z direction is used in the system to ensure good electron beam transportation. It can be noted that the effective SSP output happens only with a very limited beam height of H and decreases quickly outside this height range. There is almost no output power when the distance is smaller than 3.5 μm or larger than 4 μm. The reason for this is that the interaction between the SSP field and electron beam is not optimal as the field of this specific SSP mode is weak when this distance is very small. Also, it decays rapidly away from the metallic grating surface when the distance is very large. The output SSP power increases gradually along with the interaction waveguide length to an extent. The peak SSP power is reached around 40 W at the end of the structure with a 7.6 mm length for the operation frequency of 1 THz with 18.40 kV and a 0.4 A electron beam. Therefore, the beam height is set as this optimized value and the interaction length is chosen as 7.6 mm for its good output performance. The beam width of emissive shape is also optimized with a half duty cycle and beam width of 3.75 µm. The injected electron beam current is 0.4 A. The interaction length here is largely extended compared to previous studies on the open space metallic grating; thus, it is possible to realize long-distance excitation, provided that the other parameters are properly set [36].

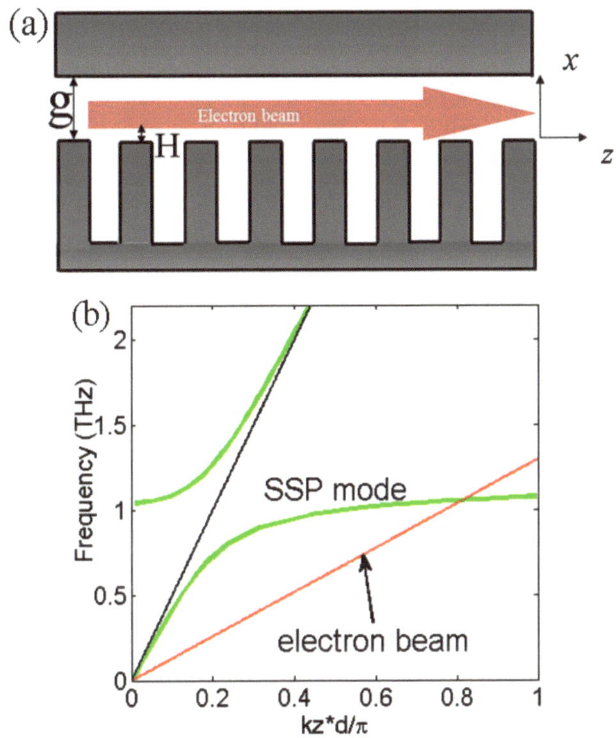

Figure 3. (**a**) Schematic diagrams of the proposed THz SSP excitation on the roofed metallic grating by an injected electron beam on the *x*-*z* plane. The gap size of the metal plate above the grating is marked by *g*. The height of the electron beam is marked by *H*. (**b**) The dispersion lines of the SSP mode and electron beam in the forward wave regime.

In order to further illustrate the SSP excitation mechanism induced by the electron beam, which is dependent on the SSP dispersion and electron beam energy, the influences of beam energy and the gap size of *g* on the generated SSP power are specifically studied and analyzed in simulations. Figure 5a presents the SSP output power spectrums with different electron beam voltages ranging from 18.10 to 18.70 kV with a 0.1 kV step. The SSP power spectrum, which is plotted by the dependency between the output power and operation frequency, is directly calculated from SSP without an input driving signal. The SSP power spectrum width is about 1 GHz within a 3 dB output level. The obvious red shift of the generated SSP power spectrum is observed with the increased beam voltage. This is caused by the increased electron beam velocity with an increased beam voltage; thus, the optimized operation frequency also decreases according to the above-mentioned beam voltage formulism. The beam current is 0.4 A and the other parameters are kept the same. In the middle of the red line with 18.40 kV, the optimized operation frequency is 1 THz. In addition, the peak value of the generated SSP power slightly decreases along with the increased beam voltage. Thus, the electron beam voltage can provide a simple tuning method to tune the generated SSP power spectrum freely. The SSP power is generated from the decreased electron beam energy across the interaction length. Figure 5b plots the electron beam energy variation along with the interaction length for three different beam voltages of 18.1 (blue line), 18.4 (red line) and 18.7 kV (green line), respectively. It can be clearly seen that the electron beam energy decreases gradually along with the interaction length; thus, the generated SSP power increases at its corresponding optimized operation frequency increases, i.e., 0.9995 THz, 1.0 THz and 1.0005 THz, respectively. In addition, the maximum nest decreased electron beam energy for these three different beam voltages at the end of the structure is about $U * I = 0.3 \times 0.4 \times 1000 = 120$ W; I is the injected electron beam current. Due to some diffraction losses and ohmic attenuations, the decreased electron beam energy cannot be fully converted to SSP power. So, the conversion efficiency from the injected electron beam energy to the generated S

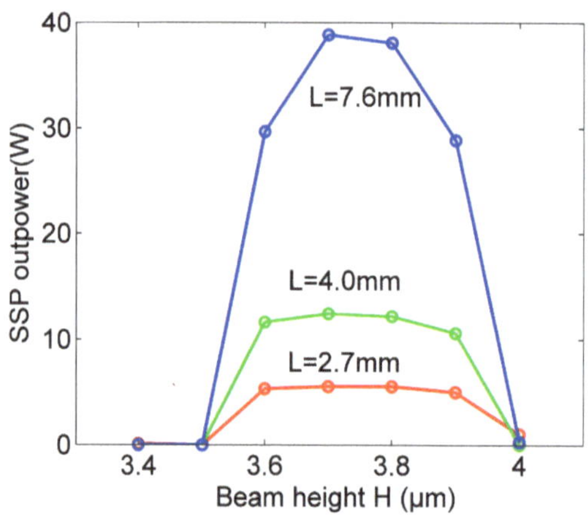

Figure 4. Output SSP power distributions along the electron beam height of H with different interaction lengths of 2.7, 4.0 and 7.6 mm from Figure 3, respectively. Beam voltage and current are 18.40 kV and 0.4 A, respectively.

SP power is about $40/(18{,}400 \times 0.4) = 0.54\%$.

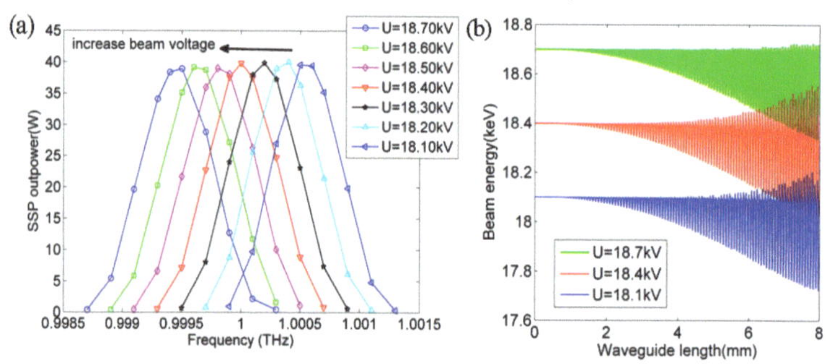

Figure 5. (**a**) Output SSP power spectrums with different electron beam voltages ranging from 18.10 kV to 18.70 kV with a beam current of 0.4 A. The interaction length is 7.6 mm. (**b**) The electron beam energy variations along with the interaction length for different beam voltages of 18.1, 18.4 and 18.7 kV at its corresponding optimized operation frequency, respectively.

4. Influence of Gap Size on the SSP Excitation

The above studies indicate that the interaction between the SSP mode and injected electron beam is similar to that on the open space grating [36]. To demonstrate its difference on the roofed metallic grating, the gap size of g in Figure 3 is specifically considered for its transmission property and the generated SSP power spectrum with the other conditions kept constant. Figure 6 gives the simulated transmission spectrums with different gap sizes of the roofed metallic grating ranging from 10 (blue line) to 50 μm (green line). The inset shows the model in the simulations. The black line and red line are the results for g = 20 and 30 μm, respectively. The band-stop property is observed for this periodical roofed metallic grating. In addition, the band-stop window within −2 dB becomes larger as the gap size decreases gradually, as indicated by the black arrow. This distinct transmission property agrees well with the dispersion characteristic variation in Figure 2a. The band-stop window in the transmission is caused by the spectrum gap between the asymptotical frequency of the bounded SSP mode and cutoff frequency of the leaky mode. Obviously, the leaky SSP mode dispersion change mainly contributes to this band-stop window shift as the gap size poses a weak influence on the bounded SSP mode, as shown in Figure 2a.

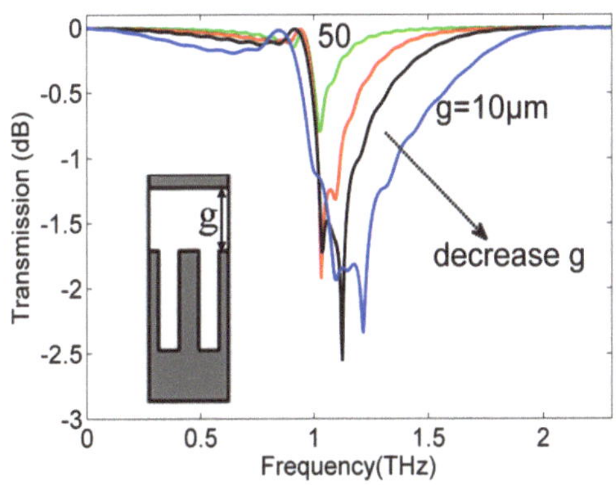

Figure 6. The transmission magnitude distributions of the roofed metallic grating with different gap sizes between the metallic grating and metal plate ranging from 10 to 50 μm. The inset is the calculated model in the simulations.

To demonstrate the effect of gap size on the generated SSP power spectrums with the roofed metallic grating, three different gap sizes of g = 25, 30 and 35 μm were employed to illustrate its interaction mechanism. The other conditions were set the same as each other. Figure 7 provides the simulated results of the generated SSP power spectrums with different gap sizes of g = 25, 30 and 35 μm taken from Figure 3, respectively. The electron beam parameters are 18.40 kV and 0.4 A. The metallic grating length is 7.6 mm. It can be concluded that the generated SSP power spectrum shifts gradually to a lower frequency band as the gap size decreases with the same injected electron beam. This variation agrees well with previous dispersion characteristic and transmission property analysis with different gap sizes of the structure. In addition, the peak value of the generated SSP power spectrum decreases along with the decreased gap size. This is because transmission becomes worse as shown in Figure 6 as the gap size decreases; thus, the peak power of SSP also slightly decreases. Changing the gap size between the metal plate and grating can tune the output SSP power spectra and thus should be considered carefully.

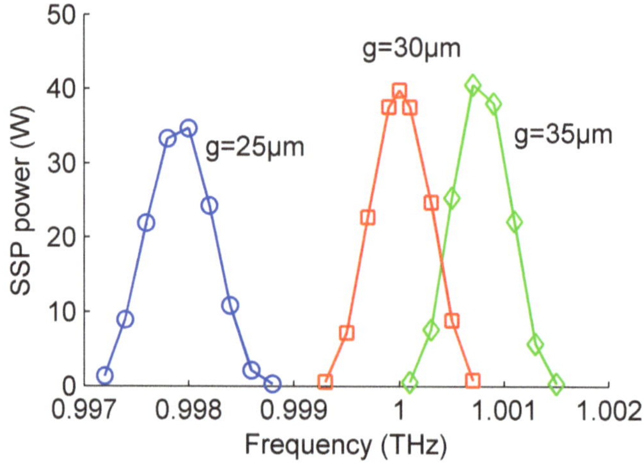

Figure 7. The generated SSP power spectrums with different gap sizes of g = 25, 30 and 35 μm taken from Figure 3, respectively. The electron beam parameters are 18.40 kV and 0.4 A. The metallic grating length is 7.6 mm.

5. Conclusions

In this paper, a compact and high-power THz radiation source based on the excitation of the SSP mode on a roofed metallic grating by an electron beam is numerically demonstrated and investigated. The effect of the fundamental dispersion theory on the structure is presented analytically and also compared with that on the conventional bare periodical structure. The excitation of the SSP mode on the structure is also investigated and analyzed by particle-in-cell simulations with various interaction conditions such as different beam energies, height and waveguide lengths, etc. The effects of gap size between the metallic grating and metal plate on the SSP transmission and its generated power spectrum are also examined and studied. The presented studies can provide a new path to develop compact and high-power THz radiation sources induced by an electron beam on the roofed metallic plasmonic waveguide.

Author Contributions: Conceptualization, Y.L.; methodology, H.C.; software, H.C. and X.Z.; validation, Y.W.; formal analysis, Y.Z.; investigation, Y.W.; resources, Y.Z.; data curation, L.L.; writing—original draft preparation, Y.L.; writing—review and editing, Y.L.; visualization, J.S.; supervision, J.S.; project administration, L.L. and J.S.; funding acquisition, Y.L. All authors have read and agreed to the published version of the manuscript.

Funding: This research was funded by open funding from the National Key Laboratory of Scattering and Radiation.

Data Availability Statement: The data presented in this study are available on request from the corresponding author.

Conflicts of Interest: The authors declare no conflicts of interest.

References

1. Zhang, X.; Xu, Q.; Xia, L.; Li, Y.; Gu, J.; Tian, Z.; Ouyang, C.; Zhang, W. Terahertz surface plasmonic waves: A review. *Adv. Photon.* **2020**, 2, 014001. [CrossRef]
2. Garcia-Vidal, F.J.; Fernández-Domínguez, A.I.; Martin-Moreno, L.; Zhang, H.C.; Tang, W.; Peng, R.; Cui, T.J. Spoof surface plasmon photonics. *Rev. Mod. Phys.* **2022**, *94*, 025004. [CrossRef]
3. Matsui, T. A brief review on metamaterial-based vacuum electronics for terahertz and microwave science and technology. *J. Infrared Millim. Terahertz Waves* **2017**, *38*, 1140–1161. [CrossRef]
4. Liu, P.-K.; Huang, T.-J. Terahertz surface plasmon polaritons and their applications. *J. Infrared Millim. Waves* **2020**, *39*, 169–190.

5. Tang, W.X.; Zhang, H.C.; Ma, H.F.; Jiang, W.X.; Cui, T.J. Concept, theory, design, and applications of spoof surface plasmon polaritons at microwave frequencies. *Adv. Opt. Mater.* **2018**, *7*, 1800421. [CrossRef]
6. Liu, Y.-Q.; Li, L.; Yin, H. Long-range spoof surface plasmons (LRSSP) on the asymmetric double metal gratings. *IEEE Photonics J.* **2021**, *13*, 4800209. [CrossRef]
7. Liu, X.; Zhu, L.; Wu, Q.; Feng, Y. Highly-confined and low-loss spoof surface Plasmon polaritons structure with periodic loading of trapezoidal grooves. *AIP Adv.* **2015**, *5*, 077123. [CrossRef]
8. Han, C.; Chu, Y.; Wang, Z.; Zhao, X. Spoof surface plasmonic waveguide devices with compact length and low-loss. *J. Appl. Phys.* **2017**, *122*, 123301. [CrossRef]
9. Lu, S.; Xu, K.-D.; Guo, Y.-J.; Chen, Q. Compact Spoof Surface Plasmon Polariton Waveguides with Simple Configurations and Good Performance. *IEEE Trans. Plasma Sci.* **2021**, *49*, 3786–3792. [CrossRef]
10. Tian, L.; Liu, J.; Zhou, K.; Gao, Y.; Liu, S. Investigation of mechanism: Spoof SPPs on periodically textured metal surface with pyramidal grooves. *Sci. Rep.* **2016**, *6*, 32008. [CrossRef]
11. Gao, Z.-H.; Li, X.-S.; Mao, M.; Sun, C.; Liu, F.-X.; Zhang, L.; Zhao, L. Ultra-Compact Low-Pass Spoof Surface Plasmon Polariton Filter Based on Interdigital Structure. *Micromachines* **2023**, *14*, 1687. [CrossRef] [PubMed]
12. Valerio, G.; Sipus, Z.; Grbic, A.; Quevedo-Teruel, Q. Nonresonant modes in plasmonic holey metasurfaces for the design of artificial flat lenses. *Opt. Lett.* **2017**, *42*, 2026–2029. [CrossRef] [PubMed]
13. Yang, Y.; Wang, J.; Li, Y.; Pang, Y.; Meng, Y.; Cheng, Q.; Cui, T.; Qu, S. 2D Achromatic Flat Focusing Lens Based on Dispersion Engineering of Spoof Surface Plasmon Polaritons: Broadband and Profile-Robust. *J. Phys. D* **2018**, *51*, 045108. [CrossRef]
14. Guo, Y.; Zhang, Z.; Pu, M.; Huang, Y.; Li, X.; Ma, X.; Luo, X. Spoof plasmonic metasurfaces with catenary dispersion for two-dimensional wide-angle focusing and imaging. *iScience* **2019**, *21*, 145–156. [CrossRef] [PubMed]
15. Huang, T.J.; Liu, J.Y.; Yin, L.Z.; Han, F.Y.; Liu, P.K. Superfocusing of terahertz wave through spoof surface plasmons. *Opt. Express* **2018**, *26*, 22722–22732. [CrossRef] [PubMed]
16. Yang, Z.-B.; Guan, D.-F.; Zhang, Q.; You, P.; Hou, X.-X.; Xu, S.-D.; Yong, S.-W. A hybrid substrate-integrated waveguide and spoof surface plasmon-polariton one-layer dual bandpass filter formed by resonant tunneling effect. *Appl. Phys. Express* **2018**, *11*, 114101. [CrossRef]
17. Zhu, H.; Zhang, Y.; Ye, L.; Li, Y.; Xu, Y.; Xu, R. On-chip terahertz bandpass filter based on substrate integrated plasmonic waveguide. *Results Phys.* **2021**, *27*, 104553. [CrossRef]
18. Wu, J.; Yang, X.; Su, P.; Yu, W.; Zheng, L. Defects Detection Method Based on Programmable Spoof Surface Plasmon Polaritons in Non-Metallic Composites. *Micromachines* **2023**, *14*, 756. [CrossRef]
19. Khan, B.; Kamal, B.; Ullah, S.; Lah, A.; Ali, H.; Ullah, R. Asymmetric polarization converting metasurface for microwave applications. *Opt. Mater. Express* **2022**, *12*, 3403–3415. [CrossRef]
20. Wang, C.; Li, Y.; Feng, M.; Wang, J.; Ma, H.; Zhang, J.; Qu, S. Frequency-selective structure with transmission and scattering deflection based on spoof surface plasmon polariton modes. *IEEE Trans. Antennas Propag.* **2019**, *67*, 6508–6514. [CrossRef]
21. Han, Z.; Zhang, Y.; Bozhevolnyi, S.I. Spoof surface plasmon-based stripe antennas with extreme field enhancement in the terahertz regime. *Opt. Lett.* **2015**, *40*, 2533–2536. [CrossRef]
22. Zhang, Y.; Zhang, Q.; Chan, C.H.; Li, E.; Wang, H. Emission of orbital angular momentum based on spoof localized surface plasmons. *Opt. Lett.* **2019**, *44*, 5735–5738. [CrossRef]
23. Zhang, H.C.; Liu, S.; Shen, X.; Chen, L.H.; Li, L.; Cui, T.J. Broadband amplification of spoof surface plasmon polaritons at microwave frequencies. *Laser Photonics Rev.* **2015**, *9*, 83–90. [CrossRef]
24. Sun, W.; He, Q.; Sun, S.; Zhou, L. High-efficiency surface plasmon meta-couplers: Concept and microwave-regime realizations. *Light Sci. Appl.* **2016**, *5*, e16003. [CrossRef] [PubMed]
25. Liu, Y.-Q.; Liu, P.-K. Excitation of surface plasmon polaritons by electron beam with graphene ribbon arrays. *J. Appl. Phys.* **2017**, *121*, 113104. [CrossRef]
26. Meng, Y.Y.; Ma, H.; Wang, J.F.; Lv, Y.G.; Feng, M.D.; Li, Z.Q.; Qu, S.B. Dispersion engineering of metasurfaces for supporting both TM and TE spoof surface plasmon polariton. *J. Phys. D Appl. Phys.* **2018**, *51*, 045109. [CrossRef]
27. Zhang, Y.; Zhou, Y.J.; Cai, J.; Jiang, J.H. Amplification of spoof localized surface plasmons on active plasmonic metamaterials. *J. Phys. D Appl. Phys.* **2018**, *51*, 295304. [CrossRef]
28. Wang, H.; Shen, T.; Liu, J.; Zhu, Y.; Li, H.; Wang, T. Enhancement of Terahertz Emission by Silver Nanoparticles in a Liquid Medium. *Micromachines* **2023**, *14*, 1593. [CrossRef] [PubMed]
29. Fang, M.; Niu, K.; Huang, Z.; Sha, W.; Wu, X.; Koschny, T.; Scoukoulis, C.M. Investigation of broadband terahertz generation from metasurface. *Opt. Exp.* **2018**, *26*, 14241–14250. [CrossRef]
30. Talebi, N. Interaction of electron beams with optical nanostructures and metamaterials: From coherent photon sources towards shaping the wave function. *J. Opt.* **2017**, *19*, 103001. [CrossRef]
31. Liu, Y.-Q.; Du, C.-H.; Liu, P.-K. Terahertz Electronic Source Based on Spoof Surface Plasmon on the Doubly Corrugated Metallic Waveguide. *IEEE Trans. Plasma Sci.* **2016**, *44*, 3288–3294. [CrossRef]
32. Mann, K.L.; Sajal, V.; Panwar, A.; Sharma, N.K. Excitation of terahertz radiation by parametric mixing of four waves in magnetized plasma. *Optik* **2019**, *186*, 182–186. [CrossRef]
33. Srivastav, R.K.; Panwar, A. Cherenkov terahertz surface magnetoplasmons excitation by an electron beam. *Phys. Plasmas* **2023**, *30*, 023104. [CrossRef]

34. Okajima, A.; Matsui, T. Electron-beam induced terahertz radiation from graded metallic grating. *Opt. Exp.* **2014**, *22*, 17490–17496. [CrossRef]
35. Kong, L.-B.; Huang, C.-P.; Du, C.-H.; Liu, P.-K.; Yin, X.-G. Enhancing spoof surface-plasmons with gradient metasurfaces. *Sci. Rep.* **2015**, *5*, 8772. [CrossRef] [PubMed]
36. Liu, Y.-Q.; Kong, L.-B.; Du, C.-H.; Liu, P.-K. A terahertz electronic source based on the spoof surface plasmon with sub-wavelength metallic grating. *IEEE Trans. Plasma Sci.* **2016**, *44*, 930–937. [CrossRef]
37. Zhu, J.-F.; Du, C.-H.; Bao, L.-Y.; Liu, P.-K. Regenerated amplification of terahertz spoof surface plasmon radiation. *New J. Phys.* **2019**, *21*, 033021. [CrossRef]
38. Davids, P.S.; Intravaia, F.; Dalvit, D.A.R. Spoof polariton enhanced modal density of states in planar nanostructured metallic cavities. *Opt. Exp.* **2014**, *22*, 12424–12437. [CrossRef]
39. Liu, Y.-Q.; Kong, L.-B.; Du, C.-H.; Liu, P.-K. Analysis on dispersion characteristics of rectangular metal grating based on spoof surface plasmons. *Acta Phys. Sin.* **2015**, *64*, 174102.
40. Joy, S.R.; Erementchouk, M.; Mazumder, P. Spoof surface plasmon resonant tunneling mode with high quality and Purcell factors. *Phys. Rev. B* **2017**, *95*, 075435. [CrossRef]
41. Yang, J.; Zhao, M.; Liu, L.; Xiang, H.; Han, D. Analysis of the Symmetric and Anti-Symmetric Modes in Spoof–Insulator–Spoof Waveguides. *J. Phys. Soc. Jpn.* **2017**, *86*, 064401. [CrossRef]
42. Zhao, W.; Ju, D.; Jiang, Y. Efficient localization of terahertz waves within a gradient dielectric-filled metallic grating. *Appl. Phys. Express* **2014**, *7*, 124301. [CrossRef]
43. Islam, M.; Kumar, G. Terahertz surface plasmons propagation through periodically tilted pillars and control on directional properties. *J. Phys. D Appl. Phys.* **2016**, *49*, 435104. [CrossRef]
44. Liu, Y.-Q.; Kong, L.-B.; Du, C.-H.; Liu, P.-K. Spoof surface plasmon modes on doubly corrugated metal surfaces at terahertz frequencies. *J. Phys. D Appl. Phys.* **2016**, *49*, 235501. [CrossRef]
45. Tehranian, A.; Ahmadi-Boroujeni, M.; Abbaszadeh, A. Achieving subwavelength field confinement in sub-terahertz regime by periodic metallo-dielectric waveguides. *Opt. Express* **2019**, *27*, 4226–4237. [CrossRef] [PubMed]
46. Liu, Y.-Q.; Sun, J.; Li, L.; Yin, H. Asymmetric propagation of spoof surface plasmons along doubly corrugated metal surfaces. *AIP Adv.* **2020**, *10*, 045005. [CrossRef]
47. Liu, Y.-Q.; Li, L.; Yin, H. Surface plasmon dispersion and modes on the graphene metasurface with periodical ribbon arrays. *Mater. Res. Express* **2020**, *7*, 075801. [CrossRef]
48. Zhang, Y.; Feng, Y.; Zhao, J. Graphene-Enabled Tunable Phase Gradient Metasurface for Broadband Dispersion Manipulation of Terahertz Wave. *Micromachines* **2023**, *14*, 2006. [CrossRef]
49. Chen, L.; Wei, Y.; Zang, X.; Zhu, Y.; Zhuang, S. Excitation of dark multipolar plasmonic resonances at terahertz frequencies. *Sci. Rep.* **2016**, *6*, 22027. [CrossRef]
50. Zhou, J.; Liu, D.; Liao, C.; Li, Z. CHIPIC: An efficient code for electromagnetic PIC modeling and simulation. *IEEE Trans. Plasma Sci.* **2009**, *37*, 2002–2011. [CrossRef]
51. Chen, Z.; Wang, J.; Wang, Y. Efficiency improvement of THz overmoded surface wave oscillator by circular spoof surface plasmon polaritons coupler. *Phys. Plasmas* **2017**, *24*, 023118. [CrossRef]
52. Joe, J.; Louis, L.J.; Scharer, J.E.; Booske, J.H.; Basten, M.A. Experimental and theoretical investigations of a rectangular grating structure for low-voltage traveling wave tube amplifiers. *Phys. Plasmas* **1997**, *4*, 2707–2715. [CrossRef]
53. Carlsten, B.E. Modal analysis and gain calculations for a sheet electron beam in a ridged waveguide slow-wave structure. *Phys. Plasmas* **2002**, *9*, 5088–5096. [CrossRef]
54. Li, R.J.; Ruan, C.J.; Fahad, A.K.; Zhang, C.Y.; Li, S.S. Broadband and high-power terahertz radiation source based on extended interaction klystron. *Sci. Rep.* **2019**, *9*, 4584. [CrossRef]
55. Cao, M.; Liu, W.; Wang, Y.; Li, K. Dispersion characteristics of three-dimensional dielectric-loaded grating for terahertz Smith-Purcell radiation. *Phys. Plasmas* **2014**, *21*, 023116. [CrossRef]
56. Zhang, X.; Liu, S.; Zheng, X.; Liao, K. A high-performance mode multiplexer based on hybrid spoof surface plasmon polariton and substrate integrated waveguide. *Int. J. RF Microw. Comput.-Aided Eng.* **2022**, *32*, e23475. [CrossRef]

micromachines

Communication

Increasing the Light Extraction Efficiency of Organic Light-Emitting Devices by Electrochemically Corroded Patterned Substrates

Yang Wang *, Zhonghao Li, Yu Bai and Yingzhi Wang

School of Electronic Information Engineering, Changchun University of Science and Technology, Changchun 130022, China
* Correspondence: wyang@cust.edu.cn

Abstract: A substrate with microstructure can increase the light extraction efficiency of OLEDs. However, the present preparation methods for micro- and nanostructures are not suited for broad-area manufacturing. In this research, we suggested an electrochemical etching approach to patterning Si substrates and effectively generated a vast area of micro-/nanostructures on the surface of Si. We created OLEDs using this patterned substrate. It was discovered through this study that when the current density is 100 mA/cm^2, the brightness increases by 1.67 times and the efficiency increases by 1.43 times, over a planar equivalent. In the future, this electrochemical etching process for patterned silicon substrates might give rise to a new approach to the large-scale manufacture of microstructured silicon substrates.

Keywords: OLED; silicon substrate; electrochemically corroded

Citation: Wang, Y.; Li, Z.; Bai, Y.; Wang, Y. Increasing the Light Extraction Efficiency of Organic Light-Emitting Devices by Electrochemically Corroded Patterned Substrates. *Micromachines* **2024**, *15*, 67. https://doi.org/10.3390/mi15010067

Academic Editor: Jinn-Kong Sheu

Received: 29 November 2023
Revised: 20 December 2023
Accepted: 27 December 2023
Published: 29 December 2023

1. Introduction

Organic light-emitting diodes have numerous distinct advantages, such as rapid reaction time, great contrast, low power consumption, and so on [1–4]. Because they can be applied to TVs, mobile phones, and other full-color flat-panel displays and solid-state lighting sources, they have received a great deal of attention from academics and the business community. For a long time, research on improving the performance of OLED devices has mostly concentrated on the basic topic of how to effectively enhance device luminosity efficiency. Enhancing the light extraction efficiency of OLED devices, in addition to enhancing their internal quantum efficiency, is an efficient technique to enhance the luminosity efficiency of OLED devices [5–10]. Periodic micro-/nanostructure substrates may significantly increase the light removal efficiency of OLED devices [11,12]. However, many of the present technologies for creating micro-/nanostructure substrates are not suited for large-area preparation, which slows down the development of microstructure substrate OLEDs. Although electrochemical corrosion may be used to produce a vast range of microstructure patterns, it is often isotropic, making it difficult to regulate the direction of corrosion [13–16]. In this paper, we describe a method of electrochemically corroding silicon wafers that allows us to regulate the direction of corrosion to obtain the desired microstructure, which can be utilized to build a silicon-based microstructure substrate over a broad area.

2. Experiment

Figure 1 depicts the microstructure substrate manufacturing device, an electrolytic cell with a side opening. The working electrode (WE) silicon was placed in the opening position by a sealing gasket, and was directly linked to the potentiostat via the ohm contact layer. The electrode (CE) is made of a 25 cm^2 Pt wire, the reference electrode (RE) is made of a saturated calomel electrode, and the three-electrode system is controlled by a potentiostat

(PARSTAT 2273 type). The etching solution is a HF acid solution that has been diluted with deionized water and has had a surfactant added. The thermostat bath keeps the temperature of the solution at room temperature (25 °C).

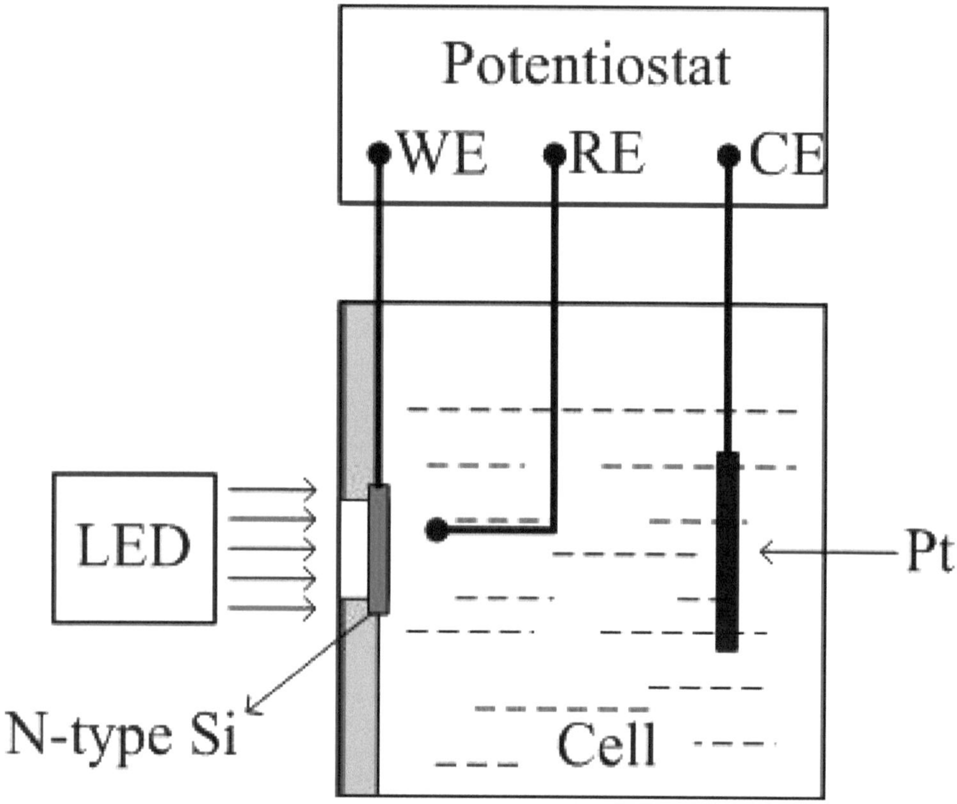

Figure 1. The microstructure substrate fabrication device.

After receiving the patterned substrate, we cleaned the substrate and then utilized the thermal evaporation technique to grow 80 nm of Ag on the patterned substrate as the anode. Following that, a 60 nm N,N′-Bis(naphthalen-1-yl)-N,N′-bis(phenyl)benzidine (NPB) acting as the hole transport layer, a 70 nm Tris(8-hydroxyquinolinato)aluminum (Alq) acting as the emitting layer, and a 20 nm LiF and Ag layer acting as the composite cathode were evaporated [17–22]. All of the layers were prepared by thermal evaporation in a high-vacuum system (Shenyang Sida Vacuum Technology Research Institute, SD400B) with a pressure of less than 4×10^{-4} Pa. The deposition rate of the electrode was about 1 nm/s, and the deposition rate of the organic material was 1–2 Å/s. The deposition rate was monitoring by a film thickness monitor (SHANGHAI TAIYAO VACUUM TECHNOLOGY CO., LTD., FTM-V, Shanghai, China) The effective luminous area of the device was 1 mm^2. After completing the OLED preparation of the patterned microstructure substrate, the current–voltage–brightness (J-V-L) test was completed using a computer-controlled Keithley2611 digital source table and a Konica Minolta CS-100A photometer. All of the tests were carried out at room temperature.

3. Results and Discussion

Because holes are minority carriers in n-type silicon, they must be generated by light or a stronger electric field during the electrochemical corrosion process. To produce the holes in the study, an LED was employed as the light source. Distinct wavelengths of light stimulate silicon to create distinct photocurrents under the same light power circumstances, indicating that the spectral response varies with wavelength.

The associated electrochemical corrosion photocurrent measured under different LED wavelengths is shown in Figure 2a. To decrease the dark current, the solution concentration was kept low (1 wt%) throughout the measurement, and the anionic surfactant was utilized. Figure 2b depicts the predicted relative spectral response curve of silicon photoelectrochemical corrosion. The spectral response rises dramatically with increasing wavelength, as illustrated in Figure 2b. For this experiment, the junction created between the silicon and the HF solution in photoelectrochemical corrosion of the n-type silicon was equal to a reverse-biased PN junction. When backlighting occurs, the created holes diffuse to the interface, and the silicon corrodes to make a photocurrent. In the experiment, the LED light source was placed on the back of the silicon, far away from the PN junction contact (the silicon wafer thickness was 400 m). Because the majority of the photogenerated holes are recombined and do not contribute to the current, the photocurrent value was minimal.

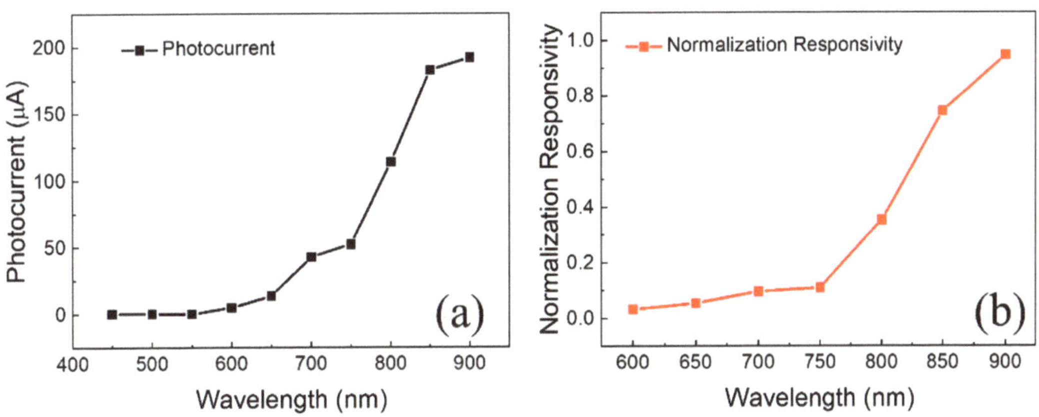

Figure 2. The wavelength responsivity curves of the electrochemical corrosion experiment. (**a**) Photocurrent wavelength characteristics, and (**b**) normalization responsivity wavelength characteristics.

It is generally known that the absorption of silicon to light decreases as the wavelength of the light increases. As a result, the greater the wavelength of light, the deeper the depth of penetration into silicon. Because the LED is placed on the back of the silicon in this experiment, the longer wavelength light may penetrate deeper, bringing the holes created by excitation closer to the Si/HF contact and improving the spectrum response. We selected 850 nm as the LED emission wavelength after careful analysis.

Initially, we investigated the effect of the corrosion voltage on the patterning of the silicon substrate. To begin, the corrosion voltage was set to 0.6 V, 1 V, 1.5 V, 3 V, 5 V, and 9 V, with the corrosion current density set to 10 mA/cm^2 and the corrosion time set to 1 h. The front of the sample was then ground and polished. As shown in Figure 3, the cross-sectional morphology of the hole was obtained using a metallographic microscope (JEOL, JSM-7500F, Beijing, China). As can be seen in Figure 3, when the voltage increases, the cross-sectional form of the hole progressively changes from square to round. When the voltage was increased to 3 V, the shape transformed into a four-pointed star. When the voltage was 5 V, the hole forks, and when the voltage was 9 V, it split into several smaller holes. The phenomenon is generated by the pores' spatial layout and the unequal distribution of the electric field at the pores' tips. The holes are first gathered near the tip in a process of macroporous silicon electrochemical corrosion, and the variation in the spacing around the tip leads to a difference in the supply of holes. The stronger the electric field, on the other hand, the more unequal the distribution of the holes at the tip. The combination of the two results in the experimental phenomena is depicted in Figure 3.

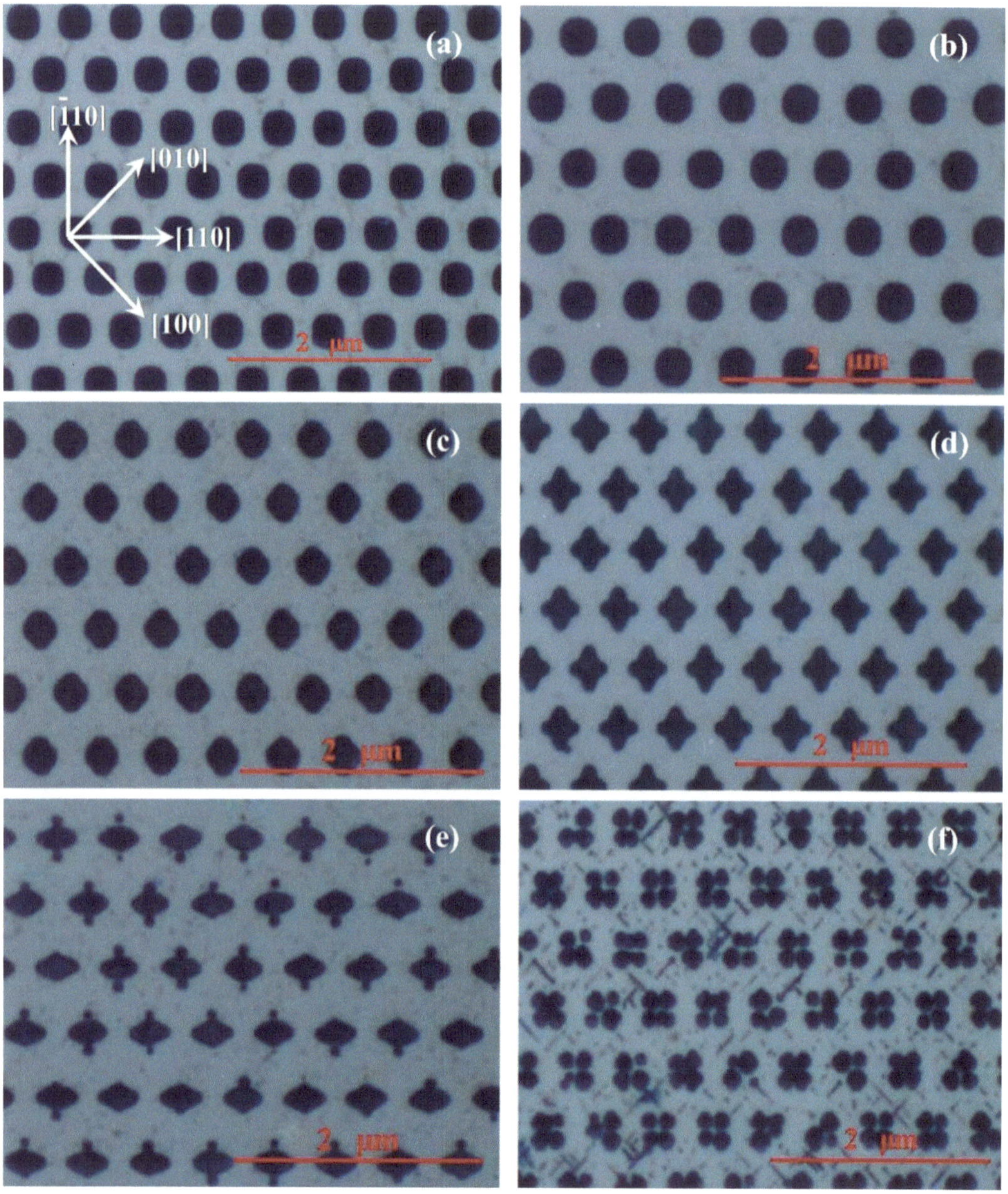

Figure 3. Microscopic photos of hole cross-sections under different voltages. (**a**) 0.6 V, (**b**) 1 V, (**c**) 1.5 V, (**d**) 3 V, (**e**) 5 V, and (**f**) 9 V.

The deeper the hole that produces the pattern on the silicon substrate, the longer the corrosion time. The link between hole depth and corrosion time is seen in Figure 4. When the corrosion voltage is fixed, the corrosion time mostly impacts the particle transport process in the solution. It had an effect on the transit of F-ions in this study. A sufficiently high corrosion voltage was utilized in this study to exclude the impact of F-ion concentration distribution on the experiment and meet the polarization requirement that the electrochemical reaction is exclusively affected by mass transfer. The corrosion depth of the hole was monitored every minute. As can be seen in Figure 4, the growth rate of the hole depth decreases as the corrosion duration rises. F-ions are primarily used in the electrochemical corrosion of silicon substrates to operate on the tip of the Si/HF interface, causing silicon atoms to mix with fluorine and form soluble substances, thereby enlarging the hole. The electrochemical reaction is mainly limited by mass transfer transport in the case of sufficient holes, and the F-ions in the solution are continuously transported to the tip of the hole under the action of the concentration gradient, but the reaction product is transported from the tip of the hole to the solution, which will hinder the transport of F-ions. As the hole deepens, more reaction products accumulate in the hole, and the transport of F-ions will produce greater obstacles. As a result, the corrosion process slows.

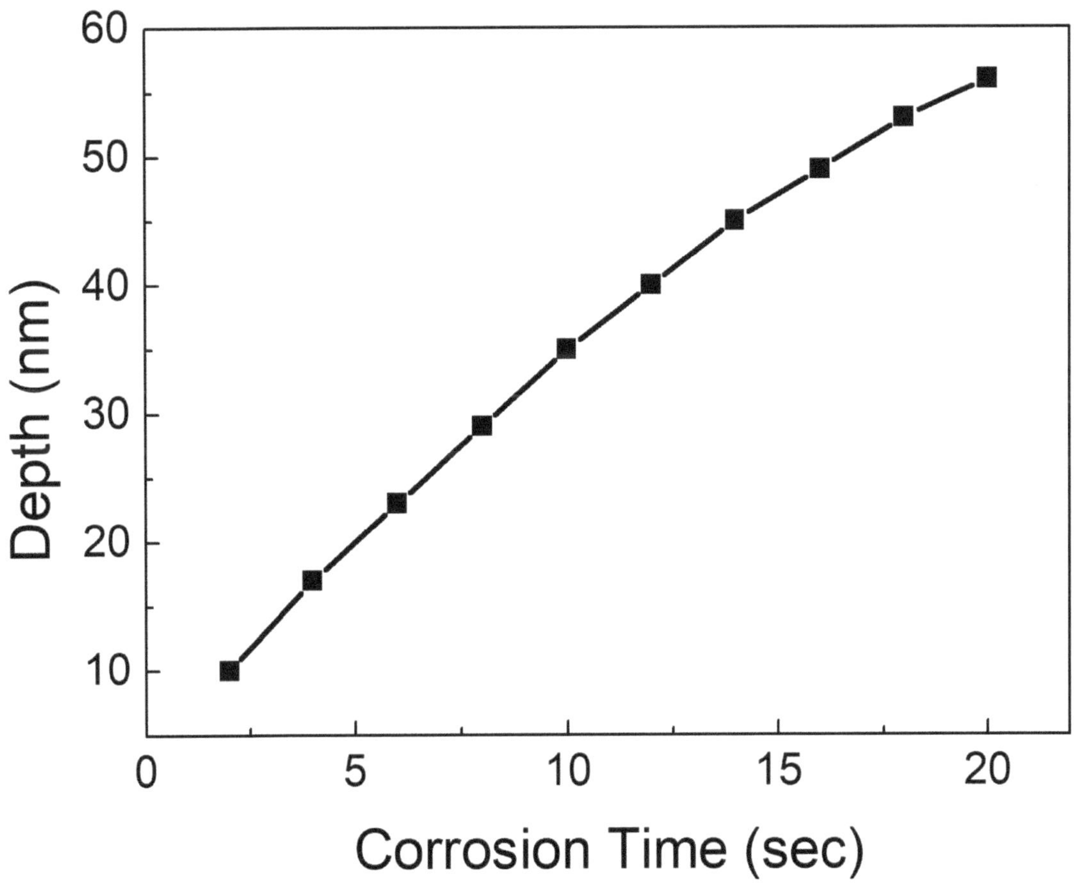

Figure 4. The relationship between the hole depth and the corrosion time.

Patterns of varying shapes and depths may be produced on silicon substrates by manipulating the corrosion voltage and corrosion duration. The patterned silicon substrate may significantly increase OLED light extraction efficiency. We used the finite–difference time–domain approach to simulate light transmission in the patterned OLED to determine which design had the highest light extraction efficiency. We employed the Drude model to describe the dielectric coefficient of metals and materials, and the refractive index of the substance was fitted into the Drude model parameters using ellipsometer data (J. A. Woollam, M-2000UI, Shanghai, China). During the simulation, we used a periodic boundary condition for directions perpendicular to the pattern and a perfect matching layer (PML) to cut out other boundaries. We employed a modulated Gauss impulse with a center frequency inside the visible light spectrum of interest for the incident wave. At the completion of the computation, we isolated the transmission and reflection light components and utilized the Poynting vector ratio to represent transmission, reflection, and absorption. Figure 5 depicts the simulated spectra of the patterned OLED with the hole depth set at 20 nm. Figure 3 shows that as the voltage exceeds 5 V, the surface of the microstructure begins to reveal more flaws, making it unsuitable as an OLED substrate. As a result, we did not model these two categories.

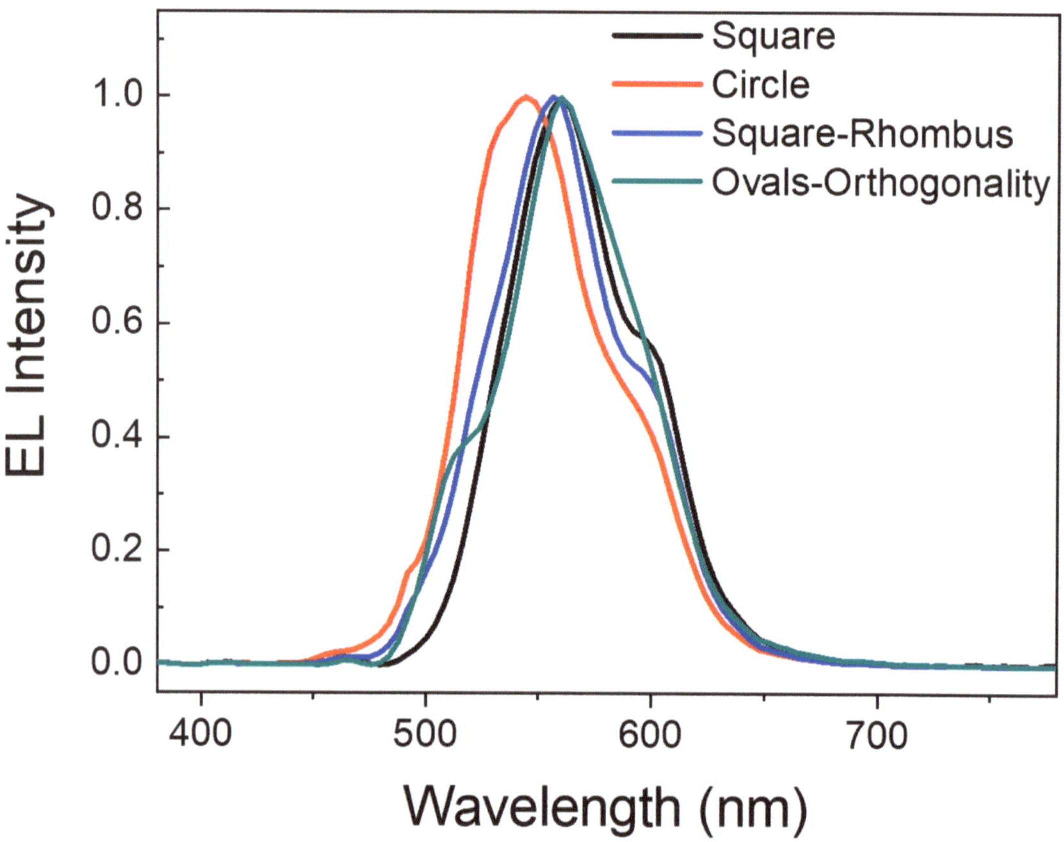

Figure 5. The simulation spectra of the OLEDs with a pattern of square, circle, square rhombus and oval orthogonality, respectively.

Using the OLED simulation spectra with the circular pattern as an example, we can observe that there are three additional emission peaks at 493 nm, 506 nm, and 599 nm in addition to the Alq emission peak. To determine the cause of these extra peaks, we simulated and examined the field intensity in a circle-patterned OLED at an observation angle of 0°, as shown in Figure 6. Figure 6a,b show the 493 nm and 506 nm peak intensity distributions in TM mode. We can observe that the maximum field intensity occurs in the xz direction at the electrode/organic layer contact and in directions along the interface. This explains why the 493 nm and 506 nm peaks occur when the grating ignites surface plasma [12,20–24]. The waveguide mode, as shown in Figure 6c, has the maximum field strength in the xz directions, hence the 599 nm peak is created because the waveguide mode is stimulated. This suggests that the efficiency of light extraction from patterned OLEDs may be greater than that of planar OLEDs. Because the circumstances are the same for patterned OLEDs and planar OLEDs, except for the substrate pattern shape, we can anticipate which substrate pattern shape will provide superior performance based on the shape of the spectrum and the strength of the extra peaks. As a result, we can conclude that with the circular design, OLEDs will have a higher light extraction efficiency.

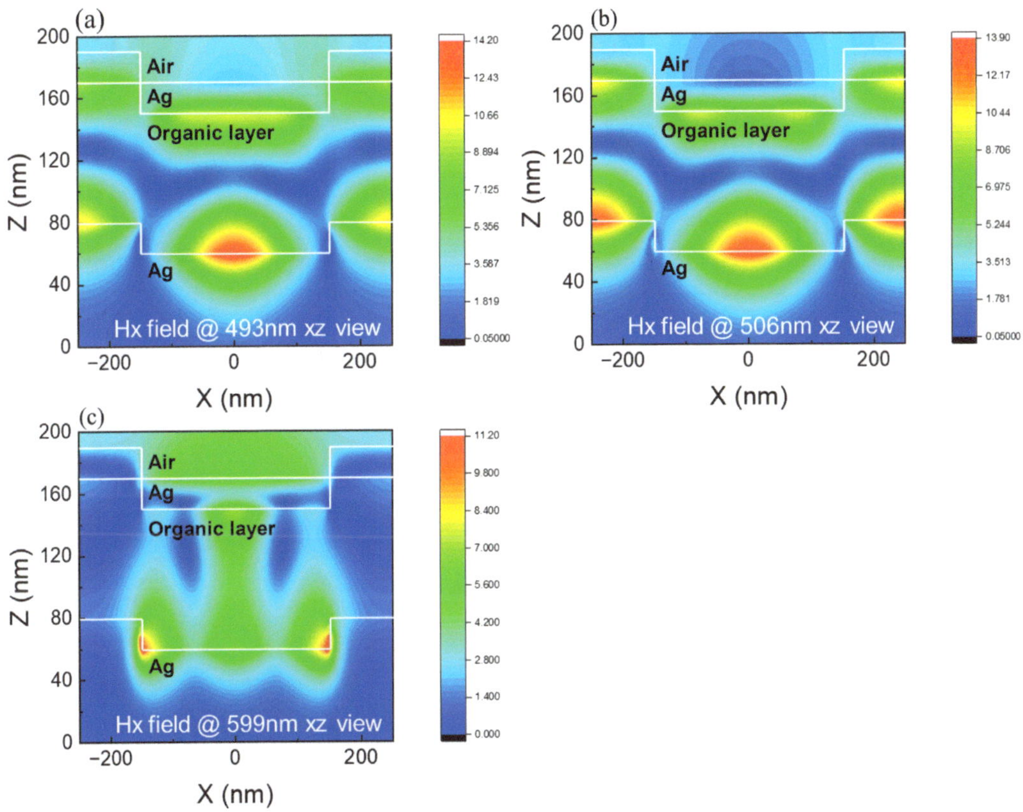

Figure 6. Distribution of magnetic field intensity in the corrugated OLEDs at wavelengths of incident polarized light of 493 nm (**a**), 506 nm (**b**), and 599 nm (**c**) in the XZ view, respectively.

To put our theory to the test, we created planar and patterned OLEDs with varying hole depths. The substrate's microstructures can be totally transmitted from the anode to the cathode. Despite the fact that the depths of the holes vary, and the area of the microstructures is considerably smaller than the luminous area of the OLED, the anode reflectivity of the patterned OLED is still extremely high overall and will not become diffuse owing to the existence of the microstructures. Figure 7 depicts the device performance curve. As seen in Figure 7, the current density of the patterned OLED rose greatly when compared to the control flat-panel device. This is because the pattern microstructure improved the device's effective area. In addition, patterned OLEDs have superior brightness and efficiency when compared to flat-panel devices. The patterned OLED offers the best performance when the patterned hole depth is 30 nm, according to the performance comparison. At a current density of 100 mA/cm^2, the patterned OLED has 1.67 times the brightness and 1.43 times the efficiency of a standard flat-panel device.

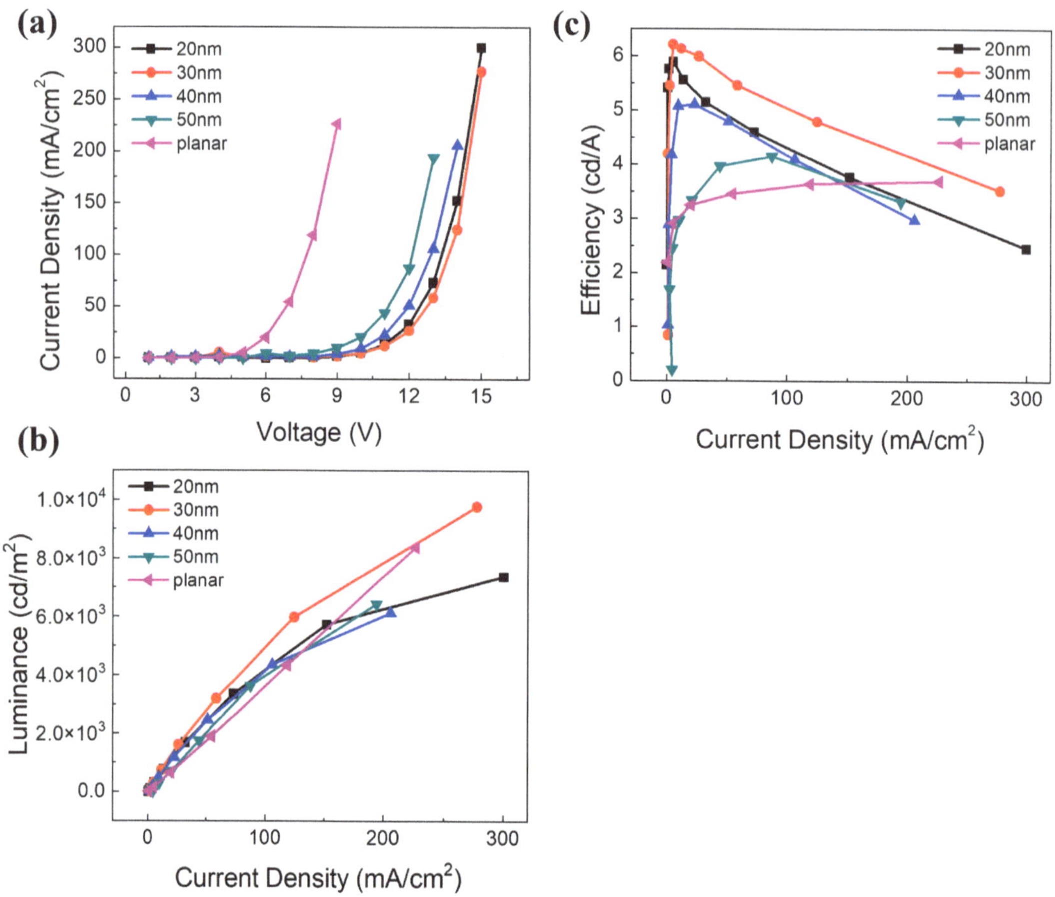

Figure 7. EL performance of the different corrugated-height and planar OLEDs. (**a**) Current density–voltage characteristics, (**b**) luminance–current density characteristics, and (**c**) efficiency–current density characteristics.

4. Conclusions

In this article, we apply the process of electrochemical corrosion to build a patterned silicon substrate. When the corrosion voltage intensities are varied, we can acquire six distinct patterned silicon substrates, and the power-on duration impacts the corrosion depth. We discovered that when the energized voltage is 1 volt and the energized time is 8 s, the resulting circle pattern silicon substrate OLED has the best performance: when the current density is 100 mA/cm^2, its brightness is 1.67 times greater than that of the ordinary flat-structure OLED, and its efficiency is 1.43 times greater than the ordinary flat-plate-structure OLED. This electrochemical corrosion approach of producing patterned silicon substrates may provide a fresh direction for the future large-scale fabrication of microstructured silicon substrates.

Author Contributions: Software, Y.B.; validation, Y.W. (Yingzhi Wang); writing—original draft, Z.L.; writing—review and editing, Y.W. (Yang Wang). All authors have read and agreed to the published version of the manuscript.

Funding: This research was funded by the Education Department of Jilin Province (JJKH20200754KJ) and the Jilin Province Science and Technology Development Project (20210101182JC).

Data Availability Statement: Data are contained within the article.

Conflicts of Interest: The authors declare no conflicts of interest.

References

1. Hobson, P.A.; Wedge, S.; Wasey, J.A.E.; Sage, I.; Barnes, W.L. Surface plasmon mediated emission from organic light-emitting diodes. *Adv. Mater.* **2002**, *14*, 1393–1396. [CrossRef]
2. Gu, G.; Garbuzov, D.Z.; Burrows, P.E.; Venkatesh, S.; Forrest, S.R.; Thompson, M.E. High-external-quantum-efficiency organic light-emitting devices. *Opt. Lett.* **1997**, *22*, 396–398. [CrossRef] [PubMed]
3. Kim, D.H.; Cho, C.O.; Roh, Y.G.; Jeon, H.; Park, Y.S.; Cho, J.; Im, J.S.; Sone, C.; Park, Y.; Choi, W.J.; et al. Enhanced light extraction from gan-based light-emitting diodes with holographically generated two-dimensional photonic crystal patterns. *Appl. Phys. Lett.* **2005**, *87*, 855. [CrossRef]
4. Xie, G.; Meng, Y.; Wu, F.; Tao, C.; Zhang, D.; Liu, M.; Xue, Q.; Chen, W.; Zhao, Y. Erratum: "very low turn-on voltage and high brightness tris-(8-hydroxyquinoline)aluminum-based organic light-emitting diodes with a moox p-doping layer" [Appl. Phys. Lett. 92, 093305 (2008)]. *Appl. Phys. Lett.* **2008**, *93*, 229902. [CrossRef]
5. Bai, Y.; Chuai, Y.; Wang, Y.; Wang, Y. Higher Light Extraction Efficiency in Organic Light-Emitting Devices by Employing 2D Periodic Corrugation. *Front. Chem.* **2022**, *9*, 807867. [CrossRef]
6. Moreland, J.; Adams, A.; Hansma, P.K. Efficiency of light emission from surface plasmons. *Phys. Rev. B Condens. Matter* **1982**, *25*, 2297–2300. [CrossRef]
7. Cai, X.L.; Xu, Y.C.; Pan, Y.; Li, L.; Pu, Y.X.; Zhuang, X.M.; Li, C.l.; Wang, Y. Solution-Processable Pure-Red Multiple Resonance-induced Thermally Activated Delayed Fluorescence Emitter for Organic Light-Emitting Diode with External Quantum Efficiency over 20%. *Angew. Chem.* **2023**, *62*, e202216473. [CrossRef]
8. Cheng, W.; Wang, L.-D.; Zhou, Y.-Y.; Bian, Z.-Q.; Tong, B.-H.; Liu, Z.-W.; Wang, S. Blue iridium(III) complexes with high internal quantum efficiency based on 4-(pyridin-3-yl)pyrimidine derivative and their electroluminescent properties. *Dye. Pigment.* **2020**, *177*, 108257. [CrossRef]
9. Usman, M.; Mushtaq, U.; Zheng, D.-G.; Han, D.-P.; Rafiq, M.; Muhammad, N. Enhanced Internal Quantum Efficiency of Bandgap-Engineered Green W-Shaped Quantum Well Light-Emitting Diode. *Appl. Sci.* **2020**, *9*, 77. [CrossRef]
10. Shim, J.; Shin, D. Measuring the internal quantum efficiency of light-emitting diodes: Towards accurate and reliable room-temperature characterization. *Nanophotonics* **2018**, *7*, 1601–1615. [CrossRef]
11. Bai, Y.; Feng, J.; Liu, Y.-F.; Song, J.-F.; Simonen, J.; Jin, Y.; Chen, Q.-D.; Zi, J.; Sun, H.-B. Outcoupling of trapped optical modes in organic light-emitting devices with one-step fabricated periodic corrugation by laser ablation. *Org. Electron.* **2011**, *12*, 1927–1935. [CrossRef]
12. Worthing, P.T.; Barnes, W.L. Efficient coupling of surface plasmon polaritons to radiation using a bi-grating. *Appl. Phys. Lett.* **2001**, *79*, 3035–3037. [CrossRef]
13. Bálint, M.; Ali, G.; Gábor, H.; Béla, P.; Ilona, F. Electrochemical corrosion and electrochemical migration characteristics of SAC-1Bi-xMn solder alloys in NaCl solution. *Corros. Sci.* **2023**, *213*, 110965.
14. Dileep, P.; Dheeraj, J.; Vikrant, S.; Behera, C.K. Electrochemical corrosion behavior of tantalum coated 316L stainless steel by D.C. Magnetron sputtering for orthopedic applications. *Appl. Surf. Sci. Adv.* **2023**, *13*, 100365.
15. Ming, T.Y.; Peng, Q.J.; Han, Y.L.; Zhang, T. Effect of water jet cavitation peening on electrochemical corrosion behavior of nickel-based alloy 600 in NaCl solution. *Mater. Chem. Phys.* **2023**, *295*, 127122. [CrossRef]
16. Wu, J.J.; Wang, W.; Shen, M.L.; Zhu, S.L. A comparative study of the electrochemical corrosion behavior between Cr;2;N and CrN coatings. *Heat Treat. Surf. Eng.* **2022**, *4*, 64–69. [CrossRef]
17. Peng, H.; Zhu, X.; Sun, J.; Xie, Z.; Xie, S.; Wong, M.; Kwok, H.-S. Efficient organic light-emitting diode using semitransparent silver as anode. *Appl. Phys. Lett.* **2005**, *87*, 913. [CrossRef]
18. Choi, H.W.; Kim, S.Y.; Kim, W.K.; Lee, J.L. Enhancement of electron injection in inverted top-emitting organic light-emitting diodes using an insulating magnesium oxide buffer layer. *Appl. Phys. Lett.* **2005**, *87*, 2469. [CrossRef]
19. Han, S.; Feng, X.; Lu, Z.; Johnson, D.; Wood, R. Transparent-cathode for top-emission organic light-emitting diodes. *Appl. Phys. Lett.* **2003**, *82*, 2715–2717. [CrossRef]
20. Van Oosten, D.; Spasenović, M.; Kuipers, L. Nanohole chains for directional and localized surface plasmon excitation. *Nano Lett.* **2010**, *10*, 286–290. [CrossRef]
21. Odom, T.W.; Gao, H.; Henzie, J. Direct evidence for surface plasmon-mediated enhanced light transmission through metallic nanohole arrays. *Nano Lett.* **2006**, *6*, 2104–2108.

22. Liu, W.C.; Tsai, D.P. Optical tunneling effect of surface plasmon polaritons and localized surface plasmon resonance. *Phys. Rev. B Condens. Matter* **2002**, *65*, 155432. [CrossRef]
23. Li, J.; Ning, C.Z. Interplay of collective excitations in quantum-well intersubband resonances. *Phys. Rev. Lett.* **2003**, *91*, 097401. [CrossRef] [PubMed]
24. Giannattasio, A.; Barnes, W. Direct observation of surface plasmon-polariton dispersion. *Opt. Express* **2005**, *13*, 428–434. [CrossRef]

micromachines

Article

Double-Strip Array-Based Metasurfaces with BICs for Terahertz Thin Membrane Detection

Yanchun Shen [1,*], Jinlan Wang [1], Hongyu Sheng [2], Xiaoming Li [1], Jing Yang [1], Hongmei Liu [1] and Dejun Liu [3,*]

1 College of Information Engineering, Guangzhou Railway Polytechnic, Guangzhou 511300, China; wangjinlan@gtxy.edu.cn (J.W.); ming00ming@126.com (X.L.); yangjing@gtxy.edu.cn (J.Y.); liuhongmei@gtxy.edu.cn (H.L.)
2 College of Robotics, Beijing Union University, Beijing 100101, China; jzw@buu.edu.cn
3 Department of Physics, Shanghai Normal University, Shanghai 200234, China
* Correspondence: shenyanchun@gtxy.edu.cn (Y.S.); dejunliu1990@shnu.edu.cn (D.L.)

Abstract: A double-strip array-based metasurface that supports the sharp quasi-bound states in the continuum (quasi-BICs) is demonstrated in terahertz regions. By tuning the structural parameters of metal strips, the conversion of BICs and quasi-BICs is controllable. The simulated results exhibit an achieved maximum Q-factor for quasi-BICs that exceeds 500, corresponding to a bandwidth that is less than 1 GHz. The optical response of quasi-BICs is mainly affected by the properties of substrates. Resonant frequencies decrease linearly with increasing refractive index. The bandwidth of quasi-BICs decreases to 0.9 GHz when n is 2.2. The sharp quasi-BICs are also sensitive to changes in material absorption. Low-loss materials show higher Q-factors. Thus, the selection of a suitable substrate material will be beneficial in achieving resonance with a high Q value. The sensitivity of DSAs for molecules is assessed using a thin membrane layer. The DSAs show high sensitivity, which achieves a frequency shift of 70 GHz when the thickness of the membrane is 10 μm, corresponding to a sensitivity of 87.5 GHz/RIU. This metasurface with sharp quasi-BICs is expected to perform well in THz sensing.

Keywords: terahertz detection; metasurface; bound states in the continuum; high quality

Citation: Shen, Y.; Wang, J.; Sheng, H.; Li, X.; Yang, J.; Liu, H.; Liu, D. Double-Strip Array-Based Metasurfaces with BICs for Terahertz Thin Membrane Detection. *Micromachines* **2024**, *15*, 43. https://doi.org/10.3390/mi15010043

Academic Editor: Muhammad Ali Butt

Received: 8 December 2023
Revised: 22 December 2023
Accepted: 23 December 2023
Published: 25 December 2023

1. Introduction

Bound states in the continuum (BICs) describe exotic localized eigenstates embedded in the continuous spectrum, which is initially proposed by von Neumann and Wigner in 1929 [1]. BICs eliminate radiation loss and thus allow for resonances with an infinite quality factor (Q-factor), showing high potential in the field of lasing [2], biosensing [3], imaging [4], and nonlinear harmonic generation [5]. In terahertz (THz) regions, BICs are widely applied in sensing applications. By changing the incident angle or structural parameters, BICs can be converted to quasi-BICs with high Q-factors. Resonances with high Q-factor show an enhanced field, benefitting light-matter interactions. Metasurfaces with periodic meta-atoms are perfect platforms to manipulate BICs [6–10]. Recently, different types of metasurfaces that consist of various materials have been proposed for BICs manipulations [11–16]. Dielectric metasurfaces are excellent candidates for high Q-factor BICs due to their low material losses [17–19]. The Q-factor of dielectric metasurfaces can be increased by up to 10^4 by breaking the symmetries of metasurfaces [20]. The Q-factor measured by THz-TDS is lower than 100. Furthermore, dielectric metasurfaces confine the enhanced field inside the structure, exhibiting weak interactions between the field and analyte layers. Metal hole array (MHA)-based metasurfaces have been reported for the detection of protein molecules because of the induced surface plasmon polaritons (SPPs) with the surface-enhanced field [21,22]. For the printed horseradish peroxidase with 500 pg/mm^2, due to the change in refractive index, the transmission dip is red-shifted. MHAs need oblique incident waves to excite sharp resonance, which is complicated and limited in THz sensing [21]. In order to address this problem, Liu, D. et al. have introduced quasi-BICs for full metal structures [14,23]. By tuning the structural parameter, a maximum Q-factor of 102 of quasi-BICs can be measured. Subsequently, Wang et al. used such kinds of BICs metasurfaces for polymer membrane sensing [24]. A sensitivity of 151 GHz/RIU for 50-μm polymer was obtained, higher than bulky dielectric substrate metamaterials. To further improve the sensitivity of quasi-BICs sensors, a metasurface based on four-hole arrays was proposed, which realizes a polarization-independent quasi-BIC and a high Q-factor exceeding 2000 after optimization [25]. For a 25 μm-thick polyimide membrane, the measured results exhibit that the frequency shift is 97.5 GHz. The full metallic structure shows high Ohmic loss, hindering the generation of high Q-factor quasi-BICs. Thus, metal-dielectric-based metasurfaces with about 200 nm thick metal layers have been demonstrated. Such metasurfaces reduce the Ohmic loss from thick metal. Y. K. Srivastava et al. reported an asymmetric split ring on a low refractive index substrate with quasi-BICs, which allows the detection of an analyte with a thickness of 7 nm at terahertz frequencies [26]. The analyte layer is coated on the metasurface, which is difficult to clean after sensed. Thus, reusable sensors based on BICs metasurfaces are in urgent need. In addition, the effect of material properties of substrates such as thickness and absorption on the Q-factor of resonances is not clear, which warrants further analysis.

Here, the quasi-BICs that are supported by the metasurface consisting of double-strip arrays are studied at THz frequencies. The transformation between BICs and quasi-BICs can be controlled by changes in the length of one metal strip. The achieved maximum Q-factor exceeds 500, corresponding to a bandwidth that is less than 1 GHz. We thoroughly analyzed the induced quasi-BICs by using the Fano fitting and electric field vectors; results show that the quasi-BICs manifests itself as a type of Fano resonance when structural symmetry breaks down. Differing from traditional resonances of WA with the same profile, the field vector direction of quasi-BICs on the strip surface is opposite, showing a phase difference of 180°. The role of substrate materials on resonant responses of quasi-BICs is thoroughly discussed. The bandwidth of quasi-BICs is influenced by the substrate thickness. The narrowest bandwidth (1.8 GHz) can be realized when the substrate thickness is 50 μm and 60 μm. Refractive indices of the substrate also dramatically change the optical response of the metasurfaces. Resonant frequencies decrease linearly with increasing refractive index. The bandwidth of quasi-BICs decreases to 0.9 GHz when n is 2.2. The sharp quasi-BICs are sensitive to the changes in materials absorption. Low-loss materials show higher Q-factors. Thus, suitable substrate material selection will benefit the achieving high Q-factor resonance. Finally, the sensitivity of proposed DSAs for molecules is assessed using the BSA layer. The DSAs show high sensitivity, achieving a frequency shift of 70 GHz when the thickness of BSA is 10 μm, corresponding to a sensitivity of 87.5 GHz/RIU. Thus, this metasurface with sharp quasi-BICs is expected to perform well in THz sensing.

2. The Design of Double-Strip Arrays

The 3D configuration of proposed metasurfaces based on double-strip arrays (DSAs) can be seen in Figure 1. Differently from the previous reference with thick quartz substrate [27], the proposed DSAs consists of two metal strips that cover a flexible ultrathin substrate. The thin substrate can reduce material absorption and improve Q-factors [28]. This flexible substrate is polyethylene (PE) (n = 1.52 at 1 THz) [29]. The thickness of the substrate is d with a value of 50 μm. Here, numerical results are carried out through CST Studio Suite. The metal conductivity strongly affects the profile of resonances when the metasurface has extremely low structural asymmetry [30]. In the simulation, the metal is set as a perfect electron conductor (PEC) with a thickness of 200 nm, and the material absorption of PE is not considered. As shown in Figure 1b, each double-strip array is considered as a unit and the period of the array is set as Λ. The metal strip has a width of w. The two strips have a respective length of L1 and L2. What follows are the detailed parameters: the period is Λ = 260 μm; the width of strips is w = 60 μm; the length of strips is L1 = 200 μm. L2 is selected as a variable parameter, which determines the structural states (symmetry or asymmetry). Such states provide a route for the conversion of perfect BICs and quasi-BICs [31]. For transverse electric (TE) modes, the electric field is transverse to the direction of propagation while the magnetic field is normal to the direction of propagation. For transverse magnetic (TM) modes, the magnetic field is transverse to the direction of propagation while the electric field is normal to the direction of propagation. In the CST simulation, the electric field of the TE and TM modes is, respectively, perpendicular to the X- and Y-axes, as shown in Figure 1a.

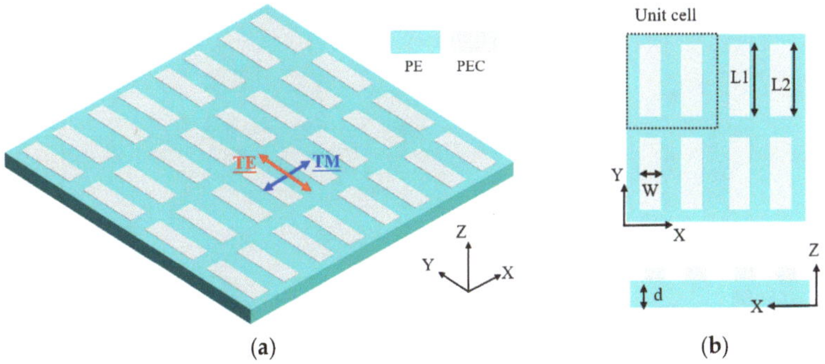

Figure 1. (**a**) Configuration of double-wire arrays; (**b**) the unit cell of double-strip arrays.

3. Simulation Results and Analysis

Figure 2 shows the transmission spectra at normal incidence of TE and TM modes for DSAs with various structural parameters. The DSAs are symmetric in the X- and Y-directions when L1 is equal to L2. For a periodic structure, the free-space frequency f_{WA} satisfies the WA condition. Thus, it shows a transmission minimum. The formulation of WA is expressed as $f_{WA} = C/(\Lambda^* n_{eff})$, in which n_{eff} is the effective refractive index. In Figure 2a, a transmission dip is found in the spectrum for TE modes, which can be termed a Wood anomaly (WA) [32,33]. Other resonance is not induced because the symmetry state of DSAs is maintained. The corresponding spectra are different from the complementary structure of double-slit arrays in the previous reference [14]. As L2 increases or decreases, a sharp resonant dip appears because the structural symmetry is broken. When L2 = 180 μm, a resonance with a sharp profile is induced at 0.49 THz (red line). As L2 increases to 220 μm, the sharp resonance shifts to 0.44 THz (blue line). Such resonance performs as a typical feature of Fano resonance with an asymmetric line shape [21,34]. But for TM modes, the structural symmetry in the X-direction is sustained, and, thus, sharp resonance is not induced. As proved in previous references [35–37], perfect BICs are a dark mode that happen only at L1 = L2, which convert to the bright mode of quasi-BICs with a high Q value when the L2 is different from L1 due to the broken structural symmetry. In other words, the changes in structural symmetry cause the emergence of quasi-BICs.

To further confirm the existence of such BICs under TE wave incidences, the map of transmission spectra of the DSAs versus different L2s is calculated and shown in Figure 3a. Clearly, the sharp resonant profile vanishes when L2 is close to the value of 200 μm, which means that the BICs are almost decoupled to the incidence waves [37]. The result shows that the ideal BIC appears at 0.485 THz, which can be adjusted by changing the material properties of substrates [31]. With L2 increases, the quasi-BICs appear and redshift. Quasi-BICs bandwidths widen, corresponding to the decreases of Q-factors. Similarly, when L2 decreases, the induced quasi-BICs move to higher frequencies. These results describe the characteristics of BICs. Such BICs are also called symmetry-protected BICs [38]. Once the structural symmetry is broken, bound states of quasi-BICs change to radiation modes [35]. Figure 3b reveals the alteration of L2 results in the changes of quasi-BICs. One can see that the resonant profiles can be modified by varying the structural parameters of L2. As L2 increases from 170 μm to 190 μm, the corresponding Q-factor shows an exponential growth.

It is clear in Figure 3b that the simulated maximum Q-factor exceeds 500, corresponding to a bandwidth less than 1 GHz. However, it is difficult to realize ultrahigh Q-factors in experiments because of the limitation of fabrication conditions for small asymmetric parameters. In addition, the low resolution of THz-TDS restricts the observation of high Q-factors. The broken symmetry also changes the volume of metasurfaces, resulting in slight changes of resonant frequencies, as shown in Figure 3b. The changes in L2 refer to a changed asymmetric factor $a = (|(L2 - L1)/L1|)$. The corresponding Q-factor satisfies the formula $Q \propto 1/a^2$ [14,21,39]. Therefore, the smaller the asymmetric factor, the larger the Q-factor [Figure 3c].

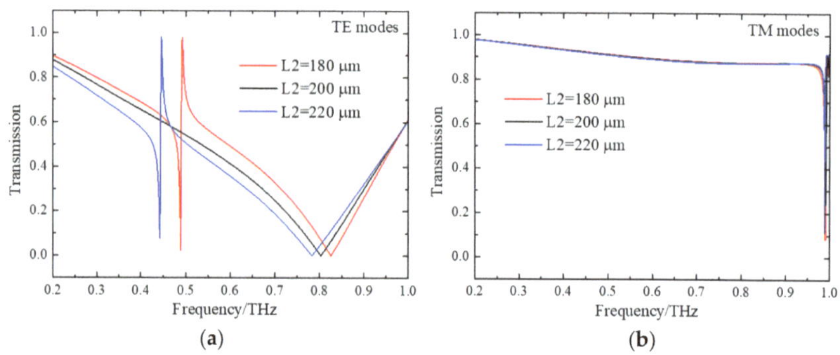

Figure 2. (**a**) For TE waves, the transmission spectra of DSAs with different L2. (**b**) For TM waves, the transmission spectra of DSAs with different L2.

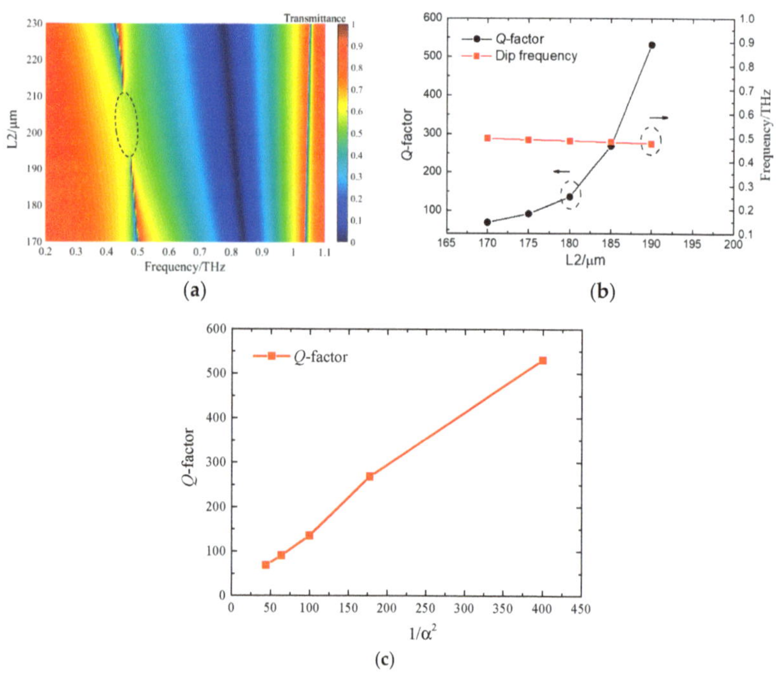

Figure 3. (**a**) The transmission spectra of double-strip arrays with different L2s. (**b**) The corresponding Q-factor and resonant frequency with L2 changes. (**c**) The relation between Q-factor and asymmetric parameter α.

In Figure 4a, a sharp quasi-BICs is depicted in the transmission spectrum, where L2 is 185 μm. Here, the structural state of DSAs is broken. Thus, the quasi-BICs with a Q-factor of 286.6 can be observed at 0.484 THz. This resonance originating from the symmetry breaking would be fit by the Fano formula [23],

$$T = \left| a_1 + ja_2 + \frac{b}{\omega - \omega_0 + j\gamma} \right|^2 \quad (1)$$

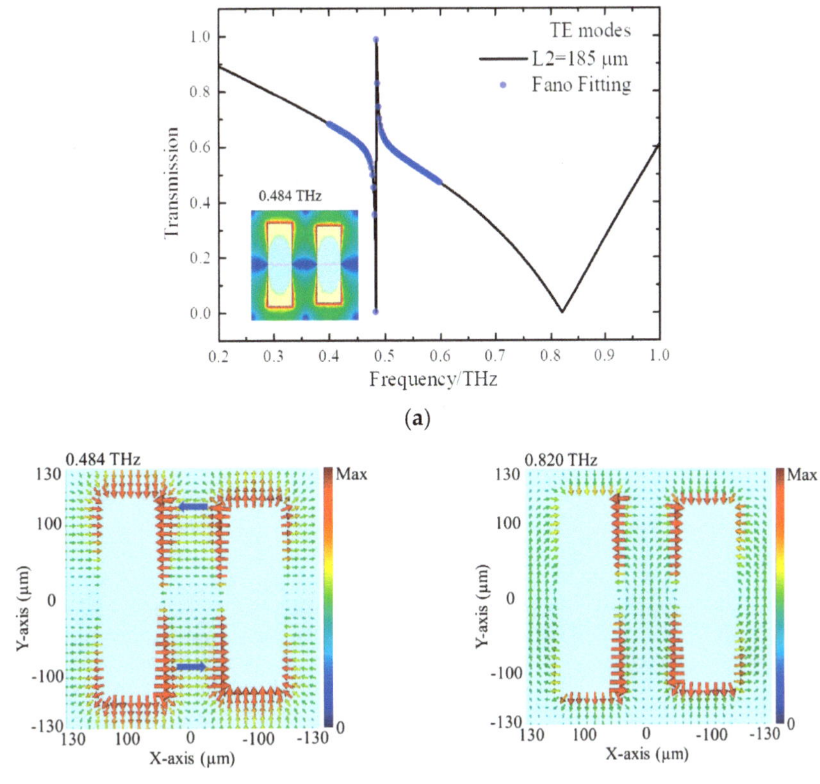

Figure 4. (**a**) The transmission spectrum of DSAs when L2 is 185 μm, where the inset figure is the field distribution of 0.484 THz; (**b**) field vectors of 0.484 THz for quasi-BICs; (**c**) field vectors of 0.820 THz for WA.

This means that the quasi-BICs manifest themselves as a type of Fano resonance as the structural symmetry is broken [40,41]. It can be seen in Figure 4a, for a field distribution at 0.484 THz, that the strong field is located at the edge of metal strips, which is sensitive to material changes. To better understand the physical mechanism of quasi-BICs, the electric field vector distribution is simulated. The resonance of WA at 0.820 THz is selected as an example for comparison. At 0.484 THz, the surface field is excited and concentrated at the upper and lower boundaries of the metal strip and the edge of the gap between the two strips. The vector directions on the surface of the two strips are opposite, showing the phase difference of 180°, which indicates the quasi-BICs are formed. But at 0.820 THz of WA, the surface vector directions are the same. Thus, the profile of quasi-BICs differs from the traditional resonances of WA.

The properties of substrate materials have a giant effect on the performance of quasi-BICs [28,42–45]. Correctly selecting the substrate with suitable parameters is advantageous for obtaining high Q-factor modes. The substrate effect appears due to the finite dielectric contrast between the superstrate and substrate claddings [42]. Here, the relationship between PE thickness and the profile of quasi-BICs is analyzed. The calculated transmission spectra with various substrate thicknesses under TE mode incidence is depicted in Figure 5a. It is clearly seen that the resonance shifts as the PE thickness, *d*, changes. In addition to the spectral shift induced by the PE, a dramatic change in resonant width is found. Figure 5b summarizes the resonant frequency and width as a function of PE thickness *d*. With the increasing PE thickness, the resonance moves toward low frequencies. For example, d = 10 μm corresponds to a resonance at 0.540 THz, shifting to 0.455 THz once d increases to 110 μm. The bandwidth of quasi-BICs is also affected by the substrate, which decreases first and then increases as the PE thickness increases. The narrowest bandwidth of 1.8 GHz can be realized when the thickness is 50 μm and 60 μm. The bandwidth increases to 2.7 GHz when the PE thickness increases to 100 μm. Thus, selecting a suitable substrate thickness benefits achieving a high Q-factor resonance.

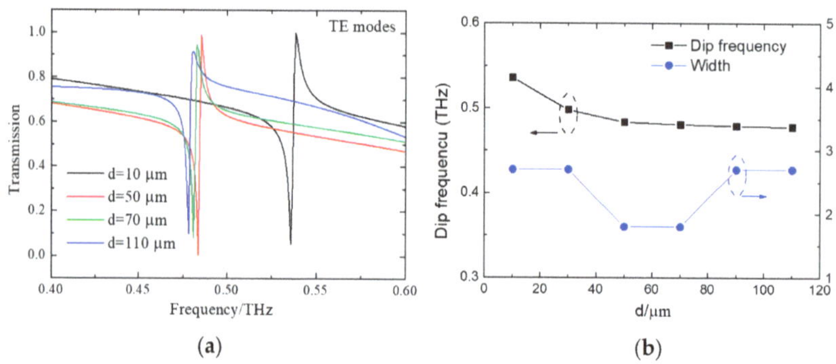

Figure 5. (**a**) The calculated transmission spectra for different PE thicknesses; (**b**) the changing trend of resonant frequency and width with the alters of PE thickness *d*.

The above results proved that this metasurface can achieve high Q-factor quasi-BICs when the structural symmetry is broken and thus its potential to be used as a biosensor [24,25,28,46]. To verify the functionality of DSA-based sensors, we have simulated the transmission spectra of DSAs with different substrate refractive indices (*n*). The thickness of the substrate is fixed as 50 μm and L2 is set as 185 μm. As shown in Figure 6a, as *n* increases, resonant peaks in the transmission spectrum gradually shift towards low frequencies. When *n* equals 1.92, the resonant dip changes to 0.415 THz. This suggests that the transmission dip position is significantly influenced by changes in the substrate's refractive index. It also means that the resonance peaks will be sensitive to biological samples with different refractive indices [11,47,48]. For a clearer visual representation of how the refractive index shifts the resonant dip, a spectral map of the refractive index and resonance frequency is present in Figure 6b. The figure clearly illustrates the resonance peak shift. Meanwhile, the resonant bandwidth is also reduced as the refractive index of substrates increases. To learn more about how the substrate refractive index affects resonance characteristics, we have plotted the curve of resonant frequency and bandwidth with different *n*. In Figure 6c, with *n* increasing, the resonant frequency shows a linear decline due to its changed dispersion. The frequency of the resonant dip appears at 0.366 THz while *n* equals 2.2. However, the quasi-BICs bandwidths remain unchanged with a value of 1.8 GHz when *n* is altered from 1.4 to 2.0. It decreases to 0.9 GHz when *n* changes to 2.2. Thus, the optical response of the metasurfaces is drastically altered by the substrate's refractive index.

Sharp quasi-BICs are sensitive to changes in materials such as absorption and refractive index [14,23,25,35]. In full metal-based metasurfaces, high conductivities show higher Q-factors [14,23,25]. To further reveal the effect of the material loss of the substrate on quasi-BICs, transmission spectra with different loss tangents have been simulated and performed. Parameters for structures are chosen to support sharp quasi-BICs. Here, L2 is selected as 180 μm and *d* is set as 50 μm. As seen in Figure 7a, the resonance becomes broad as the tanδ increases. It means that the Q-factor is reduced. Resonant frequency and bandwidth of quasi-BICs are summarized and shown in Figure 7b. Transmission spectra results show that material absorption does not significantly alter the locations of resonances [49,50]. But resonant bandwidth nonlinear decreases as the loss tangent increases. This confirms that the quasi-BICs are indeed sensitive when the loss of substrate is changed.

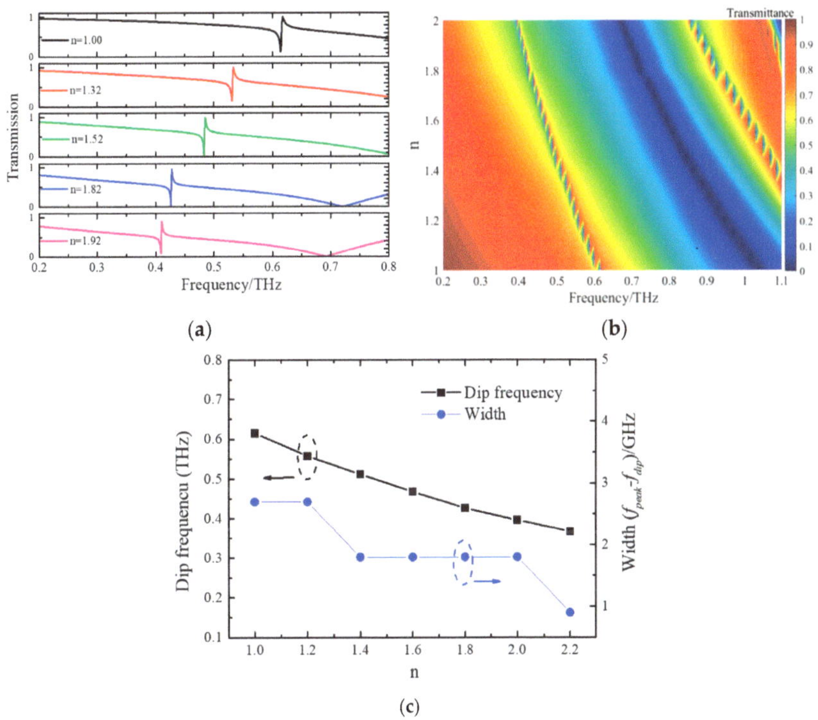

Figure 6. (**a**) The calculated spectra of DSAs for different *n*; (**b**) the spectral map for different *n*; (**c**) the relationship between *n* and the resonant frequency and bandwidth.

THz sensing technology is beneficial for thin-film detection with a small volume of samples. A serum albumin protein of bovine serum albumin (BSA) could prevent nonspecific binding sites from being used during protein–protein interactions [51]. Thus, BSA is frequently utilized as a THz biomarker for biological detection. Here, the BSA membrane is selected to assess the performance and potential of the DSAs-based sensors [52,53]. BSA has a refractive index of 1.8 in 0.2–1.2 THz [51]. Our proposed structure is proven to be sensitive to detect the molecular layer with various thicknesses. Putting the analyte layer on the metasurface without processing, there is an air gap between the analyte layer and the metasurface [24]. In reference [26], the detection material is put on the metasurface-based sensors. Thus, such metasurfaces cannot be reused. In reference [25], alcohol is used to eliminate the air gap between the metasurface and the sensed membrane. After the alcohol evaporates, the membrane sticks tightly to the metasurface. Here, in the simulation, we do not consider that air gap. As seen in Figure 8a, the calculated spectra of DSAs with various thicknesses (*st*) of the BSA layer are presented. Numerical results reveal that small analyte thickness changes can result in a clear spectral shift. A 1 μm-thick BSA layer results in a frequency shift of 34 GHz. As the BSA thickness increases, the quasi-BICs further move to lower frequencies. In Figure 8b, we have summarized the resonant frequency and frequency shift with different BSA thicknesses. The proposed metasurface is sensitive to the attached molecular layers due to the enhanced electromagnetic field on the metal surfaces [54,55]. The strong field–analyte interaction results in large resonant frequency shifts. As the BSA layer thickness increases, the frequency shift Δf shows a nonlinear trend. The frequency shift Δf is 70 GHz when the thickness of the BSA is 10 μm, which corresponds to a sensitivity ($S = \Delta f/\Delta n$) of 87.5 GHz/RIU. The realization sensitivity is higher than that of full metal structures (60.6 GHz/RIU) [25]. Such metasurfaces can be used for microfluidic sensing because of their high sensitivity [56].

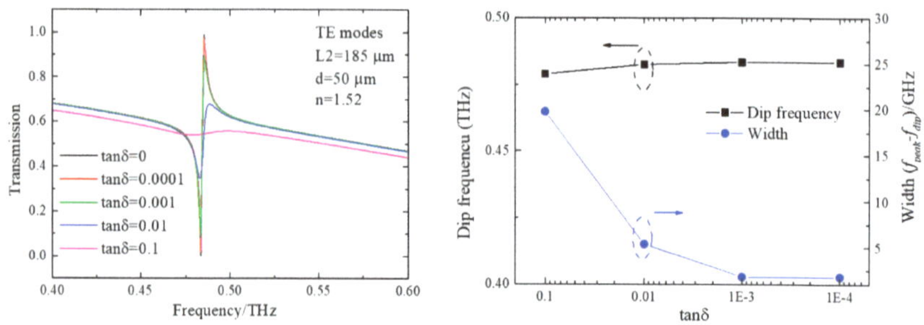

(**a**) (**b**)

Figure 7. (**a**) The calculated spectra with different loss tangent; (**b**) resonant frequency and bandwidth of quasi-BICs as a function of tanδ.

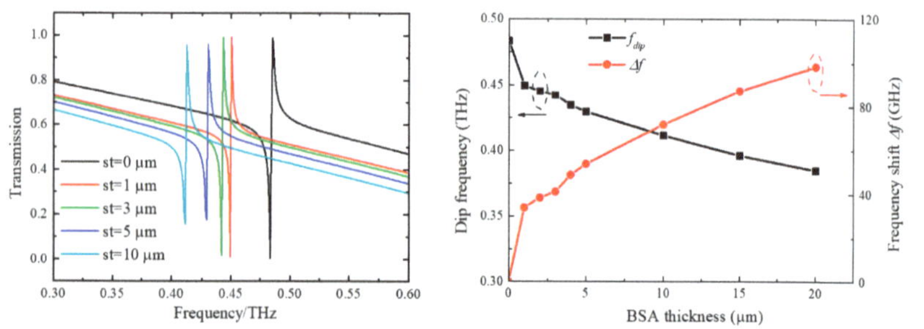

(**a**) (**b**)

Figure 8. (**a**) The calculated spectra with different BSA thicknesses (*st*); (**b**) the summary of resonant frequency and frequency shift for different thicknesses (*st*).

4. Conclusions

We have investigated the optical response of quasi-BICs in a metasurface consisting of double-strip arrays. The transformation between BICs and quasi-BICs can be controlled by changes in the length of one metal strip. The achieved maximum Q-factor exceeds 500, corresponding to a bandwidth that is less than 1 GHz. We have thoroughly analyzed the induced quasi-BICs by using the Fano fitting and electric field vectors. Results show that the quasi-BICs manifest themselves as a type of Fano resonance once the symmetry is broken. Differing from traditional resonances of WA with the same profile, the field vector direction of quasi-BICs on the strip surface is opposite, showing a phase difference of 180°. More discussion is given to the impact of substrate on the performance of quasi-

BICs. The thickness of the substrate has a great impact on the bandwidth of quasi-BICs. The narrowest bandwidth of 1.8 GHz can be realized when the substrate thickness is 50 μm and 60 μm. The optical response of the metasurfaces is also significantly altered by the substrate's refractive index. The resonant frequency shows a linear decline when n increases. The bandwidth of quasi-BICs decreases to 0.9 GHz when n is 2.2. The sharp quasi-BICs are sensitive to the changes in absorption of materials. Low-loss materials show higher Q-factors. Thus, suitable substrate material selection will benefit the achieving high Q-factor resonance. Finally, the BSA layer is used to assess DSA sensitivity. The DSAs show high sensitivity, achieving a frequency shift of 70 GHz when the thickness of BSA is 10 μm, corresponding to a sensitivity of 87.5 GHz/RIU. Thus, this metasurface with sharp quasi-BICs is expected to perform well in THz sensing.

Author Contributions: Structure design and data calculation and writing, Y.S. and D.L.; Writing, J.W., H.S., X.L., J.Y. and H.L.; funding acquisition, Y.S. All authors have read and agreed to the published version of the manuscript.

Funding: This research was funded by Guangdong Key Fields Research Programs for Universities, China (2021ZDZX1097).

Data Availability Statement: Upon reasonable request, Yanchun Shen, the corresponding author, will provide the data that support the study's conclusions.

Conflicts of Interest: The authors declare no conflict of interest.

References

1. von Neuman, J.; Wigner, E. Über merkwürdige diskrete Eigenwerte. *Physikalische Zeitschrift* **1929**, *30*, 465–467.
2. Hwang, M.S.; Lee, H.C.; Kim, K.H.; Jeong, K.Y.; Kwon, S.H.; Koshelev, K.; Kivshar, Y.; Park, H.G. Ultralow-threshold laser using super-bound states in the continuum. *Nat. Commun.* **2021**, *12*, 4135. [CrossRef] [PubMed]
3. Romano, S.; Zito, G.; Lara Yepez, S.N.; Cabrini, S.; Penzo, E.; Coppola, G.; Rendina, I.; Mocellaark, V. Tuning the exponential sensitivity of a bound-state-in-continuum optical sensor. *Opt. Express* **2019**, *27*, 18776–18786. [CrossRef] [PubMed]
4. Dong, Z.; Mahfoud, Z.; Paniagua-Dominguez, R.; Wang, H.; Fernandez-Dominguez, A.I.; Gorelik, S.; Ha, S.T.; Tjiptoharsono, F.; Kuznetsov, A.I.; Bosman, M.; et al. Nanoscale mapping of optically inaccessible bound-states-in-the-continuum. *Light Sci. Appl.* **2022**, *11*, 20. [CrossRef] [PubMed]
5. Carletti, L.; Koshelev, K.; De Angelis, C.; Kivshar, Y. Giant Nonlinear Response at the Nanoscale Driven by Bound States in the Continuum. *Phys. Rev. Lett.* **2018**, *121*, 033903. [CrossRef] [PubMed]
6. Liu, Z.; Xu, Y.; Lin, Y.; Xiang, J.; Feng, T.; Cao, Q.; Li, J.; Lan, S.; Liu, J. High-Q quasibound states in the continuum for nonlinear metasurfaces. *Phys. Rev. Lett.* **2019**, *123*, 253901. [CrossRef] [PubMed]
7. Liang, Y.; Koshelev, K.; Zhang, F.; Lin, H.; Lin, S.; Wu, J.; Jia, B.; Kivshar, Y. Bound states in the continuum in anisotropic plasmonic metasurfaces. *Nano. Lett.* **2020**, *20*, 6351–6356. [CrossRef]
8. Zhou, Y.; Guo, Z.; Zhao, X.; Wang, F.; Yu, Z.; Chen, Y.; Liu, Z.; Zhang, S.; Sun, S.; Wu, X. Dual-quasi bound states in the continuum enabled plasmonic metasurfaces. *Adv. Opt. Mater.* **2022**, *10*, 2200965. [CrossRef]
9. Tang, Y.; Liang, Y.; Yao, J.; Chen, M.K.; Lin, S.; Wang, Z.; Zhang, J.; Huang, X.G.; Yu, C.; Tsai, D.P. Chiral bound states in the continuum in plasmonic metasurfaces. *Laser Photonics Rev.* **2023**, *17*, 2200597. [CrossRef]
10. Meng, B.; Wang, J.; Zhou, C.; Huang, L. Bound states in the continuum supported by silicon oligomer metasurfaces. *Opt. Lett.* **2022**, *47*, 1549–1552. [CrossRef]
11. Romano, S.; Zito, G.; Torino, S.; Calafiore, G.; Penzo, E.; Coppola, G.; Cabrini, S.; Rendina, I.; Mocella, V. Label-free sensing of ultralow-weight molecules with all-dielectric metasurfaces supporting bound states in the continuum. *Photonics Res.* **2018**, *6*, 726–733. [CrossRef]
12. Koshelev, K.; Lepeshov, S.; Liu, M.; Bogdanov, A.; Kivshar, Y. Asymmetric metasurfaces with high-Q resonances governed by bound states in the continuum. *Phys. Rev. Lett.* **2018**, *121*, 193903. [CrossRef] [PubMed]
13. Zhao, X.; Chen, C.; Kaj, K.; Hammock, I.; Huang, Y.; Averitt, R.D.; Zhang, X. Terahertz investigation of bound states in the continuum of metallic metasurfaces. *Optica* **2020**, *7*, 1548–1554. [CrossRef]
14. Liu, D.; Yu, X.; Wu, F.; Xiao, S.; Itoigawa, F.; Ono, S. Terahertz high-Q quasi-bound states in the continuum in laser-fabricated metallic double-slit arrays. *Opt. Express* **2021**, *29*, 24779–24791. [CrossRef] [PubMed]
15. Liang, Y.; Lin, H.; Lin, S.; Wu, J.; Li, W.; Meng, F.; Yang, Y.; Huang, X.; Jia, B.; Kivshar, Y. Hybrid anisotropic plasmonic metasurfaces with multiple resonances of focused light beams. *Nano Lett.* **2021**, *21*, 8917–8923. [CrossRef] [PubMed]
16. Berghuis, A.M.; Castellanos, G.W.; Murai, S.; Pura, J.L.; Abujetas, D.R.; van Heijst, E.; Ramezani, M.; Sánchez-Gil, J.A.; Rivas, J.G. Room temperature exciton-polariton condensation in silicon metasurfaces emerging from bound states in the continuum. *Nano Lett.* **2023**, *23*, 5603–5609. [CrossRef] [PubMed]

17. Bulgakov, E.N.; Sadreev, A.F. Bloch bound states in the radiation continuum in a periodic array of dielectric rods. *Phys. Rev. A* **2014**, *90*, 053801. [CrossRef]
18. Han, S.; Cong, L.; Srivastava, Y.K.; Qiang, B.; Rybin, M.V.; Kumar, A.; Jain, R.; Lim, W.X.; Achanta, V.G.; Prabhu, S.S.; et al. All-dielectric active terahertz photonics driven by bound states in the continuum. *Adv. Mater.* **2019**, *31*, e1901921. [CrossRef]
19. Cen, W.; Lang, T.; Wang, J.; Xiao, M. High-Q Fano Terahertz resonance based on bound states in the continuum in All-dielectric metasurface. *Appl. Surf. Sci.* **2022**, *575*, 151723. [CrossRef]
20. Han, S.; Pitchappa, P.; Wang, W.; Srivastava, Y.K.; Rybin, M.V.; Singh, R. Extended bound states in the continuum with symmetry-broken terahertz dielectric metasurfaces. *Adv. Opti. Mater.* **2021**, *9*, 2002001. [CrossRef]
21. Yoshida, H.; Ogawa, Y.; Kawai, Y.; Hayashi, S.; Hayashi, A.; Otani, C.; Kato, E.; Miyamaru, F.; Kawase, K. Terahertz sensing method for protein detection using a thin metallic mesh. *Appl. Phys. Lett.* **2007**, *91*, 253901. [CrossRef]
22. Miyamaru, F.; Tanaka, M.; Hangyo, M. Effect of hole diameter on terahertz surface-wave excitation in metal-hole arrays. *Phys. Rev. B* **2006**, *74*, 153416. [CrossRef]
23. Liu, D.; Wu, F.; Yang, R.; Chen, L.; He, X.; Liu, F. Quasi-bound states in the continuum in metal complementary periodic cross-shaped resonators at terahertz frequencies. *Opt. Lett.* **2021**, *46*, 4370–4373. [CrossRef] [PubMed]
24. Wang, L.; Cao, J.; Li, X.; Zhao, Y.; Shi, H.; Fu, L.; Liu, D.; Liu, F. Quasi-BICs enabled proximity sensing based on metal complementary H-shaped arrays at terahertz frequencies. *IEEE Photonics J.* **2022**, *14*, 1–8. [CrossRef]
25. Liu, D.; Yu, X.; Wu, F.; Du, W.; Chen, L.; Liu, F.; Kuwahara, M.; Ono, S. Terahertz asymmetric metallic hole arrays with polarization-independent quasi-bound states in the continuum for membrane sensing. *Opt. Express* **2023**, *31*, 23608–23620. [CrossRef] [PubMed]
26. Srivastava, Y.K.; Ako, R.T.; Gupta, M.; Bhaskaran, M.; Sriram, S.; Singh, R. Terahertz sensing of 7 nm dielectric film with bound states in the continuum metasurfaces. *Appl. Phys. Lett.* **2019**, *115*, 151105. [CrossRef]
27. Abujetas, D.R.; van Hoof, N.; ter Huurne, S.; Gómez Rivas, J.; Sánchez-Gil, J.A. Spectral and temporal evidence of robust photonic bound states in the continuum on terahertz metasurfaces. *Optica* **2019**, *6*, 996–1001. [CrossRef]
28. Beruete, M.; Jáuregui-López, I. Terahertz sensing based on metasurfaces. *Adv. Opt. Mater.* **2019**, *8*, 1900721. [CrossRef]
29. Tuchin, V.V.; Fedulova, E.V.; Genina, E.A.; Nazarov, M.M.; Angeluts, A.A.; Meglinski, I.V.; Kitai, M.S.; Sokolov, V.I.; Shkurinov, A.P. Studying of dielectric properties of polymers in the terahertz frequency range. In Proceedings of the Saratov Fall Meeting 2011: Optical Technologies in Biophysics and Medicine XIII, Saratov, Russia, 27–30 September 2011.
30. Srivastava, Y.K.; Manjappa, M.; Cong, L.; Cao, W.; Al-Naib, I.; Zhang, W.; Singh, R. Ultrahigh-QFano Resonances in Terahertz Metasurfaces: Strong influence of metallic conductivity at extremely low asymmetry. *Adv. Opt. Mater.* **2016**, *4*, 457–463. [CrossRef]
31. Qin, M.; Duan, J.; Xiao, S.; Liu, W.; Yu, T.; Wang, T.; Liao, Q. Strong coupling between excitons and quasibound states in the continuum in bulk transition metal dichalcogenides. *Phys. Rev. B* **2023**, *107*, 045417. [CrossRef]
32. Liu, D.; Hattori, T. Sharp resonances in terahertz free-standing three-dimensional metallic woven meshes. *Opt. Express* **2020**, *28*, 30174–30186. [CrossRef] [PubMed]
33. Sarrazin, M.; Vigneron, J.-P.; Vigoureux, J.-M. Role of Wood anomalies in optical properties of thin metallic films with a bidimensional array of subwavelength holes. *Phys. Rev. B* **2003**, *67*, 085415. [CrossRef]
34. Wu, F.; Qi, X.; Luo, M.; Liu, T.; Xiao, S. Ultraslow light effect driven by quasibound states in the continuum in compound grating waveguide structures. *Phys. Rev. B* **2023**, *108*, 165404. [CrossRef]
35. Hsu, C.W.; Zhen, B.; Stone, A.D.; Joannopoulos, J.D.; Soljačić, M. Bound states in the continuum. *Nat. Rev. Mater.* **2016**, *1*, 16048. [CrossRef]
36. Yuan, L.; Lu, Y.Y. Bound states in the continuum on periodic structures: Perturbation theory and robustness. *Opt. Lett* **2017**, *42*, 4490–4493. [CrossRef] [PubMed]
37. Bogdanov, A.A.; Koshelev, K.L.; Kapitanova, P.V.; Rybin, M.V.; Gladyshev, S.A.; Sadrieva, Z.F.; Samusev, K.B.; Kivshar, Y.S.; Limonov, M.F. Bound states in the continuum and Fano resonances in the strong mode coupling regime. *Adv. Photonics* **2019**, *1*, 016001. [CrossRef]
38. Kupriianov, A.S.; Xu, Y.; Sayanskiy, A.; Dmitriev, V.; Kivshar, Y.S.; Tuz, V.R. Metasurface engineering through bound states in the continuum. *Phys. Rev. Appl.* **2019**, *12*, 014024. [CrossRef]
39. Huang, L.; Jin, R.; Zhou, C.; Li, G.; Xu, L.; Overvig, A.; Deng, F.; Chen, X.; Lu, W.; Alu, A.; et al. Ultrahigh-Q guided mode resonances in an all-dielectric metasurface. *Nat. Commun.* **2023**, *14*, 3433. [CrossRef]
40. Li, Z.; Zhang, X. Enhanced optical squeezing from quasi-bound states in the continuum and Fano resonances without nonlinearity. *New J. Phys.* **2019**, *21*, 123050. [CrossRef]
41. Lee, S.G.; Kim, S.H.; Kee, C.S. Creation of Fano resonances and bound states in the continuum in metallic metasurface superlattices. *Opt. Express* **2021**, *29*, 21492–21501. [CrossRef]
42. Mousavi, S.H.; Khanikaev, A.B.; Shvets, G. Optical properties of Fano-resonant metallic metasurfaces on a substrate. *Phys. Rev. B* **2012**, *85*, 155429. [CrossRef]
43. Sadrieva, Z.F.; Sinev, I.S.; Koshelev, K.L.; Samusev, A.; Iorsh, I.V.; Takayama, O.; Malureanu, R.; Bogdanov, A.A.; Lavrinenko, A.V. Transition from optical bound states in the continuum to leaky resonances: Role of substrate and roughness. *ACS Photonics* **2017**, *4*, 723–727. [CrossRef]
44. Kim, K.H.; Kim, I.P. Quasi-bound states in the continuum with high Q-factors in metasurfaces of lower-index dielectrics supported by metallic substrates. *RSC Adv.* **2022**, *12*, 1961–1967. [CrossRef] [PubMed]

45. Chorsi, H.T.; Lee, Y.; Alu, A.; Zhang, J.X.J. Tunable plasmonic substrates with ultrahigh Q-factor resonances. *Sci. Rep.* **2017**, *7*, 15985. [CrossRef] [PubMed]
46. Luchansky, M.S.; Bailey, R.C. High-Q optical sensors for chemical and biological analysis. *Anal. Chem.* **2012**, *84*, 793–821. [CrossRef] [PubMed]
47. Fan, F.; Chen, S.; Wang, X.-H.; Wu, P.-F.; Chang, S.-J. Terahertz Refractive Index Sensing Based on Photonic Column Array. *IEEE Photonics Technol. Lett.* **2015**, *27*, 478–481. [CrossRef]
48. Shi, C.; Hu, J.; Liu, X.; Liang, J.; Zhao, J.; Han, H.; Zhu, Q. Double-layer symmetric gratings with bound states in the continuum for dual-band high-Q optical sensing. *Beilstein J. Nanotechnol.* **2022**, *13*, 1408–1417. [CrossRef]
49. Yang, C.; Sang, T.; Li, S.; Wang, Y.; Cao, G.; Hu, L. Tailoring the light absorption of monolayer graphene via accidental quasi-bound states in the continuum. *J. Opt. Soc. Am. B* **2022**, *39*, 2531–2539. [CrossRef]
50. Zhang, X.; Sang, T.; Pian, C.; Wang, Y.; Wang, J. Dual-band polarization-independent high quality factor Fano resonances using a twisted tetrameric nanohole slab. *J. Opt. Soc. Am. B* **2023**, *40*, 1652–1660. [CrossRef]
51. Sun, Y.; Du, P.; Lu, X.; Xie, P.; Qian, Z.; Fan, S.; Zhu, Z. Quantitative characterization of bovine serum albumin thin-films using terahertz spectroscopy and machine learning methods. *Biomed. Opt. Express* **2018**, *9*, 2917–2929. [CrossRef]
52. Yoneyama, H.; Yamashita, M.; Kasai, S.; Kawase, K.; Ueno, R.; Ito, H.; Ouchi, T. Terahertz spectroscopy of native-conformation and thermally denatured bovine serum albumin (BSA). *Phys. Med. Biol.* **2008**, *53*, 3543–3549. [CrossRef] [PubMed]
53. Li, X.; Fu, X.; Liu, J.; Du, Y.; Hong, Z. Investigation of thermal denaturation of solid bovine serum albumin by terahertz dielectric spectroscopy. *J. Mol. Struct.* **2013**, *1049*, 441–445. [CrossRef]
54. Feng, G.; Chen, Z.; Wang, Y.; Liu, X.; Liu, Y.; Liu, X.; Sun, F.; Yang, Y.; Chen, S. Enhanced Fano resonance for high-sensitivity sensing based on bound states in the continuum. *Chin. Opt. Lett.* **2023**, *21*, 031202. [CrossRef]
55. Zhou, R.; Wang, C.; Huang, Y.; Huang, K.; Wang, Y.; Xu, W.; Xie, L.; Ying, Y. Label-free terahertz microfluidic biosensor for sensitive DNA detection using graphene-metasurface hybrid structures. *Biosens. Bioelectron.* **2021**, *188*, 113336. [CrossRef]
56. Xu, J.; Liao, D.; Gupta, M.; Zhu, Y.; Zhuang, S.; Singh, R.; Chen, L. Terahertz microfluidic sensing with dual-torus toroidal metasurfaces. *Adv. Opt. Mater.* **2021**, *9*, 2100024. [CrossRef]

micromachines

Article

Progress of Edge-Emitting Diode Lasers Based on Coupled-Waveguide Concept

Lili Han [1], Zhaowei Wang [1,*], Nikita Yu. Gordeev [2,*], Mikhail V. Maximov [3], Xiansheng Tang [1], Artem A. Beckman [2], Grigoriy O. Kornyshov [3], Alexey S. Payusov [2], Yuri M. Shernyakov [2], Alexey E. Zhukov [4], Kuilong Li [1], Ruizhan Zhai [1], Zhongqing Jia [1], He Yang [5] and Wei Zhang [1]

1 Laser Institute, Qilu University of Technology (Shandong Academy of Sciences), Jinan 250104, China
2 Ioffe Institute, St. Petersburg 194021, Russia
3 Nanophotonics Laboratory, Alferov University, St. Petersburg 194021, Russia
4 International Laboratory of Quantum Optoelectronics, National Research University Higher School of Economics, St. Petersburg 190121, Russia
5 School of Instrumentation and Optoelectronic Engineering, Beihang University, Beijing 100191, China
* Correspondence: zw.wang@qlu.edu.cn (Z.W.); gordeev@switch.ioffe.ru (N.Y.G.)

Abstract: Semiconductor lasers have developed rapidly with the steady growth of the global laser market. The use of semiconductor laser diodes is currently considered to be the most advanced option for achieving the optimal combination of efficiency, energy consumption, and cost parameters of high-power solid-state and fiber lasers. In this work, an approach for optical mode engineering in planar waveguides is investigated. The approach referred to as Coupled Large Optical Cavity (CLOC) is based on the resonant optical coupling between waveguides and allows the selection of high-order modes. The state-of-art of the CLOC operation is reviewed and discussed. We apply the CLOC concept in our waveguide design strategy. The results in both numerical simulation and experiment show that the CLOC approach can be considered a simple and cost-efficient solution for improving diode laser performance.

Keywords: diode lasers; CLOC; waveguide; high power

Citation: Han, L.; Wang, Z.; Gordeev, N.Y.; Maximov, M.V.; Tang, X.; Beckman, A.A.; Kornyshov, G.O.; Payusov, A.S.; Shernyakov, Y.M.; Zhukov, A.E.; et al. Progress of Edge-Emitting Diode Lasers Based on Coupled-Waveguide Concept. *Micromachines* **2023**, *14*, 1271. https://doi.org/10.3390/mi14061271

Academic Editors: Niall Tait and Jinn-Kong Sheu

Received: 28 March 2023
Revised: 30 May 2023
Accepted: 16 June 2023
Published: 20 June 2023

1. Introduction

The global laser market steadily grows. A significant part of the market is semiconductor lasers, most of which are represented by high-power edge-emitting lasers due to their compactness, reliability, high efficiency, and high level of technology development. Currently, the development of solid-state laser technology is largely determined by progress in improving the performance of pumping laser diodes. Even in high-power industrial laser systems, more reliable and compact diode-pumped lasers are intensively replacing lamp-pumped lasers. High-power diode lasers that appeared in the late 80s allowed the creation of compact diode-pumped units with output power as high as 10 kW. The high quantum efficiency and narrow spectral line allow selective pumping into the absorption line of the active element with optical efficiency exceeding 50%. Thus, the use of semiconductor laser diodes is presently considered the most advanced option for achieving an optimal combination of efficiency, energy consumption, and cost parameters for high-power solid-state and fiber lasers.

Cost is one of the main factors affecting the use of high-power diode lasers. One watt of the laser output optical power (dollars-per-watt, $/W) in the typical 9xx nm waveband cost more than US$100 20 years ago and now approaches US$1 [1]. A prime cost is fixed by the manufacturing technologies of both the laser chips themselves and the laser modules. The maximum continuous wave (CW) optical power delivered by commercially available single lasers lies in the range of 10–20 W for a 100 μm light emission aperture, depending on the emitting wavelength. One method to increase the power of diode laser modules is to combine several lower-power oscillators to create a higher-power beam. While that strategy does increase the overall power, it does not reduce $/W. A natural way to reduce $/W and increase the market attractiveness of high-power diode lasers is to enhance the maximal output optical power of a laser diode while keeping the laser cost unchanged. In principle, it can be performed by waveguide engineering, which is an effective approach for simultaneously improving optical, electrical, and thermal characteristics. Particularly it makes it possible to increase the output power by enlarging the laser aperture.

In this paper, we review our approach for designing transverse waveguides of broad-area lasers, which are referred to as coupled large optical cavities (CLOC). We briefly discuss existing trends in developing laser waveguides, present our general waveguide design strategy, describe the physical principles of the CLOC operation, show the implementation of the concept, and finally discuss some aspects related to the CLOC application for other types of diode lasers.

2. Trends in Designing Edge-Emitting Laser Waveguides

There are thermal and non-thermal limitations on the optical power of diode lasers. The diode laser parameters affect the output performance in a rather complicated manner. Fundamentally, the output optical power of an edge-emitting laser depends on the electrical-to-optical power conversion efficiency (PCE), which determines how much of the electrical power is wasted on heat. The main Joule heat comes from the active region and is dissipated into a heatsink. Semiconductor layers located between the active region and the laser surface have higher thermal resistance, which prevents effective heat extraction and can cause significant overheating. The self-heating decreases the laser PCE and the output optical power, therefore. Other heat-induced effects are wavelength redshift, thermal lensing, and filamentation. The latter two processes significantly worsen the laser beam quality and can trigger catastrophic optical mirror damage (COMD). The temperature-induced red-shift in GaAs-based lasers has a rate of 0.3~04 nm/K (it depends in particular on the central wavelength). When lasers are used as pumping sources, a temperature-induced wavelength mismatch between laser and pumped media can dramatically decrease the pumping efficiency. High series resistance contributes to self-heating and, at the same time, decreases the laser PCE due to the parasitic voltage drop. The largest portion of up to 40% of the total diode laser electric resistance is considered to come from p-claddings [2]. The nonradiative recombination also contributes to the unwanted power loss mainly associated with heating. The internal efficiency reflects how effectively the injected current is converted into photons. Internal efficiency depends on the laser parameters in a complicated manner. The most influential ones are the crystal quality of the laser wafer, active region band structure, and optical waveguide design. Improperly designed active regions and waveguides can cause parasitic recombination [3]. The internal optical loss is associated mainly with the free-carrier absorption and increases, therefore, when the optical mode significantly overlaps with highly doped layers, especially p-doped one's [2]. A current-induced accumulation of carriers in the initially undoped waveguide layers can also significantly increase internal optical loss. High internal optical loss increases the threshold current, contributes to the self-heating and therefore significantly reduces the total output optical power.

In developing high-power diode lasers, one should remember that besides the output optical power, the other critical parameters are the beam divergence and quality, spectral width, reliability, and manufacturing cost.

In summary, the strategy of increasing the maximal output optical power of diode lasers should be focused on the following.

- Improvement of the heat dissipation from the active region of diode lasers due to the use of heatsinks with high thermal conductivity and reducing the thermal resistance by means of reducing the distance from the active region to the heatsink surface.
- Reducing the electric series resistance by using thinner p-claddings and undoped layers.

- Reducing the laser internal optical loss mainly by decreasing overlap of the optical mode with highly doped layers.
- Protection of laser mirrors from COMD by applying dielectric coatings, passivation and reducing the optical power density.
- Increase the focusing efficiency of the laser beam by reducing its divergence.
- Effective suppressing of the high-order lateral modes that may arise due to diode overheating and deteriorated lateral beam quality.

In 2004, a team from Ioffe Institute (St. Petersburg, Russia) [4] used a broadened optical waveguide in 1.06 μm high-power lasers, which allowed effective confinement of the laser mode within the undoped layers and, thus, reduced the optical loss down to 0.34 cm^{-1}. A quantum well active region was shifted from the waveguide center. Due to this shift, the second-order vertical optical mode has a reduced optical confinement factor and does not contribute to the lasing. CW optical power is as large as 16 W was obtained from 100 μm-wide stripes. The broadened waveguide reduced the intensity at the facet and avoided the COMD damage. The disadvantage of this approach is that the design is limited in terms of the further waveguide broadening and is too sensitive to current and temperature changes. However, the general idea of broadening the optical waveguide has evolved into the Super Large Optical Cavity (SLOC) [5] design intensively developed by Ferdinand-Braun-Institute FBH (Berlin, Germany). The approach allows increasing the laser waveguide thickness up to 10 μm providing the vertical beam divergence as narrow as 15 deg. This waveguide was originally a multimode waveguide. To suppress high-order modes, the laser claddings have reduced thickness as thin as 0.4 μm. As a result, high-order modes couple to the highly-doped contact layer and the substrate and experience extra optical losses. The fundamental mode has a low optical loss and allows making the cavity length as long as 8 mm to provide high optical power. The fundamental mode profile has a Gaussian shape which positively affects the laser beam quality. The approach still has some disadvantages. Due to the thermal and current-induced changes of the waveguide and claddings refractive indices, in some cases, high-order modes may contribute to the lasing. In 2012 the FBH team showed the maximal CW optical power of 25 W presented in [5]. The lasers have shown moderate operation time of fewer than 5000 h. In [5], they have also discussed limitations to peak power. At high current densities, band-bending in the p-side waveguide leads to a very large concentration of carriers accumulated in this region. This effect results in power saturation caused by the current-induced optical loss, which can be considered a major drawback of SLOC lasers. The fundamental issues of this phenomenon were theoretically studied in [3]. The authors proposed a strongly asymmetric laser waveguide allowing shifting of the active region toward the p-cladding and, thus, dramatically reducing the optical loss [6]. Since that time, a trend for using asymmetric, not very broad waveguides has been intensively developed.

The FBH team has also developed a similar concept which has evolved into the structures termed by the authors as Extreme Double Asymmetric (EDAS) [7]. The name reflects that the structure design possesses a highly asymmetric cladding layer composition and waveguide thickness. The thin p-side waveguide provides low series resistance, low internal optical losses, and low bias-driven leakage currents. Low optical losses are attributed to the small overlap of the fundamental vertical mode with the p-waveguide [8]. The team has reported CW optical power exceeding 15 W. Further improvement of the concept has resulted in the extreme-triple-asymmetric (ETAS) laser design [9,10]. A third asymmetry in the graded profile of the refractive index for the layers on either side of the quantum well allows fine-tuning of the optical field. The ETAS lasers having 100-μm wide aperture emitting at 940 nm have shown up to 63% power conversion efficiency at 14 W CW optical output power [9]. It is important to note that maximal output optical power can be increased by using a larger laser-emitting aperture [1]. A team from Fujikura Ltd. (Chiba, Japan) has reported 30 W optical power in lasers with ~200 μm aperture [2]. Such a large aperture is possible due to the effective heat dissipation resulting from using an Asymmetric Decoupled Confinement Heterostructure (ADCH), another variation of asymmetric waveguides.

The world leaders in manufacturing high-power diode lasers are Lumentum (Milpitas, CA, USA), known as JDSU before 2015, NLIGHT (Vancouver, WA, USA), and IPG (Oxford, MI, USA). Driven by the increasing power of fiber lasers, diode lasers with both high power and high brightness are being intensely developed to achieve high performance and reduced manufacturing costs. Lumentum has reported 27 W CW optical power from 100 μm aperture 4 mm long lasers emitting in 910–980 nm wavelength range [11]. The company associates high output power with improved PCE. Better heat dissipation reduces thermal lensing and decreases the blooming of the far-field divergence at high pumping currents. In 2017, IPG showed CW optical power as high as 30 W [12] from 976 nm lasers having a 5-mm cavity length. The company claims that reduced ohmic and thermal resistances, combined with improved linearity of light-current characteristics, have improved the PCE with the current. NLIGHT has introduced a new design of broad-area lasers and shown 25 W diodes for coupling into 105 μm fiber [13]. These companies have revealed no detailed information on the laser designs they used. However, in their publications, they reported to have used modified laser wafer designs and referred to the papers describing asymmetric-waveguide approaches, which is indirect evidence of using this concept.

The main disadvantages of the strongly asymmetric waveguide relate to a very low refractive index cladding-waveguide contrast at the n-side. It is made intentionally to provide high-order mode selection, namely for promoting the fundamental mode lasing. However, due to this low contrast, the fundamental mode has a large exponential decay on the n-side and its profile can be extremely sensitive to the pumping currents and temperature rise caused by the laser self-heating. Transverse mode profile instability can result in blooming the vertical divergence and in worsening the beam quality. One more drawback is that the epitaxial growth of low Al material is rather challenging because of the difficulty of precisely controlling so small Al mole fractions (a few percent). Additionally, this approach is hardly applicable to Al-free materials. On the opposite p-side, the waveguide contrast is rather high, which should prevent the lasing mode from penetrating into the highly doped p-cladding. High contrast is usually provided by higher Al fraction in AlGaAs claddings, which in turn worsens the structure's thermal resistance. The high Al fraction may cause difficulties in effective p-doping. No information is available about the application of this method to 808 nm laser structures. In summary, despite the fact that asymmetric waveguides make it possible to obtain high optical power, the approach has significant drawbacks and can hardly be adapted for lasers of a different spectral range.

3. Single and Multimode Waveguides

Generally, a laser waveguide consists of a high refractive index core surrounded by two low refractive index claddings. Thicknesses and refractive indices of the core and claddings define the guided optical modes and their effective refractive indices (Figure 1a) [14] p. 50: $N \equiv \beta/k$, where β is the propagation constants, and k is the wavenumber in a vacuum. Figure 1b–d shows a typical evolution of the fundamental mode parameters against the waveguide thickness.

When the waveguide is relatively narrow, the fundamental mode has large exponential decay in the claddings resulting in the non-Gaussian intensity profile accompanied by large mode overlap with lossy, highly doped claddings [15]. With increasing the thickness, the fundamental mode becomes more localized within the waveguide core, so the mode-cladding overlap substantially reduces (Figure 1c). Starting from a certain waveguide thickness, the mode size is almost equal to it, follows this value (Figure 1d), and does not depend on the waveguide index steps. Thereby temperature-induced and current-induced refractive index variations would only slightly affect the fundamental mode size and, correspondingly, its far-field pattern behavior. In the simulated waveguide, its thickness range favorable for the fundamental mode stability lies above approximately 1.5 μm. Having looked at the dispersion curves in Figure 1a, one can see that the waveguide of this thickness guides several optical modes. This consideration is relevant to asymmetric multimode waveguides as well. Our waveguide design strategy is built upon ensuring fundamental mode lasing by eliminating high-order modes in naturally multimode waveguides [15]. We should note that "multimode waveguide" does not necessarily mean a broadened one. It can be rather narrow.

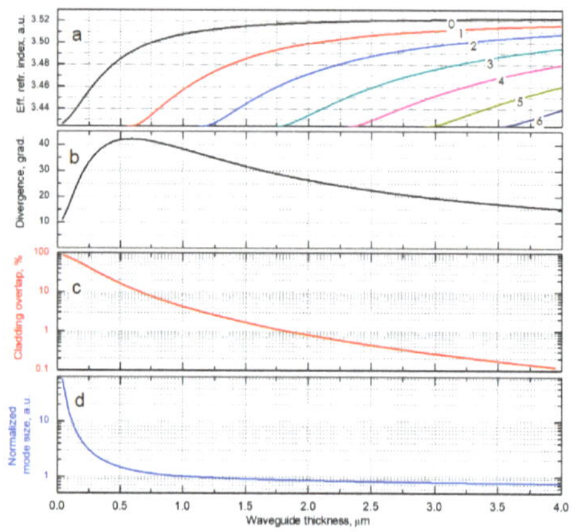

Figure 1. Dependence of the effective refractive indices (**a**), the fundamental mode beam divergence FWHM (**b**), mode-claddings overlap (**c**), and normalized mode size at the 1/e2 level (**d**) on the waveguide thickness. The modeled waveguide is symmetric with GaAs core and $Al_{0.15}Ga_{0.85}As$ claddings. The wavelength is 980 nm.

4. CLOC Operation

Our waveguide design strategy has been realized by using the CLOC concept. It is called after the physical effect that it exploits, namely, optical coupling. Let us consider two identical codirectional optical single-mode waveguides, A and B (Figure 2). If the distance between the waveguides is rather large, then each operates independently. Their guided modes are identical (Figure 2a) and characterized by effective refractive indices N_A and N_B. The optical coupling takes place if the distance between the waveguides decreases (Figure 2b) because the electromagnetic field of one waveguide is perturbed by another waveguide. As a result, in this waveguide system, new optical modes, often referred to as composite, are formed. In the considered identical A and B waveguides, these odd and even modes have different field profiles and, consequently, different effective refractive indices but almost identical intensity distributions. It is important to note that each of the modes is equally distributed between the waveguides. It means that if, for instance, a solitary waveguide A has a guided mode with intensity I_0, then in the coupled waveguide A, this mode transforms and has two times lower intensity $I_0/2$. Coupling between the waveguides is substantial if the optical modes in the solitary waveguides have equal or nearly equal effective refractive indices $N_A = N_B$ [16].

Since these indices are the mode parameters, we can construct a system containing a multi-mode waveguide WG_A optically coupled with a single-mode one WG_B in such a way that the index equality is fulfilled for a high-order eigenmode of the WG_A and the only eigenmode of the WG_B (Figure 3). As a result, two composite modes are formed, and each of them has half the intensity of the high-order eigenmode. This idea is employed in our CLOC laser concept so that the multi-mode waveguide contains an active region, and a coupled single-mode waveguide is passive and doped to pass electric current. A selected high-order mode of the active waveguide tunnels into the passive one forms two composite modes. The CLOC lasers make use of two mode-suppression mechanisms. The first one is the reduced mode intensity and gamma factor in the active waveguide; the second one is the high optical loss originating from the mode overlap with the doped passive waveguide core.

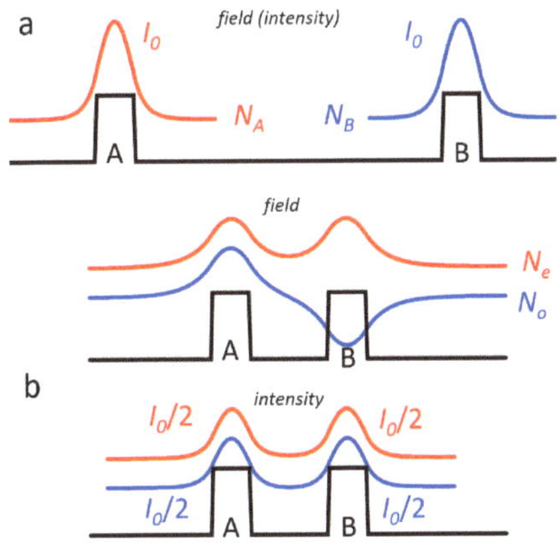

Figure 2. Mode profiles in identical codirectional waveguides: (**a**) two solitary waveguides; (**b**) two coupled waveguides.

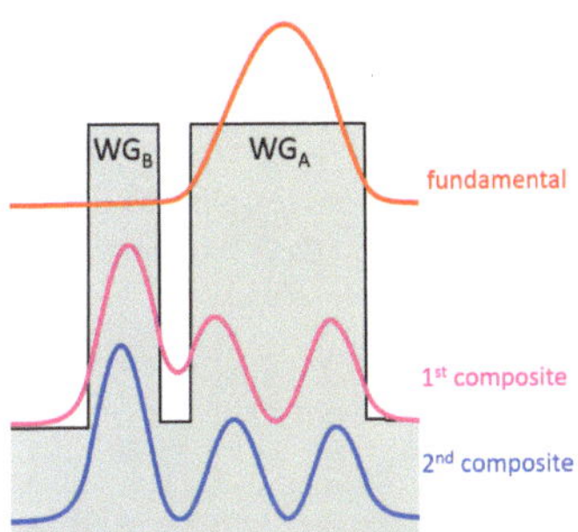

Figure 3. Intensity profiles of the fundamental and two composite modes in two coupled waveguides having different thicknesses.

Initially, the CLOC concept has been proposed for broadening transverse laser waveguide via suppressing parasitic high-order mode. In our proof-of-concept experiment [17], we have compared CLOC lasers and reference ones. Both InGaAs ($\lambda \approx 1.04$ µm) quantum well lasers had identical 2.5 µm-thick undoped GaAs waveguides (Figure 4). The only difference is that the CLOC structure had an extra n-GaAs passive waveguide separated from the active waveguide with an n-AlGaAs layer. Fundamental modes in both structures have identical profiles. The major difference between the two lasers is the absence of a second-order mode in the CLOC structure. Both CLOC and reference lasers showed similar parameters (threshold current, efficiency, spectra). However, over the entire pumping current range, the CLOC lasers have shown stable single-mode emission (Figure 4b), while the reference lasers emitted on high-order modes (Figure 4a).

The CLOC structures can have a more advanced design, "1 + 2", where two different passive single-mode waveguides are placed by either side of the active waveguide (Figure 5a). In this case, each parasitic high-order mode tunnels into the corresponding passive waveguide. We used this approach in designing a waveguide as broad as 4.8 µm [18]. The partially doped $Al_{0.1}Ga_{0.9}As$ active waveguide containing an InGaAs quantum well active region was optically coupled with two GaAs passive waveguides with the thicknesses of 310 nm and 375 nm, which allowed for suppressing the first and second transverse modes. Measurements of the transverse far-field patterns in the CW regime have shown that the single mode remains stable throughout the entire pumping current range (Figure 5b). The obtained divergence is 13 ± 1 deg. FWHM was very close to the diffraction limit for the 4.8 μm emitting aperture. The stability of far-field patterns was measured in the temperature range from 20 to 90 °C in the pulsed mode. No pronounced changes in the beam divergence were observed. These measurements confirm that in multimode-broadened CLOC waveguides, temperature- and current-induced refractive index variations weakly alter the fundamental mode size and, correspondingly, its divergence [19].

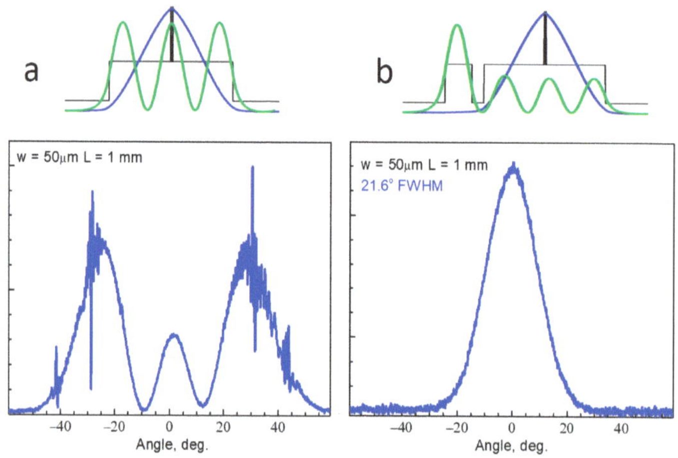

Figure 4. Mode intensity profiles and transverse far-field patterns of the high-order mode and fundamental mode for the reference (**a**) and CLOC (**b**) lasers, respectively.

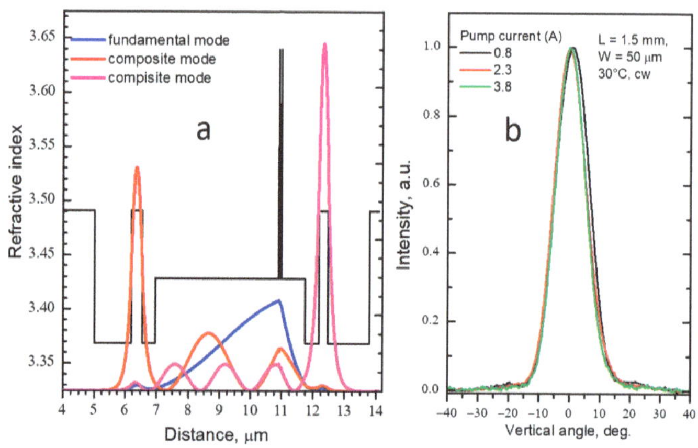

Figure 5. The "1 + 2" CLOC laser: (**a**)—refractive index profile and simulated intensity distributions of the fundamental mode; (**b**)—far-field patterns measured in CW regime at different currents.

For high-power lasers [20–22], broadened waveguides have an advantage since they reduce optical power density on the laser facets and, thus, may prevent COMD. However, the COMD threshold also can be successfully increased by using coating and passivation for the facets [23]. The narrower vertical divergence associated with the broadened waveguide is not crucial for edge-emitting lasers. We should note that using broadened waveguides in high-power lasers has some weaknesses. A larger p-side waveguide core increases optical loss and series electrical resistance. These problems can be tackled by shifting the active region toward the p-cladding. However, this approach results in the natural reduction of the optical confinement factor and sometimes promotes high-order mode lasing. The former would require using a multilayer active region.

In connection with the above, the CLOC concept can be further updated for high-power lasers. The basic idea is that the laser waveguide is multimode and rather narrow (1–1.5 μm), and the active region is extremely shifted toward the p-cladding. Parasitic high-order modes are eliminated by means of the CLOC. This approach has some benefits:

- The multimode waveguide ensures strong fundamental mode localization within the core, which reduces the optical losses in the claddings;
- For the same reason, a thinner p-cladding can be used, which decreases thermal and electrical resistances;
- Both rather narrow waveguides and shifted active regions reduce optical loss in the p-side of the waveguide core and result in lower thermal and electrical resistances.

We have experimentally realized this concept in CLOC lasers possessing the GaAs waveguide with a thickness of 1.35 μm [24]. The wafer was grown by MOCVD on an n + -GaAs substrate. Besides the active waveguide, it contains (Figure 6) a passive one with a thickness of 0.55 μm introduced to suppress first-order mode. The active region based on two InGaAs QWs (wavelength ~1 μm) is shifted by 380 nm from the active waveguide center toward the p-$Al_{0.25}Ga_{0.75}As$ cladding. The latter has a reduced thickness of 0.5 μm. In this regard, it should be noted that a thinner p-cladding layer significantly improves the electrical resistance of the laser heterostructure. For example, considering an AlGaAs cladding layer with a hole concentration of about 5×10^{17} cm^{-3} and corresponding hole mobility of about 90 $cm^2/(V \times s)$ [25], a decrease of the layer thickness from 2 to 0.5 μm results in a decrease of the resistance from 2.8×10^{-5} $\Omega \times cm^2$ (while the total electrical resistance of the whole structure is typically 10^{-4} $\Omega \times$ cm or slightly less) to 0.7×10^{-5} $\Omega \times cm^2$.

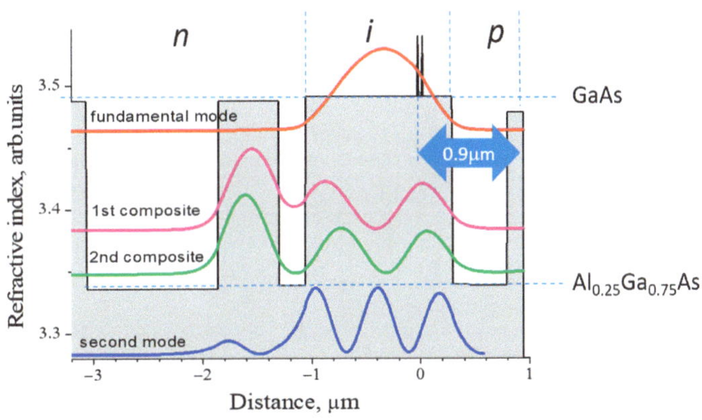

Figure 6. Refractive index and optical mode profiles of the CLOC laser with shifted active region and thin p-cladding.

The active region is located near the minimum of the second-order mode, which prevents its lasing. The laser wafer is completed by a p-GaAs contact layer (0.15 μm), so the active region is situated at a depth of only 0.92 μm from the wafer surface. The design allowed us to achieve an internal optical loss as low as 0.4 cm^{-1}, and shallow active region location has resulted in low thermal resistance of 6.0 (K/W) × mm. The lasers have shown a divergence of 34 deg. Almost unchangeable in the temperature range of 20–80 °C and the cw output power exceeding 12 W [19].

It is worth emphasizing that the feasibility of the CLOC concept is significantly supported by the high optical gain achievable by the active laser region. This is because shifting the mode position towards the p-type doped cladding layer, in combination with the extended mode profile, strongly decreases the optical confinement factor so that higher material gain is required to balance the total optical loss. An excellent candidate for exploitation in the CLOC lasers is so-called quantum well-dots, which represent very dense arrays of InGaAs quantum islands formed by deposition on slightly misoriented GaAs substrates [26]. The maximal material gain was shown to exceed 10^4 cm^{-1}. As an example, Figure 7 demonstrates the modal gain G as a function of the injection current density J for the fivefold stacked array of InGaAs quantum well-dots. The gain data were evaluated as the sum of the output loss, and the internal loss measured for the stripe lasers with cavity lengths varied from 100 to 2000 μm. The experimental data can be fitted well with the following equation: $G = G_{sat}[1 - \exp(-\chi(J - J_0)/J_0)]$, where G_{sat} stands for the saturated gain, J_0-transparency current density, χ-non-linearity parameter. The saturated gain, i.e., the maximal achievable, is about 190 cm^{-1}, which corresponds to approximately 40 cm^{-1} per single InGaAs plane in the active laser region.

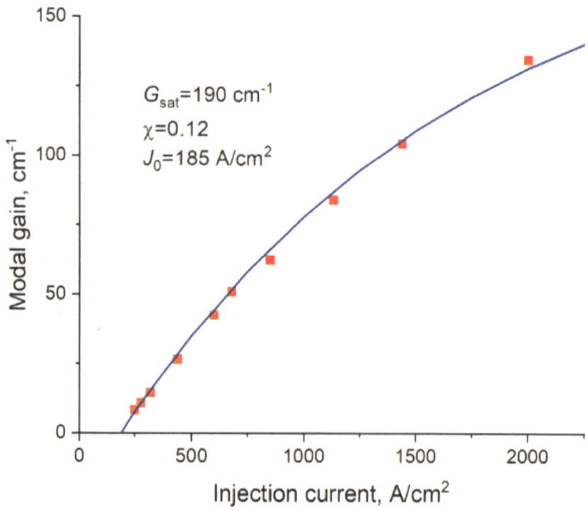

Figure 7. Modal gain against injection current for 5-times stacked InGaAs quantum well-dots: symbols-experiment, line-fit with an exponential-based function.

5. Expansion of CLOC Approach

In this section, we discuss some aspects concerning expanding the CLOC concept beyond transverse waveguides in broad-area edge-emitting lasers, namely:

- narrow-ridge lasers;
- edge-emitting lasers based on lateral CLOC waveguide;
- microdisk lasers.

Narrow-ridge lasers are usually designed to provide single-mode emission both in transverse and lateral directions. Typical lateral index-guided waveguides are formed by etching through the p-cladding. To reduce the optical loss and improve the laser reliability, the etching is stopped at a certain distance from the waveguide core. The residual cladding thickness provides an effective refractive index step required for the lateral mode guiding. The larger the optical mode penetration into the cladding, the higher the lateral effective refractive index step. The fundamental transverse mode is better localized within the core in comparison with the high-order modes, so the latter ones may have better lateral guiding (Figure 8), which promotes their lasing. This general consideration is relevant to the CLOC waveguides [27]. We have processed a CLOC wafer into broad-area lasers with 100 μm stripes and into 4 μm wide ridge lasers. The former ones showed stable emission on the fundamental transverse mode, while the latter evidently operated on the high-order modes (Figure 9). So, multimode or broadened transverse waveguides seem to be not so advantageous for use in spatial single-mode lasers. Designing such devices requires advanced 2D or even 3D simulations and should allow for a complete set of the optical modes, thermal and current-induced effects and the device imperfections caused by the epitaxial growth and the following processing.

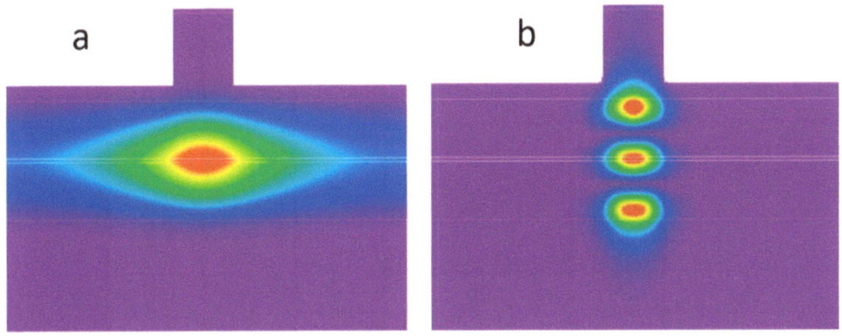

Figure 8. Calculated intensity contours for fundamental (**a**) and second-order (**b**) transverse modes in the $Al_{0.15}Ga_{0.85}As/GaAs$ waveguide (ridge width 4 μm, waveguide thickness 1.7 μm, residual cladding thickness 50 nm). The active region locates in the center of the waveguide. Lateral and transverse sizes are in different scales.

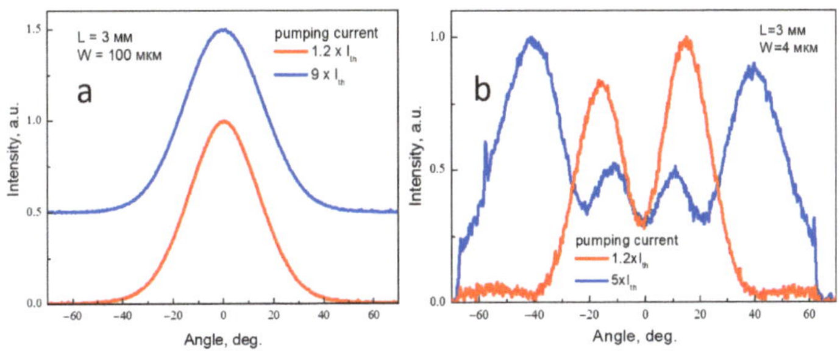

Figure 9. Transverse fundamental mode emission (**a**) of the broad-area laser and high-order mode emission (**b**) of the narrow-stripe laser processed from the same CLOC wafer.

Up to now, we have discussed the CLOC concept being used for transverse waveguides where differences in the effective refractive indices for neighbor modes are defined by the waveguide contrast and can be as high as 0.05. This value is much higher than temperature- or current-induced refractive index variations. As a result, the CLOC ensures stable fundamental transverse mode operation. The same waveguide coupling formalism known as coupled mode theory [14] can be applied to transverse and lateral waveguides. Among the latter ones, directional couplers [14] and laterally coupled diode lasers are known [28]. As often as not for these devices, identical or nearly identical waveguides are employed. However, as we mentioned for the transverse CLOC basics, an effective coupling can be obtained for different waveguides if their modes have close effective refractive indices. In this sense, the CLOC concept has a potentiality for lateral mode engineering, for instance, for increasing ridge width in single-mode lasers by means of suppressing high-order lateral modes. Broader ridges would allow for obtaining higher output optical power. We have numerically treated the lateral CLOC design [29,30]. The lateral waveguide (Figure 10) consisted of a 10 μm active ridge and an optically coupled 4 μm passive single-mode ridge. The former guides two optical modes, but the first-order one resonantly tunnels into the passive waveguide and two composite modes are formed while the fundamental mode remains unchanged. As a result, this lateral CLOC waveguide system should support spatial single-mode lasing.

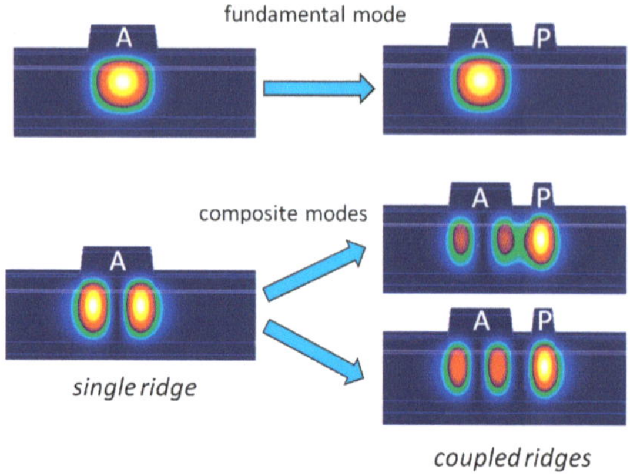

Figure 10. Simulated fundamental mode and composite modes in the two-stripe diode laser, where A—active ridge, P—passive ridge.

There were some attempts to implement similar ideas both numerically and experimentally. In [31], the authors have theoretically treated triple-ridge waveguide semiconductor lasers aimed at increasing the single-mode optical power. In these lasers, a broadened stripe waveguide is accompanied by a pair of lossy auxiliary waveguides, which in fact, reproduces the CLOC concept [30]. A theoretically obtained mode discrimination is as high as 10 cm^{-1}. In work [31], the authors report the experimental results on suppressing lateral first-order mode in a stripe laser by neighboring single-mode passive waveguide. The obtained emission is quite far from the true single mode and worsens with increasing the pumping current. The major drawback of the lateral CLOC and similar designs [31,32] relates to the very low lateral waveguide contrast, which typically does not exceed 0.001. It means that the waveguide operation is very sensitive to the post-growth processing conditions and is highly affected by temperature- and current-induced refractive index changes. Thus, implementation of the lateral coupled-waveguide concept requires further extensive theoretical and experimental studies.

We believe that the effective use of the CLOC waveguide is not limited by broad-area lasers. For instance, transverse waveguides similar to those used in edge-emitting lasers are employed in microdisk lasers [33]. Due to the small diameter of 20–50 μm, the lasers operate at high injection current densities and have a rather large thermal resistance. This hard regime requires good heat dissipation, and any reduction of thermal resistance does matter. We have used the CLOC waveguide initially designed for edge-emitting lasers for microdisk ones [33]. The wafer comprises a 1.36-μm-thick undoped GaAs active waveguiding layer separated with a 0.25-μm-thick AlGaAs optical barrier from a 0.55-μm-thick GaAs passive waveguide. The p-side consists of a 0.5-μm-thick AlGaAs cladding and a 0.15-μm-thick GaAs contact layer. An active region containing five layers of InGaAs QDs is located within the waveguide core at a distance of 0.22 μm from the p-cladding. The total thickness of the semiconductor layers between the active region and the surface of the wafer is as low as 0.95 μm. The CLOC-based microlasers were bonded p-side down onto the Si boards and measured under the cw pumping current. To compare the key parameters, a set of reference microdisk lasers based on a standard transverse waveguide has been processed and measured. In the CLOC microlasers with a diameter of 31 μm, the thermal resistance was reduced to 0.32 K/mW in comparison with 0.59 K/mW measured for the standard microdisk lasers of the same diameter. Better heat dissipation allowed for increasing maximal optical powers from 0.83 mW to 1.92 mW. For the CLOC microdisk lasers tested, the peak power increased by 2.2–2.8 times.

Microdisk lasers based on a laser heterostructure with the CLOC waveguide demonstrate high uniformity of their characteristics and, in particular, high reproducibility of the injection current corresponding to the peak power. In microlasers, the maximum power is limited by the self-heating effect, and the current of the peak power depends on many device parameters that affect heat generation and its dissipation, such as threshold current, series resistance, turn-on voltage, thermal impedance, etc. For a group of 24 tested CLOC-microlasers with a diameter of 30 μm, we have found that the standard deviation of the peak current is only 11% of the mean value, As shown in Figure 11.

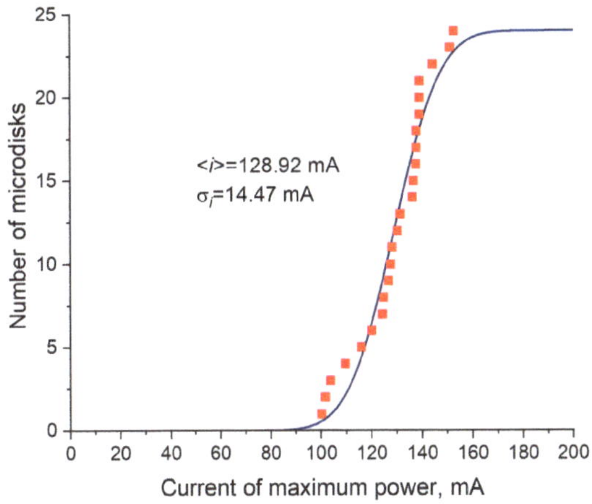

Figure 11. Number of microdisks among 24 devices with peak power current below the specified value: symbols-experiment, line-fit by a normal distribution.

6. Conclusions

To conclude, we have reviewed our CLOC concept proposed for optical mode engineering in laser planar waveguides. The approach is based on a well-known optical coupling phenomenon and allows for suppressing high-order parasitic modes and localizing the fundamental mode within the waveguide. The concept exploits the fact that only those optical modes of two waveguides interact intensively, the effective refractive indices of which are close to each other. The concept is often used in combination with shifting the active region toward the cladding layer. In view of this, a sufficient optical modal gain should be provided by the active region. The CLOC technology has been effectively employed for improving the characteristics of transverse waveguides in broad-area lasers. It enables the broadening of laser waveguides and significantly extends the range of allowed active region locations within the waveguide. Numerous results have shown that the CLOC approach leads to very stable fundamental mode lasing accompanied by reduced optical loss, as well as decreased thermal and electrical resistances (the latter is achieved owing to the possibility of decreasing the thickness of the *p*-type doped cladding layer). These parameters are crucial for high-power operations. Due to the strong localization of the fundamental transverse mode, the concept has limited applications for engineering lateral modes; however, this would be a subject for further research. Besides the edge-emitting lasers, we have successfully applied the CLOC waveguide for microdisk lasers, which has allowed reducing their thermal resistance and increasing the optical power by a factor of two at least. We believe that the CLOC approach is not limited to AlGaAs/GaAs-based lasers investigated and can be easily extended for other semiconductor compounds, which makes it rather simple and cost-efficient for the optical mode engineering and development of advanced high-power lasers.

Author Contributions: Conceptualization, L.H. and A.E.Z.; methodology, Z.W. and N.Y.G.; software, N.Y.G.; validation, L.H., N.Y.G. and M.V.M.; formal analysis, Z.W.; investigation, X.T.; resources, R.Z. and Z.J.; data curation, L.H. and Z.W.; writing—original draft preparation, L.H., N.Y.G. and M.V.M.; writing—review and editing, X.T., A.A.B., Z.W., A.E.Z. and W.Z.; visualization, G.O.K., A.S.P. and Y.M.S.; supervision, H.Y. and K.L.; project administration, G.O.K.; funding acquisition, L.H., Z.W., A.E.Z., R.Z. and Z.J. All authors have read and agreed to the published version of the manuscript.

Funding: This research was funded by National Key R&D Program of China (No. 2021YFB3201904), the Key R&D Program of Shandong Province (No. 2022CXGC020105), Natural Science Foundation of Shandong Province (Grant No. ZR2022QF142, ZR2020QF098, ZR2022QF115), Innovation Team Program of Jinan (Grant No. 2020GXRC032, 2021GXRC037), Program from Qilu University of Technology (Grant No. 2022JBZ01-04, 2022GH001, 2022PX066), Ministry of Science and Higher Education of the Russian Federation FSRM-2023-0010 (G.O.K. and M.V.M.) and FTI-LISDAS.

Data Availability Statement: The data that support the findings of this study are available from the corresponding author upon reasonable request.

Acknowledgments: A.E.Z. is thankful to the Fundamental Research Program of the HSE University for supporting optical experiments.

Conflicts of Interest: The authors declare no conflict of interest.

References

1. Zediker, M.S.; Zucker, E. High-power diode laser technology XX: A retrospective on 20 years of progress. In *High-Power Diode Laser Technology*; SPIE: Bellingham, WA, USA, 2022. [CrossRef]
2. Kaifuchi, Y.; Yoshida, K.; Yamagata, Y.; Nogawa, R.; Yamada, Y.; Yamaguchi, M. Enhanced power conversion efficiency in 900-nm range single emitter broad stripe laser diodes maintaining high power operability. In *High-Power Diode Laser Technology*; SPIE: Bellingham, WA, USA, 2019. [CrossRef]
3. Ryvkin, B.S.; Avrutin, E.A. Effect of carrier loss through waveguide layer recombination on the internal quantum efficiency in large-optical-cavity laser diodes. *J. Appl. Phys.* **2005**, *97*, 113106. [CrossRef]
4. Pikhtin, N.A.; Slipchenko, S.O.; Sokolova, Z.N.; Stankevich, A.L.; Vinokurov, D.A.; Tarasov, I.S.; Alferov, Z.I. 16 W continuous-wave output power from 100 µm-aperture laser with quantum well asymmetric heterostructure. *Electron. Lett.* **2004**, *40*, 1413. [CrossRef]
5. Crump, P.; Wenzel, H.; Erbert, G.; Tränkle, G. Progress in increasing the maximum achievable output power of broad area diode lasers. In *High-Power Diode Laser Technology and Applications*; SPIE: Bellingham, WA, USA, 2012. [CrossRef]
6. Ryvkin, B.S.; Avrutin, E.A.; Kostamovaara, J.T. Narrow versus broad asymmetric waveguides for single-mode high-power laser diodes. *J. Appl. Phys.* **2013**, *114*, 013104. [CrossRef]
7. Crump, P.; Erbert, G.; Wenzel, H.; Frevert, C.; Schultz, C.M.; Hasler, K.-H.; Staske, R.; Sumpf, B.; Maassdorf, A.; Bugge, F.; et al. Efficient High-Power Laser Diodes. *IEEE J. Sel. Top. Quantum Electron.* **2013**, *19*, 1501211. [CrossRef]
8. Kaul, T.; Erbert, G.; Maaßdorf, A.; Knigge, S.; Crump, P. Suppressed power saturation due to optimized optical confinement in 9xx nm high-power diode lasers that use extreme double asymmetric vertical designs. *Semicond. Sci. Technol.* **2018**, *33*, 035005. [CrossRef]
9. Kaul, T.; Erbert, G.; Maaßdorf, A.; Martin, D.; Crump, P. Extreme triple asymmetric (ETAS) epitaxial designs for increased efficiency at high powers in 9xx-nm diode lasers. In *High-Power Diode Laser Technology*; SPIE: Bellingham, WA, USA, 2018. [CrossRef]
10. Kauk, T.; Erbert, G.; Crump, P. Diode Laser with Improved Mode Profile Patent. WO2018138209A1. 2018. Available online: https://patents.google.com/patent/DE102017101422B4/en (accessed on 27 March 2023).
11. Skidmore, J.; Peters, M.; Rossin, V.; Guo, J.; Xiao, Y.; Cheng, J.; Shieh, A.; Srinivasan, R.; Singh, J.; Wei, C.; et al. Advances in high-power 9XXnm laser diodes for pumping fiber lasers. In *High-Power Diode Laser Technology and Applications*; SPIE: Bellingham, WA, USA, 2016. [CrossRef]
12. Gapontsev, V.; Moshegov, N.; Berezin, I.; Komissarov, A.; Trubenko, P.; Miftakhutdinov, D.; Berishev, I.; Chuyanov, V.; Raisky, O.; Ovtchinnikov, A. Highly-efficient high-power pumps for fiber lasers. In *High-Power Diode Laser Technology XV*; Zediker, M.S., Ed.; SPIE: Bellingham, WA, USA, 2017. [CrossRef]
13. Kanskar, M.; Bai, C.; Bao, L.; Biekert, N.; Chen, Z.; DeFranza, M.; DeVito, M.; Fortier, K.; Grimshaw, M.; Guan, X.; et al. High brightness diodes and 600W and 60% efficient fiber-coupled packages enabled by reduced-mode (REM) diodes. In *High-Power Diode Laser Technology XVII*; Zediker, M.S., Ed.; SPIE: Bellingham, WA, USA, 2019; p. 14. [CrossRef]
14. Okamoto, K. Coupled mode theory. In *Fundamentals of Optical Waveguides*; Elsevier: Amsterdam, The Netherlands, 2006; pp. 159–207. [CrossRef]
15. Gordeev, N.Y.; Maximov, M.V.; Zhukov, A.E. Transverse mode tailoring in diode lasers based on coupled large optical cavities. *Laser Phys.* **2017**, *27*, 086201. [CrossRef]
16. Osgood, R.; Meng, X. *General Introduction to Coupled-Mode Theory*; Springer: Cham, Switzerland, 2021; pp. 79–94. [CrossRef]

17. Gordeev, N.Y.; Payusov, A.S.; Shernyakov, Y.M.; Mintairov, S.A.; Kalyuzhnyy, N.A.; Kulagina, M.M.; Maximov, M.V. Transverse single-mode edge-emitting lasers based on coupled waveguides. *Opt. Lett.* **2015**, *40*, 2150. [CrossRef] [PubMed]
18. Serin, A.; Gordeev, N.; Payusov, A.; Shernyakov, Y.; Mintairov, S.; Maximov, M. Edge-emitting lasers based on coupled large optical cavity with high beam stability. *J. Phys. Conf. Ser.* **2017**, *929*, 012077. [CrossRef]
19. Zhukov, A.E.; Gordeev, N.Y.; Shernyakov, Y.M.; Payusov, A.S.; Serin, A.A.; Kulagina, M.M.; Mintairov, S.A.; Kalyuzhnyi, N.A.; Maksimov, M.V. Power Characteristics and Temperature Dependence of the Angular Beam Divergence of Lasers with a Near-Surface Active Region. *Tech. Phys. Lett.* **2018**, *44*, 675–677. [CrossRef]
20. Vasanth, S.; Muthuramalingam, T.; Surya Prakash, S.; Shriman Raghav, S. Investigation of SOD control on leather carbonization in diode laser cutting. *Mater. Manuf. Process.* **2023**, *38*, 544–553. [CrossRef]
21. Vasanth, S.; Muthuramalingam, T. Application of Laser Power Diode on Leather Cutting and Optimization for Better Environmental Quality Measures. *Arch. Civil Mech. Eng.* **2021**, *212*, 54. [CrossRef]
22. Vasanth, S.; Muthuramalingam, T.; Prakash, S.S.; Raghav, S.S.; Logeshwaran, G. Experimental Investigation of PWM laser standoff distance control for power diode based LBM. *Opt. Laser Technol.* **2023**, *158*, 108916. [CrossRef]
23. Harder, C. Pump diode lasers. In *Optical Fiber Telecommunications V A*; Elsevier: Amsterdam, The Netherlands, 2008; pp. 107–144. [CrossRef]
24. Zhukov, A.E.; Gordeev, N.Y.; Shernyakov, Y.M.; Payusov, A.S.; Serin, A.A.; Kulagina, M.M.; Mintairov, S.A.; Kalyuzhnyy, N.A.; Maximov, M.V. Reduction of Internal Loss and Thermal Resistance in Diode Lasers with Coupled Waveguides. *Semiconductors* **2018**, *52*, 1462–1467. [CrossRef]
25. Liu, W.C. Investigation of electrical and photoluminescent properties of MBE-grown $Al_xGa_{1-x}As$ layers. *J. Mater. Sci.* **1990**, *25*, 1765. [CrossRef]
26. Maximov, M.V.; Nadtochiy, A.M.; Mintairov, S.A.; Kalyuzhnyy, N.A.; Kryzhanovskaya, N.V.; Moiseev, E.I.; Gordeev, N.; Shernyakov, Y.M.; Payusov, A.S.; Zubov, F.I.; et al. Light emitting devices based on quantum well-dots. *Appl. Sci.* **2020**, *10*, 1038. [CrossRef]
27. Gordeev, N.Y.; Payusov, A.S.; Shernyakov, Y.M.; Mintairov, S.A.; Kalyuzhnyy, N.A.; Kulagina, M.M.; Serin, A.A.; Gordeev, S.N.; Maximov, M.V.; Zhukov, A.E. Transverse mode competition in narrow-ridge diode lasers. *Laser Phys.* **2019**, *29*, 025003. [CrossRef]
28. Vaughan, M.; Susanto, H.; Li, N.; Henning, I.; Adams, M. Stability Boundaries in Laterally-Coupled Pairs of Semiconductor Lasers. *Photonics* **2019**, *6*, 74. [CrossRef]
29. Payusov, A.S.; Serin, A.A.; Shernyakov, Y.M.; Rybalko, D.A.; Kulagina, M.M.; Maximov, M.V.; Gordeev, N.Y. Lateral mode engineering in diode lasers based on coupled ridges. *J. Phys. Conf. Ser.* **2018**, *1124*, 041043. [CrossRef]
30. Payusov, A.S.; Shernyakov, Y.M.; Kulagina, M.M.; Serin, A.A.; Maximov, M.V.; Zhukov, A.E.; Gordeev, N.Y. Resonance inhibiting of high-order lateral modes in few-stripe diode lasers. In Proceedings of the 2018 International Conference Laser Optics (ICLO), Saint Petersburg, Russia, 4–8 June 2018; IEEE: New York, NY, USA, 2018; p. 140. [CrossRef]
31. Zhao, X.; Zeng, S.; Sweatt, L.; Zhu, L. High-power single-mode triple-ridge waveguide semiconductor laser based on supersymmetry. *AIP Adv.* **2021**, *11*, 095216. [CrossRef]
32. Şeker, E.; Şengül, S.; Dadashi, K.; Olyaeefar, B.; Demir, A. Single-mode operation of electrically pumped edge-emitting lasers through cavity coupling of high order modes. In *High-Power Diode Laser Technology XX*; Zediker, M.S., Zucker, E.P., Eds.; SPIE: Bellingham, WA, USA, 2022; p. 9. [CrossRef]
33. Zubov, F.I.; Moiseev, E.I.; Nadtochiy, A.M.; Fominykh, N.A.; Ivanov, K.A.; Makhov, I.S.; Dragunova, A.S.; Maximov, M.V.; Vorobyev, A.A.; Mozharov, A.M.; et al. Improvement of thermal resistance in InGaAs/GaAs/AlGaAs microdisk lasers bonded onto silicon. *Semicond. Sci. Technol.* **2022**, *37*, 075010. [CrossRef]

micromachines

Article

Broadband All-Optical THz Modulator Based on Bi_2Te_3/Si Heterostructure Driven by UV-Visible Light

Yayan Xi, Yixuan Zhou *, Xueqin Cao, Jing Wang, Zhen Lei, Chunhui Lu, Dan Wu, Mingjian Shi, Yuanyuan Huang and Xinlong Xu *

Shaanxi Joint Laboratory of Graphene, State Key Laboratory of Photon-Technology in Western China Energy, International Collaborative Center on Photoelectric Technology and Nano Functional Materials, Institute of Photonics & Photon-Technology, School of Physics, Northwest University, Xi'an 710069, China
* Correspondence: yxzhou@nwu.edu.cn (Y.Z.); xlxuphy@nwu.edu.cn (X.X.)

Abstract: All-optical terahertz (THz) modulators have received tremendous attention due to their significant role in developing future sixth-generation technology and all-optical networks. Herein, the THz modulation performance of the Bi_2Te_3/Si heterostructure is investigated via THz time-domain spectroscopy under the control of continuous wave lasers at 532 nm and 405 nm. Broadband-sensitive modulation is observed at 532 nm and 405 nm within the experimental frequency range from 0.8 to 2.4 THz. The modulation depth reaches 80% under the 532 nm laser illumination with a maximum power of 250 mW and 96% under 405 nm illumination with a high power of 550 mW. The mechanism of the largely enhanced modulation depth is attributed to the construction of a type-II Bi_2Te_3/Si heterostructure, which could promote photogenerated electron and hole separation and increase carrier density dramatically. This work proves that a high photon energy laser can also achieve high-efficiency modulation based on the Bi_2Te_3/Si heterostructure, and the UV-Visible control laser may be more suitable for designing advanced all-optical THz modulators with micro-level sizes.

Keywords: THz modulator; Bi_2Te_3/Si heterostructure; modulation depth; photogenerated carrier

Citation: Xi, Y.; Zhou, Y.; Cao, X.; Wang, J.; Lei, Z.; Lu, C.; Wu, D.; Shi, M.; Huang, Y.; Xu, X. Broadband All-Optical THz Modulator Based on Bi_2Te_3/Si Heterostructure Driven by UV-Visible Light. *Micromachines* **2023**, *14*, 1237. https://doi.org/10.3390/mi14061237

Academic Editors: He Yang, Xinyang Su and Yizhong Huang

Received: 28 April 2023
Revised: 30 May 2023
Accepted: 8 June 2023
Published: 12 June 2023

1. Introduction

Emerging terahertz (THz) technology has attracted considerable attention due to its unique applications in various fields, such as non-invasive detection [1], imaging [2], sensing [3], and wireless communication [4]. In particular, THz technology plays a significant role in the sixth-generation (6G) system with respect to meeting the unprecedented requirements for high-speed and large bandwidth in wireless communication [5]. However, there is still a huge bottleneck concerning developing high-efficiency and low-cost THz components, including THz sources, detectors, modulators, etc [6]. In particular, the THz modulator is an essential component in communication applications with respect to manipulating THz waves with a certain high frequency. The reported control methods include electrical [7,8], optical [9,10], thermal [11], and mechanical [12] modulation. Among them, electrical and optical modulation receive the most attention from researchers because the former is the best match to the current electrical systems, and the latter is the most suitable for future all-optical systems. Compared with the electrically driven THz modulators whose modulation speed is limited by the resistance-capacitance time constant [13], all-optical motivated THz modulators possess carrier lifetime determined modulation speed, as well as reduced power consumption and minimized cross-talk [6], which makes them available for certain attractive applications such as signal processing and optical communication systems [14].

The modulation performance of optical THz modulators is determined by the optical and electrical properties of the material, such as carrier mobility, carrier lifetime, and absorption coefficient [15]. In recent years, two-dimensional (2D) materials have prompted tremendous interest in designing novel optical THz modulators due to their unique optical and electrical properties, such as high carrier mobility, tunable carrier concentration, and various bandgaps [16,17]. However, the atomical thickness of 2D material limits the optical–matter interaction strength [16], hindering their modulation performance as the core material alone. One solution is to bond them with THz metasurfaces, which can improve the modulation depth greatly at resonance frequencies [18,19]. However, this method is generally locked in several narrow frequency regions. The other widely spread method is constructing heterostructure with classical photoresponsive semiconductor materials such as Si [20] and Ge [21], which has been proven to be effective in improving the modulation performance dramatically. From the start of the graphene/Si all-optical THz modulator being verified in 2012 [9], the majority of 2D material-based heterostructures, such as WS_2/Si [22,23], $MoTe_2$/Si [24], $PtSe_2$/Si [25], and MoS_2/Si [26], have been reported in the past seven years. The modulation depth is increased from 70% to 99.9% under the continuous wave (CW) laser illumination at several frequently used wavelengths of ~800 nm, ~1064 nm, and ~450 nm. However, the role of these different illumination wavelengths in the modulation performance has mostly been ignored. Note that the wavelength-dependent penetration depth of the control light has a significant change in the Si layer, which determines the device size. Understanding of the control wavelength-determined modulation mechanism and the advantages of short wavelength control light is essential for the development of micro-based compact all-optical THz modulators.

Bi_2Te_3 is a V–VI compound topological material with unique electrical and optical properties, including high carrier mobility, tunable bandgap, high conductivity, and single Dirac cone [27–29]. Recently, it has received enormous attention in photoelectric applications such as ultrafast laser photonics, optical communications, infrared sensors, and biomedical imaging [30–32]. In the THz region, 2D Bi_2Te_3 has proven to be a promising candidate for broadband photodetection [33], chiral THz emitter [34], and high-speed spintronic devices [35]. More importantly, exploring all-optical THz modulators based on Bi_2Te_3 is significant for expanding the THz application of this material, but relevant research has not been reported.

Herein, an all-optical controlled THz modulator based on Bi_2Te_3/Si heterostructure is investigated under 532 nm and 405 nm CW laser illumination. The Bi_2Te_3 film is prepared via the physical vapor deposition (PVD) method. The heterostructure achieves an adequate broadband modulation to the THz wave ranging from 0.8 THz to 2.4 THz. With the same laser power of 250 mW, the modulation depth of Bi_2Te_3/Si reaches almost the same values of ~80% and ~76% under 532 nm and 405 nm excitation, respectively. When further increasing the power to 550 mW at 405 nm, the modulation depth can reach as high as ~96%. The largely enhanced modulation performance is ascribed to the improved photogenerated electron and hole separation as well as the carrier lifetime and density driven by the type-II band alignment of the Bi_2Te_3/Si heterostructure. This work demonstrates the excellent THz modulation performance of the Bi_2Te_3/Si heterostructure controlled via UV-Visible light, which paves a promising route for designing novel all-optical micro-functional devices.

Characterization of Sample and Experiment Setup

The Bi_2Te_3 films are prepared via the PVD method on sapphire and Si (>2000 Ω) substrates according to our previous works [36,37]. The schematic of the growth process is illustrated in Figure 1a. A total of 5 mg Bi_2Te_3 powder (99.99%, Alfa Aesar) as a precursor is put in the middle of the quartz tube, which is located in the heating zone of the furnace. Several sapphire and Si substrates are placed in a quartz boat behind the precursor at a distance of ~15 cm. Then, the temperature zone is heated to 550 °C with a rising rate of 15 min and maintained at the target temperature for 5 min. During the whole process, the atmosphere inside the quartz tube is protected by the Ar flow with a rate of 20 sccm.

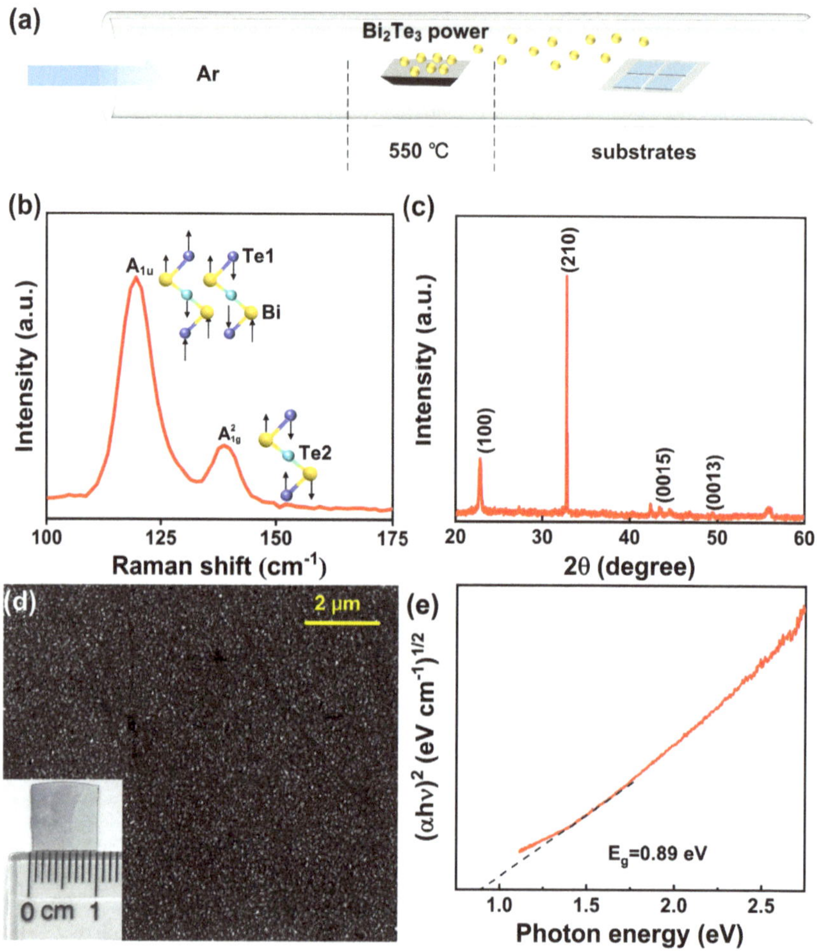

Figure 1. (**a**) Schematic of the PVD growth process of Bi_2Te_3 films on different substrates. (**b**) Raman spectrum of the Bi_2Te_3 film under a 532 nm laser excitation. The inset is the schematic of vibration modes. (**c**) XRD of the Bi_2Te_3 film. (**d**) SEM image of the Bi_2Te_3 film. The inset is the optical image. (**e**) Tauc plots ($(\alpha h\nu)^{1/2}$ versus $h\nu$) give the band gap of Bi_2Te_3.The red line and black dotted line represent the absorption edge and tangent line, respectively.

Raman spectroscopy (SmartRaman confocal-micro-Raman module, Institute of Semiconductors, Chinese Academy of Science) of a Bi_2Te_3 film measured under a 532 nm CW laser excitation is presented in Figure 1b. The peaks located at 118.5 cm^{-1} and 134 cm^{-1} correspond to the infrared active mode A_{1u} and the Raman active mode A^2_{1g}, respectively [29]. The X-ray diffraction spectroscopy (XRD, Bruker, D8 Advance) of the Bi_2Te_3 film is shown in Figure 1c. There are four diffraction peaks representing different crystalline directions, which are indexed to hexagonal Bi_2Te_3 [38]. The scan electron microscope (SEM) image (Thermo Scientific, Apreo S., Waltham, MA, USA) of the Bi_2Te_3 film is presented in Figure 1d, and the inset shows the optical image (~1 × 1 cm^2). These morphological characterization results on different scales prove that the film is uniform on a large scale. As shown in Figure 1e, the band gap of Bi_2Te_3 film is determined in the Tauc plot derived from the UV-Visible absorption spectroscopy.

The THz modulation performances of the samples are investigated via THz time-domain spectroscopy, which is illustrated in Figure 2a [39]. The THz wave is generated by exciting the ZnTe crystal with an 800 nm femtosecond laser. The THz wave is collected via a pair of off-axis parabolic mirrors, focused together with the probe laser on the detection crystal ZnTe and detected via the electro-optical sampling [40]. High-resistance Si (HR-Si) is utilized to block the femtosecond laser from the collinear THz wave. The 405 nm (or 532 nm) CW laser is then introduced with a normal incidence angle to modulate the THz transmission properties of the samples. Note that the spot diameter of CW light is about 4 mm, while the spot diameter of THz wave is approximately 3 mm. The experiment is performed with the protection of dry nitrogen. Figure 2b illustrates the transmitted THz wave modulation of a Bi_2Te_3/Si heterostructure driven by a CW laser which is geometrically expanded in the beam diameter to cover all the THz wave spots.

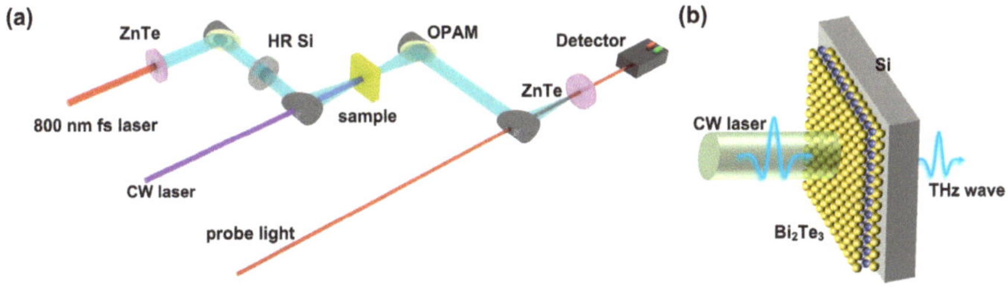

Figure 2. (**a**) Schematic of the THz time-domain spectroscopy system for the CW laser-controlled modulation experiment. (**b**) Illustration of the Bi_2Te_3/Si heterostructure THz wave modulation via CW laser.

2. Results and Discussion

Figure 3a,b show the time-domain signals of the transmitted THz waves through Bi_2Te_3/Si and Bi_2Te_3/sapphire under 532 nm CW laser illumination with different power, respectively. Note that bare sapphire is transparent in the THz region, so the Bi_2Te_3/sapphire sample only reflects the effect of Bi_2Te_3, which exhibits negligible modulation to the THz amplitude with different illumination powers, as shown in Figure 3b. This result suggests that the photocarriers generated by topological Bi_2Te_3 itself under 532 nm CW illumination could not modulate the THz amplitude. However, for the Bi_2Te_3/Si heterostructure, the transmitted THz amplitudes decrease obviously on increasing illumination power, as shown in Figure 3a. This result indicates that the CW laser illumination could generate sufficient photocarriers in Bi_2Te_3/Si heterostructure to influence the transmission of THz waves.

Similar THz wave-modulation experiments have been performed on the same Bi_2Te_3/Si and Bi_2Te_3/sapphire samples under the illumination of a CW laser in the violet region of 405 nm, as shown in Figure 3c and 3d, respectively. The results show that the Bi_2Te_3/sapphire still has negligible modulation to the THz wave, but the Bi_2Te_3 samples can absorb the THz greatly when increasing the illumination power. For the Bi_2Te_3/Si, the decline in the transmitted THz amplitudes controlled via a 405 nm laser is more significant than that via a 532 nm laser. However, this enhancement may be induced by the higher maximum power of the 405 nm laser used in the experiment, as the THz transmission under the same laser power of 532 nm and 405 nm excitation is close. Compared with Bi_2Te_3/sapphire and bare Si [39], Bi_2Te_3/Si heterostructure has a significantly improved modulation capability with respect to the THz wave. This enhancement indicates that the heterostructure has a largely optimized modulation mechanism, which is discussed in the last part of the paper.

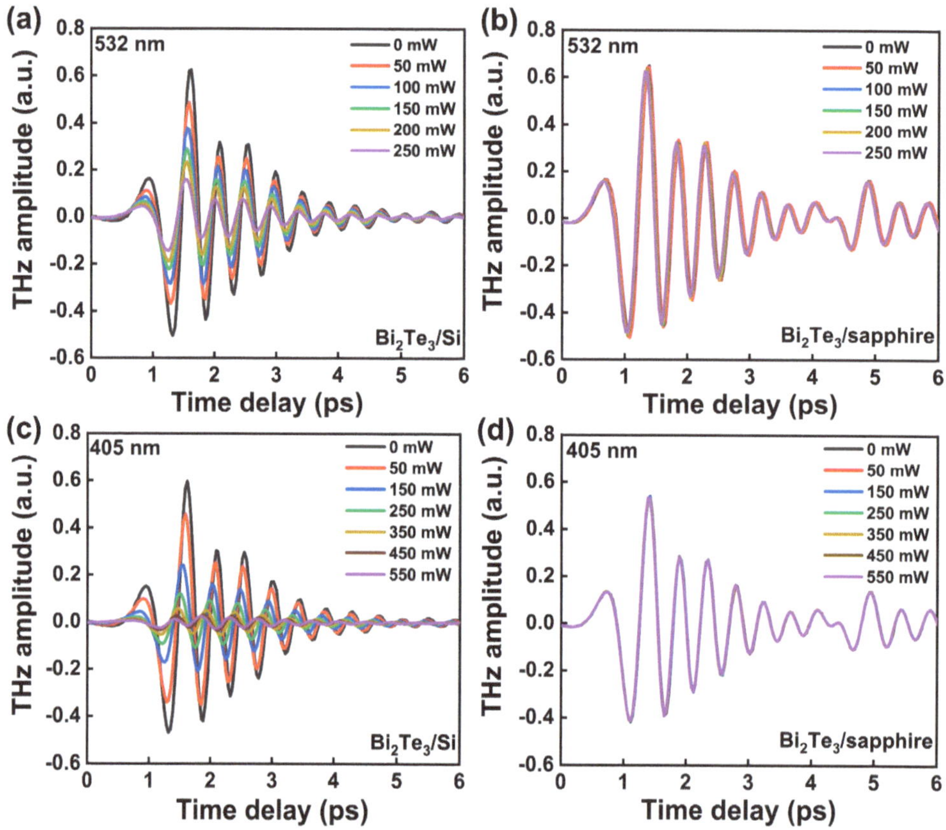

Figure 3. Time-domain signals of the transmitted THz waves through (**a**) Bi_2Te_3/Si and (**b**) Bi_2Te_3/sapphire under 532 nm CW laser excitation with different pump powers. Time-domain signals of the transmitted THz waves through (**c**) Bi_2Te_3/Si and (**d**) Bi_2Te_3/sapphire under 405 nm CW laser excitation with different pump powers.

To further evaluate the modulation performance of the heterostructure, the frequency-domain THz amplitude spectra of Bi_2Te_3/Si under 532 and 405 nm CW laser excitation with different powers are obtained via Fourier transformation, as shown in Figure 4a and 4b, respectively. The THz wave generated and detected via ZnTe crystals in this work exhibits a broadband feature that covers the 0.4–2.8 THz range. The high signal-to-noise ratio region is around 0.8–2.4 THz, and the two peaks are at ~1.4 and ~2.1 THz. Therefore, the evaluated frequency range is wider than the reported THz modulation works based on heterostructures such as WS_2/Si (0.2–1.6 THz) [23] and $PtSe_2$/Si (0.1–1.5 THz) [25]. As a result, the modulation of the Bi_2Te_3/Si heterostructure is broadband-effective and covers the whole experimental frequency region. Moreover, the THz amplitude decreases gradually with the increase in illumination power under both 532 nm and 405 nm CW laser excitation. Here, the modulation effect driven by the 405 nm laser is greater than that by the 532 nm laser.

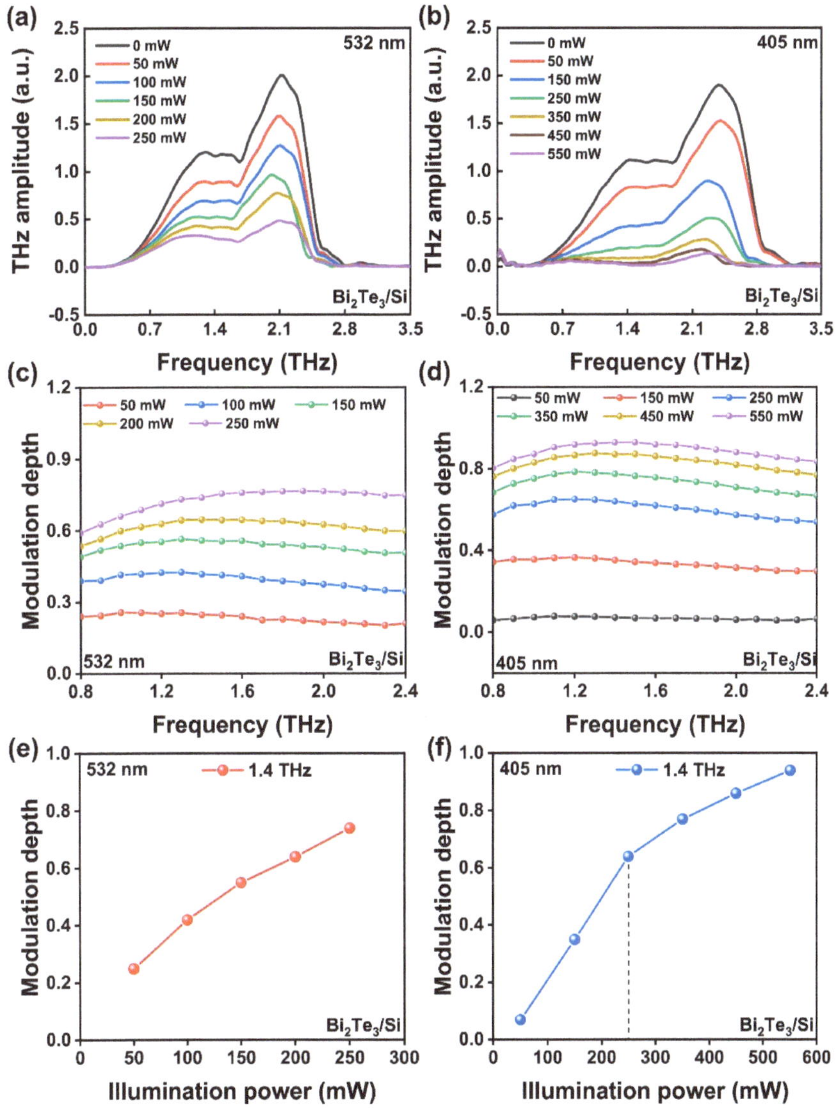

Figure 4. Frequency-domain THz amplitude spectra of Bi_2Te_3/Si under (**a**) 532 nm and (**b**) 405 nm CW laser excitation with different pump power. Modulation depth of Bi_2Te_3/Si under (**c**) 532 nm and (**d**) 405 nm CW laser excitation with different pump power. Illumination power-dependent modulation depth of Bi_2Te_3/Si at 1.4 THz under (**e**) 532 nm and (**f**) 405 nm CW laser excitation.

The modulation depth (M) that represents the ability to modulate the THz intensity of the device can be calculated quantitatively as $M = \frac{A_{off} - A_{on}}{A_{off}} \times 100\%$. Here, A_{off} and A_{on} are the THz peak-valley values without and with the CW laser illumination. Under the 532 nm and 405 nm CW laser illumination with different powers, the modulation depth of Bi_2Te_3/Si in the frequency domain is presented in Figure 4c and 4d, respectively. The heterostructure exhibits a broadband modulation feature in all the conditions as mentioned above. Taking the 1.4 THz position as an example, the modulation depth under 532 nm excitation increases gradually from 25% to 80% when increasing the illumination power from 50 to 250 mW, as shown in Figure 4e. The modulation depth increases with the enhancement of illumination power in the experimental measurement range. In comparison, under 405 nm excitation, the modulation depth of Bi_2Te_3/Si rises from 7% to 94% with the enhancement of illumination power from 50 to 550 mW (Figure 4f). Note that in the relatively low illumination power region (<250 mW) with the same photon number, the modulation depth of Bi_2Te_3/Si under 532 nm (250 mW) excitation is 76%, which is superior to the modulation depth of ~49% under 405 nm (190 mW) excitation. However, the modulation depth grows faster in the low illumination power region for the 405 nm excitation condition and becomes more significant in the higher illumination power region up to 550 mW. As a result, both the 405 nm and the 532 nm CW lasers for Bi_2Te_3/Si show excellent broadband modulation performance in the THz region. Within our experimental conditions, a 405 nm ultraviolet laser at high power shows better THz amplitude modulation ability than 532 nm visible light.

To elucidate the modulation mechanism, the modulation response properties of the Bi_2Te_3/Si heterostructure are assessed through the examination of its illumination-on/off behavior. Figure 5a shows the time response of the transmitted THz amplitude controlled via the 532 nm CW laser with a power of 250 mW. This result suggests a clear THz amplitude modulation from 0 to 74% under illumination-off to -on conditions. The photoresponsivity (R) is generally described in units of THz absorption per watt of incident laser power [41,42]. Hence, the R is defined by the analytical equation $R = \frac{A_{off} - A_{on}}{P}$, where P is the illumination power. The photoresponsivity of the Bi_2Te_3/Si heterostructure is calculated to be ~1.76 W^{-1}. Similarly, the time series of illumination-off and illumination-on for THz amplitude of Bi_2Te_3/Si is presented in Figure 5b under 405 nm CW laser illumination with the same power. Note that the THz signal is almost completely shielded under the illumination-on condition, and the photoresponsivity is calculated to be 2.4 W^{-1}, which is ~1.4 times larger than that of the 532 nm CW laser excitation condition. This indicates that the THz modulation performance is superior for the 405 nm CW laser illumination condition.

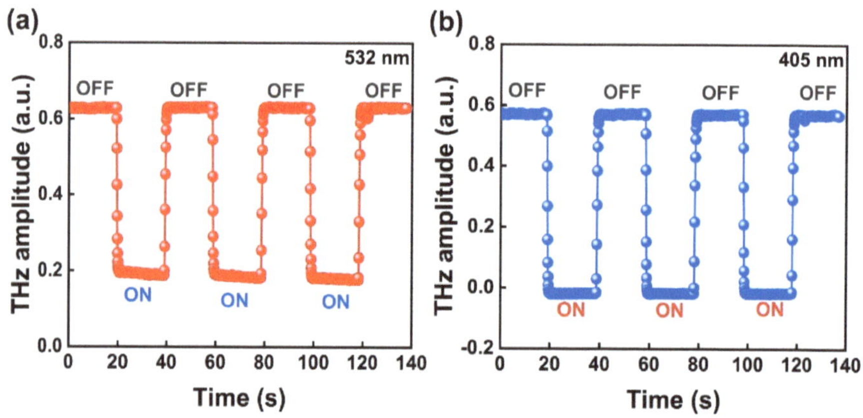

Figure 5. Time series of illumination-off and illumination-on about the THz transmission amplitude under (**a**) 532 nm and (**b**) 405 nm CW laser excitation with 250 mW.

Additionally, the response and recover time under both 532 and 405 nm laser excitation is approximately 1 s, as shown in Figure 5. However, this long time parameter is dominated by the mechanical baffle switching time which is close to 1 s instead of the real modulation speed governed by the carrier dynamic process of the sample. Therefore, one can expect a much faster modulation speed in the future when using a Bi_2Te_3/Si THz modulator controlled via choppers or other electrical control units.

Next, the modulation mechanism of the Bi_2Te_3/Si heterostructure under CW laser illumination is discussed. The THz amplitude transmission $T(\omega)$ of the Bi_2Te_3/Si heterostructure is related to its THz photoconductivity $\sigma(\omega)$ according to Tinkham's equation [43]:

$$\sigma(\omega) = \frac{(1+n_s)}{Z_0 d}\left(\frac{1}{T(\omega)} - 1\right) \tag{1}$$

where n_s is the refractive index of Si, and Z_0 and d are vacuum impedance and photoconductive layer thickness, respectively. In the THz region, complex photoconductivity of semiconductor materials generally follows the Drude model [44]:

$$\tilde{\sigma}(\omega) = \frac{{\omega_p}^2 \varepsilon_0 \tau}{1 + i\omega\tau} \tag{2}$$

where ω_p, ε_0, τ, and ω are plasma frequency, vacuum permittivity, lifetime of photogenerated carriers, and THz frequency, respectively. Further, the plasma frequency is related to the carrier density n as ${\omega_p}^2 = ne^2/m^*\varepsilon_0$, where e and m^* are electron charge and effective electron mass, respectively. Therefore, the THz transmission is related to the photoconductivity and the carrier density, which is largely affected by the material's optoelectronic properties and the illumination laser conditions. In a simplified intrinsic absorption model, the relationship between the average photogenerated charge carriers in the photoconductive layer and the carrier lifetime can be estimated with the following formula [25,45]:

$$n = \frac{PA\alpha\tau}{h\nu Sd} \tag{3}$$

where A is the absorptivity (excludes the reflection and transmission losses), α is the responsivity that represents the photocarrier-generation efficiency, $h\nu$ is the photon energy, and d is the layer thickness. Note that the photogenerated charge carriers are not uniformly distributed in real semiconductors, but this model can still reflect the influence of the key factors. For example, according to Equations (2) and (3), the carrier density is proportional to the photoconductivity, and to the illumination power. According to Equation (1), the carrier density is inversely proportional to the THz amplitude transmission with an offset, which is proportional to the opposite amont of modulation depth. Therefore, as shown in Figure 4e,f, the THz modulation depth is enhanced by increasing the laser power.

To elucidate the charge transformation mechanism of the Bi_2Te_3/Si heterostructure under CW laser excitation, the carrier separation process at the interface is discussed in the following part. The energy band alignment of the Bi_2Te_3/Si heterostructure before and after contact are shown in Figure 6a and 6b, respectively. Here, E_{vac} is the vacuum level, and Φ_1, χ_1, E_{c1}, E_{v1}, E_{F1}, and E_{g1} are work function and electron affinity, conduction band bottom, valence band maximum, Fermi level, and band gap of Bi_2Te_3, respectively. Similarly, Φ_2, χ_2, E_{c2}, E_{v2}, E_{F2}, and E_{g2} are the corresponding parameters of Si. ΔE_c and ΔE_v are defined to be the difference between the two conduction bands and valence bands, respectively. In this experiment, the Bi_2Te_3 film prepared via the PVD method is the same as the material reported in our previous work for photoelectrochemical applications, so the band alignment parameters can be found in the UV-Vis absorption spectroscopy, X-ray photoelectron spectroscopy, and ultraviolet photoelectron spectroscopy results [36]. E_{c1}, E_{v1}, E_{F1}, and E_{g1} are about 5.02 eV, 5.89 eV, 5.86 eV, and 0.87 eV, respectively. The Si substrate is slightly n-doped and we estimate the band alignment with parameters according to Meng's report [46], in which E_{c2}, E_{v2}, E_{F2}, and E_{g2} are around 4.29 eV, 5.41 eV, 4.69 eV, and 1.12 eV, respectively. Therefore, ΔE_c and ΔE_v are calculated to be ~0.73 eV and ~0.48 eV, respectively. After the contact between Bi_2Te_3 and Si, the Fermi levels reach the same value for an equilibrium state and a type-II heterostructure is formed as illustrated in Figure 6b. When the CW laser illuminates the sample, an interband transition occurs in both Bi_2Te_3 and Si. Under the built-in electric field, photogenerated electrons transfer from Bi_2Te_3 to Si while holes are in the opposite direction, as shown in Figure 6c. The separation of electrons and holes can be largely enhanced by the band alignment of the type-II heterostructure. The separation of electrons and holes significantly reduces the recombination of photocarriers, leading to an enhancement of the carrier lifetime as well as the carrier density, so the THz modulation depth is largely improved by the more sensitive photoconductivity.

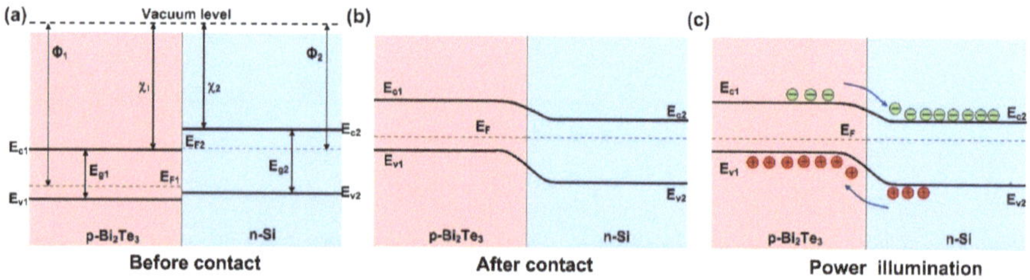

Figure 6. (**a**) Energy band structure of Bi_2Te_3 and Si before contact. (**b**) Band alignment of the Bi_2Te_3/Si heterostructure after contact. (**c**) Band alignment of the Bi_2Te_3/Si heterostructure under laser illumination.

In this work, the photon energies of the 405 nm and 532 nm lasers are larger than the band gap energies of both Bi_2Te_3 and Si. Therefore, the THz transmission properties of single Bi_2Te_3 or Si can be explained via the model mentioned above. For the Bi_2Te_3/sapphire sample, Bi_2Te_3 is the only photoconductive layer because the absorptivity of the sapphire substrate is zero. However, the absorptivity of Bi_2Te_3 is also limited by the very thin film thickness and the lifetime of photogenerated carriers is short due to electron–hole recombination. As a result, the photocarrier density in Bi_2Te_3 is very low and thus the modulation is almost negligible, as shown in Figure 3b,d. For bare Si, the absorptivity under both 405 nm and 532 nm excitation is high (only the reflection loss should be considered) because the film thickness is far longer than the penetration depth, which can be regarded as the thickness of the photoconductive layer. Hence, the modulation of bare Si can be observed in many optically controlled THz modulation works [22,39,47,48]. However, the modulation depth of bare Si is low because the recombination of photocarriers limits the lifetime and carrier density. Therefore, the effective separation of electrons and holes, which can be realized via reasonable heterostructure construction, becomes an important way of designing high-performance THz modulators.

Lastly, the modulation performances of 532 nm and 405 nm CW lasers are compared and discussed. Because both of the two wavelengths can induce interband transition in the Bi_2Te_3/Si heterostructure, the main difference could originate from the factors given in Equations (1–3). Under 532 nm and 405 nm CW laser excitation, the penetration depths of Si are ~1 and ~0.1 μm, respectively [49,50]. With the increase in illumination power, the photoconductivity and carrier concentration exhibit a growth trend in both the wavelength cases. As shown in Figure 7a,b, when the power is enhanced from 0 mW to 250 mW, the conductivity changes from 3.69×10^3 S/m to 3.25×10^4 S/m, and the carrier density varies from 7.18×10^{13} cm^{-3} to 6.34×10^{14} cm^{-3} under 532 nm excitation. In contrast to the previous statement, when exposed to 250 mW illumination power under 405 nm excitation, the photoconductivity and carrier density can reach 6.13×10^5 S/m and 1.19×10^{16} cm^{-3} (Figure 7c and d), respectively. This enhancement is mainly induced by the narrow penetration depth at 405 nm, but the modulation performance under 405 nm and 532 nm excitation has no significant difference because the d factor is eliminated in Equation (1). When the illumination power further increases to 550 mW, the conductivity and carrier density of the photoconductive layer can reach 5.84×10^6 S/m and 1.14×10^{17} cm^{-3}, respectively. Note that the laser power dependence of the carrier density measured in the experiment is not linear as predicted in Equation (3), especially in the high-power region. This could be attributed the enhancement of the carrier lifetime induced by the heterostructure, which is ignored in the calculation. These results may provide a meaningful reference from the perspective of microdevice design. According to previous reports [49,50], the penetration depth values of Si for several typical laser wavelengths of 405 nm, 532 nm, 800 nm, and 1064 nm are ~0.1 μm, ~1 μm, ~10 μm, and ~900 μm, respectively. This shows that the penetration depth changes dramatically with the excitation wavelength. Therefore, the laser with higher photon energy may make sufficient utilization of the carrier separation capacity at the finite heterostructure region. In addition, the modulation devices could be fabricated smaller and more compactly when controlled via short-wavelength lasers. For example, Si films with submicron thicknesses could be sufficient for the THz modulator design under 405 nm excitation, which can also achieve a modulation depth close to 100% as verified in this work.

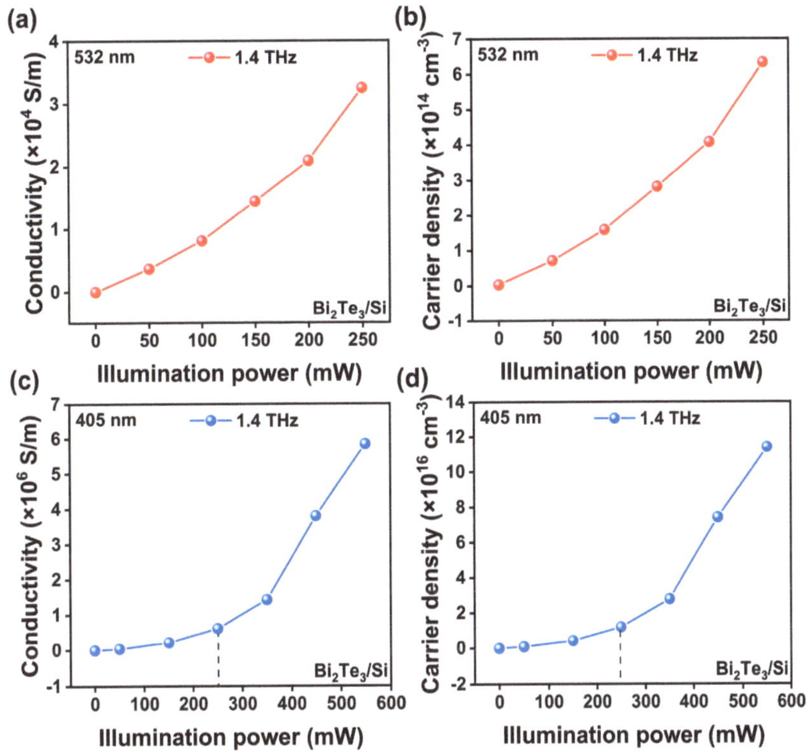

Figure 7. (**a**) Photoconductivity and (**b**) carrier concentration of Bi_2Te_3/Si photoconductive layer under 532 nm excitation with different illumination powers. (**c**) Photoconductivity and (**d**) carrier concentration of Bi_2Te_3/Si photoconductive layer under 405 nm excitation with different illumination powers.

3. Conclusions

An all-optical THz modulator based on Bi_2Te_3/Si heterostructure is investigated in this work. The 2D Bi_2Te_3 is prepared via the PVD method on sapphire and Si substrates directly, and these samples are measured via THz time-domain spectroscopy. Under the control of CW lasers at 532 nm and 405 nm, broadband modulation performance is observed in the 0.8–2.4 THz region. The modulation depth can reach ~80% under 532 nm CW laser illumination with a power of 250 mW, and ~96% under 405 nm illumination with a higher power of 550 mW. The significantly enhanced modulation capacity of the Bi_2Te_3/Si compared to that of Bi_2Te_3 is attributed to the highly improved photocarrier separation at the type-II heterostructure interface, and the modulation mechanism is discussed with the help of the Drude model and a simplified light absorption model. This work provides a useful reference for the design of compact all-optical THz modulators with no loss of performance.

Author Contributions: Funding acquisition, Y.Z., Y.H. and X.X.; Investigation, Y.X.; Methodology, C.L., D.W. and M.S.; Visualization, Y.Z.; Writing—original draft, Y.X., Y.Z. and X.C.; Writing—review & editing, Y.X., Y.Z., X.C., J.W., Z.L., C.L. and X.X. All authors have read and agreed to the published version of the manuscript.

Funding: This work was supported by National Natural Science Foundation of China (No. 11974279, 12074311, 12004310, 12261141662), and Natural Science Foundation of Shaanxi Province (2019JM-236), Natural Science Foundation of Education Department of Shaanxi Province (Grants No. 21JK0917).

Data Availability Statement: Not applicable.

Conflicts of Interest: The authors declare no conflict of interest.

References

1. Yu, L.; Hao, L.; Meiqiong, T.; Jiaoqi, H.; Wei, L.; Jinying, D.; Xueping, C.; Weiling, F.; Yang, Z. The medical application of terahertz technology in non-invasive detection of cells and tissues: Opportunities and challenges. *RSC Adv.* **2019**, *9*, 9354–9363. [CrossRef] [PubMed]
2. Afsah-Hejri, L.; Hajeb, P.; Ara, P.; Ehsani, R.J. A Comprehensive Review on Food Applications of Terahertz Spectroscopy and Imaging. *Compr. Rev. Food Sci. Food Saf.* **2019**, *18*, 1563–1621. [CrossRef] [PubMed]
3. Naftaly, M.; Vieweg, N.; Deninger, A. Industrial Applications of Terahertz Sensing: State of Play. *Sensors* **2019**, *19*, 4203. [CrossRef] [PubMed]
4. Abohmra, A.; Khan, Z.U.; Abbas, H.T.; Shoaib, N.; Imran, M.A.; Abbasi, Q.H. Two-Dimensional Materials for Future Terahertz Wireless Communications. *IEEE Antennas Propag.* **2022**, *3*, 217–228. [CrossRef]
5. Serghiou, D.; Khalily, M.; Brown, T.W.C.; Tafazolli, R. Terahertz Channel Propagation Phenomena, Measurement Techniques and Modeling for 6G Wireless Communication Applications: A Survey, Open Challenges and Future Research Directions. *IEEE Commun. Surv. Tut.* **2022**, *24*, 1957–1996. [CrossRef]
6. Gopalan, P.; Sensale-Rodriguez, B. 2D Materials for Terahertz Modulation. *Adv. Opt. Mater.* **2020**, *8*, 1900550. [CrossRef]
7. Kleine-Ostmann, T.; Dawson, P.; Pierz, K.; Hein, G.; Koch, M. Room-temperature operation of an electrically driven terahertz modulator. *Appl. Phys. Lett.* **2004**, *84*, 3555–3557. [CrossRef]
8. Wang, X.B.; Cheng, L.; Wu, Y.; Zhu, D.P.; Wang, L.; Zhu, J.-X.; Yang, H.; Chia, E.E.M. Topological-insulator-based terahertz modulator. *Sci. Rep.* **2017**, *7*, 13486. [CrossRef]
9. Weis, P.; Garcia-Pomar, J.L.; Höh, M.; Reinhard, B.; Brodyanski, A.; Rahm, M. Spectrally Wide-Band Terahertz Wave Modulator Based on Optically Tuned Graphene. *ACS Nano* **2012**, *6*, 9118–9124. [CrossRef]
10. Zhou, J.; Zhou, T.; Yang, D.; Wang, Z.; Zhang, Z.; You, J.; Xu, Z.; Zheng, X.; Cheng, X.-a. Optically Controlled Extraordinary Terahertz Transmission of Bi_2Se_3 Film Modulator. *Photon. Sens.* **2019**, *9*, 268–276. [CrossRef]
11. Wang, T.; He, J.; Guo, J.; Wang, X.; Feng, S.; Kuhl, F.; Becker, M.; Polity, A.; Klar, P.J.; Zhang, Y. Thermally switchable terahertz wavefront metasurface modulators based on the insulator-to-metal transition of vanadium dioxide. *Opt. Express* **2019**, *27*, 20347–20357. [CrossRef] [PubMed]
12. Cheng, L.; Jin, Z.; Ma, Z.; Su, F.; Zhao, Y.; Zhang, Y.; Su, T.; Sun, Y.; Xu, X.; Meng, Z.; et al. Mechanical Terahertz Modulation Based on Single-Layered Graphene. *Adv. Opt. Mater.* **2018**, *6*, 1700877. [CrossRef]
13. Chan, W.L.; Chen, H.-T.; Taylor, A.J.; Brener, I.; Cich, M.J.; Mittleman, D.M. A spatial light modulator for terahertz beams. *Appl. Phys. Lett.* **2009**, *94*, 213511. [CrossRef]
14. Yang, H.; Wang, Y.; Tiu, Z.C.; Tan, S.J.; Yuan, L.; Zhang, H. All-Optical Modulation Technology Based on 2D Layered Materials. *Micromachines* **2022**, *13*, 92. [CrossRef]
15. Degl'Innocenti, R.; Kindness, S.J.; Beere, H.E.; Ritchie, D.A. All-integrated terahertz modulators. *Nanophotonics* **2018**, *7*, 127–144. [CrossRef]
16. Khan, K.; Tareen, A.K.; Aslam, M.; Wang, R.; Zhang, Y.; Mahmood, A.; Ouyang, Z.; Zhang, H.; Guo, Z. Recent developments in emerging two-dimensional materials and their applications. *J. Mater. Chem. C* **2020**, *8*, 387–440. [CrossRef]
17. Sangwan, V.K.; Hersam, M.C. Electronic Transport in Two-Dimensional Materials. *Annu. Rev. Phys. Chem.* **2018**, *69*, 299–325. [CrossRef]
18. Fan, Y.; Shen, N.-H.; Zhang, F.; Zhao, Q.; Wei, Z.; Zhang, P.; Dong, J.; Fu, Q.; Li, H.; Soukoulis, C.M. Photoexcited Graphene Metasurfaces: Significantly Enhanced and Tunable Magnetic Resonances. *ACS Photon.* **2018**, *5*, 1612–1618. [CrossRef]
19. Li, Q.; Tian, Z.; Zhang, X.; Xu, N.; Singh, R.; Gu, J.; Lv, P.; Luo, L.-B.; Zhang, S.; Han, J.; et al. Dual control of active graphene–silicon hybrid metamaterial devices. *Carbon* **2015**, *90*, 146–153. [CrossRef]
20. Wu, X.; Pan, X.; Quan, B.; Wang, L. Optical modulation of terahertz behavior in silicon with structured surfaces. *Appl. Phys. Lett.* **2013**, *103*, 121112. [CrossRef]

21. Li, Y.; Wen, T. Optical Wavelength-dependent THz Wave Modulation based on Graphene Coated Germanium. *J. Phys. Conf. Ser.* **2021**, *1750*, 012040. [CrossRef]
22. Fan, Z.; Geng, Z.; Lv, X.; Su, Y.; Yang, Y.; Liu, J.; Chen, H. Optical Controlled Terahertz Modulator Based on Tungsten Disulfide Nanosheet. *Sci. Rep.* **2017**, *7*, 14828. [CrossRef] [PubMed]
23. Yang, D.-S.; Jiang, T.; Cheng, X.-A. Optically controlled terahertz modulator by liquid-exfoliated multilayer WS_2 nanosheets. *Opt. Express* **2017**, *25*, 16364–16377. [CrossRef] [PubMed]
24. Qiao, J.; Wang, S.; Wang, Z.; He, C.; Zhao, S.; Xiong, X.; Wang, S.; Zhang, X.; Tao, X. Ultrasensitive and Broadband All-Optically Controlled THz Modulator Based on $MoTe_2$/Si van der Waals Heterostructure. *Advanced Optical Materials* **2020**, *8*, 2000160. [CrossRef]
25. Jakhar, A.; Kumar, P.; Moudgil, A.; Dhyani, V.; Das, S. Optically Pumped Broadband Terahertz Modulator Based on Nanostructured $PtSe_2$ Thin Films. *Adv. Opt. Mater.* **2020**, *8*, 1901714. [CrossRef]
26. Chen, S.; Fan, F.; Miao, Y.; He, X.; Zhang, K.; Chang, S. Ultrasensitive terahertz modulation by silicon-grown MoS_2 nanosheets. *Nanoscale* **2016**, *8*, 4713–4719. [CrossRef]
27. Ao, D.-W.; Liu, W.-D.; Chen, Y.-X.; Wei, M.; Jabar, B.; Li, F.; Shi, X.-L.; Zheng, Z.-H.; Liang, G.-X.; Zhang, X.-H.; et al. Novel Thermal Diffusion Temperature Engineering Leading to High Thermoelectric Performance in Bi_2Te_3-Based Flexible Thin-Films. *Adv. Sci.* **2022**, *9*, 2103547. [CrossRef]
28. Liu, J.; Wang, H.; Li, X.; Chen, H.; Zhang, Z.; Pan, W.; Luo, G.; Yuan, C.; Ren, Y.; Lei, W. Ultrasensitive flexible near-infrared photodetectors based on Van der Waals Bi_2Te_3 nanoplates. *Appl. Surf. Sci.* **2019**, *484*, 542–550. [CrossRef]
29. Goncalves, L.M.; Couto, C.; Alpuim, P.; Rolo, A.G.; Völklein, F.; Correia, J.H. Optimization of thermoelectric properties on Bi_2Te_3 thin films deposited by thermal co-evaporation. *Thin Solid Films* **2010**, *518*, 2816–2821. [CrossRef]
30. Zhang, L.; Liu, J.; Li, J.; Wang, Z.; Wang, Y.; Ge, Y.; Dong, W.; Xu, N.; He, T.; Zhang, H.; et al. Site-Selective Bi_2Te_3-$FeTe_2$ Heterostructure as a Broadband Saturable Absorber for Ultrafast Photonics. *Laser Photon. Rev.* **2020**, *14*, 1900409. [CrossRef]
31. Pei, J.; Cai, B.; Zhuang, H.-L.; Li, J.-F. Bi_2Te_3-based applied thermoelectric materials: Research advances and new challenges. *Natl. Sci. Rev.* **2020**, *7*, 1856–1858. [CrossRef] [PubMed]
32. Ma, Y.; Zhang, D.-Y.; Peng, Z.; Guan, S.; Zhai, J. Delivery of Platinum(IV) Prodrugs via Bi_2Te_3 Nanoparticles for Photothermal-Chemotherapy and Photothermal/Photoacoustic Imaging. *Mol. Pharmaceutics* **2020**, *17*, 3403–3411. [CrossRef] [PubMed]
33. Yao, J.; Shao, J.; Wang, Y.; Zhao, Z.; Yang, G. Ultra-broadband and high response of the Bi_2Te_3-Si heterojunction and its application as a photodetector at room temperature in harsh working environments. *Nanoscale* **2015**, *7*, 12535–12541. [CrossRef] [PubMed]
34. Chen, X.; Wang, H.; Wang, C.; Ouyang, C.; Wei, G.; Nie, T.; Zhao, W.; Miao, J.; Li, Y.; Wang, L.; et al. Efficient Generation and Arbitrary Manipulation of Chiral Terahertz Waves Emitted from Bi_2Te_3-Fe Heterostructures. *Adv. Photon. Res.* **2021**, *2*, 2000099. [CrossRef]
35. Chen, X.; Wang, H.; Liu, H.; Wang, C.; Wei, G.; Fang, C.; Wang, H.; Geng, C.; Liu, S.; Li, P.; et al. Generation and Control of Terahertz Spin Currents in Topology-Induced 2D Ferromagnetic Fe_3GeTe_2 | Bi_2Te_3 Heterostructures. *Adv. Mater.* **2022**, *34*, 2106172. [CrossRef]
36. Luo, M.; Lu, C.; Liu, Y.; Han, T.; Ge, Y.; Zhou, Y.; Xu, X. Band alignment of type-I SnS_2/Bi_2Se_3 and type-II SnS_2/Bi_2Te_3 van der Waals heterostructures for highly enhanced photoelectric responses. *Sci. China Mater.* **2022**, *65*, 1000–1011. [CrossRef]
37. Lu, C.; Luo, M.; Dong, W.; Ge, Y.; Han, T.; Liu, Y.; Xue, X.; Ma, N.; Huang, Y.; Zhou, Y.; et al. Bi_2Te_3/Bi_2Se_3/Bi_2S_3 Cascade Heterostructure for Fast-Response and High-Photoresponsivity Photodetector and High-Efficiency Water Splitting with a Small Bias Voltage. *Adv. Sci.* **2023**, *10*, 2205460. [CrossRef]
38. Zhang, H.T.; Luo, X.G.; Wang, C.H.; Xiong, Y.M.; Li, S.Y.; Chen, X.H. Characterization of nanocrystalline bismuth telluride (Bi_2Te_3) synthesized by a hydrothermal method. *J. Cryst. Growth* **2004**, *265*, 558–562. [CrossRef]
39. Yao, Z.; Huang, Y.; Du, W.; He, C.; Zhu, L.; Zhang, L.; Xu, X. Interface-Induced Enhancement of THz Generation and Modulation in Hexagonal Boron Nitride/Si Mixed-Dimensional Van Der Waals Heterostructure. *IEEE Trans. Terahertz Sci. Technol.* **2020**, *10*, 101–106. [CrossRef]
40. Wu, Q.; Zhang, X.C. Free-space electro-optic sampling of terahertz beams. *Appl. Phys. Lett.* **1995**, *67*, 3523–3525. [CrossRef]
41. Wei, M.; Zhang, D.; Zhang, L.; Jin, L.; Zhang, H. High-Performance Multifunctional Photodetector and THz Modulator Based on Graphene/TiO_2/p-Si Heterojunction. *Nanoscale Res. Lett.* **2021**, *16*, 134. [CrossRef] [PubMed]
42. Zhai, Z.-H.; Zhu, H.-F.; Shi, Q.; Chen, S.-C.; Li, J.; Li, Z.-R.; Schneider, H.; Zhu, L.-G. Enhanced photoresponses of an optically driven VO_2-based terahertz wave modulator near percolation threshold. *Appl. Phys. Lett.* **2018**, *113*, 231104.
43. Neu, J.; Regan, K.P.; Swierk, J.R.; Schmuttenmaer, C.A. Applicability of the thin-film approximation in terahertz photoconductivity measurements. *Appl. Phys. Lett.* **2018**, *113*, 233901.
44. Wang, M.; Vajtai, R.; Ajayan, P.M.; Kono, J. Electrically tunable hot-silicon terahertz attenuator. *Appl. Phys. Lett.* **2014**, *105*, 141110. [CrossRef]
45. Cao, Y.; Gan, S.; Geng, Z.; Liu, J.; Yang, Y.; Bao, Q.; Chen, H. Optically tuned terahertz modulator based on annealed multilayer MoS_2. *Sci. Rep.* **2016**, *6*, 22899.
46. Tao, M.; Agarwal, S.; Udeshi, D.; Basit, N.; Maldonado, E.; Kirk, W.P. Low Schottky barriers on n-type silicon (001). *Appl. Phys. Lett.* **2003**, *83*, 2593–2595. [CrossRef]
47. Jakhar, A.; Kumar, P.; Husain, S.; Dhyani, V.; Das, S. Integration of Nanometer-Thick 1T-TaS_2 Films with Silicon for an Optically Driven Wide-Band Terahertz Modulator. *ACS Appl. Nano Mater.* **2020**, *3*, 10767–10777. [CrossRef]

48. Lai, W.; Ge, C.; Yuan, H.; Dong, D.; Fang, Y. NIR Light Driven Terahertz Wave Modulator with a Large Modulation Depth Based on a Silicon-PEDOT: PSS-Perovskite Hybride System. *Adv. Mater. Technol.* **2020**, *5*, 1901090. [CrossRef]
49. Binetti, S.; Le Donne, A.; Rolfi, A.; Jäggi, B.; Neuenschwander, B.; Busto, C.; Frigeri, C.; Scorticati, D.; Longoni, L.; Pellegrino, S. Picosecond laser texturization of mc-silicon for photovoltaics: A comparison between 1064nm, 532nm and 355nm radiation wavelengths. *Appl. Surf. Sci.* **2016**, *371*, 196–202. [CrossRef]
50. Uhd Jepsen, P.; Schairer, W.; Libon, I.H.; Lemmer, U.; Hecker, N.E.; Birkholz, M.; Lips, K.; Schall, M. Ultrafast carrier trapping in microcrystalline silicon observed in optical pump–terahertz probe measurements. *Appl. Phys. Lett.* **2001**, *79*, 1291–1293. [CrossRef]

micromachines

Article

Antireflection Structures for VIS and NIR on Arbitrarily Shaped Fused Silica Substrates with Colloidal Polystyrene Nanosphere Lithography

David Schmelz [1,*], Guobin Jia [2], Thomas Käsebier [1], Jonathan Plentz [2] and Uwe Detlef Zeitner [3,4]

1 Institute of Applied Physics, Abbe Center of Photonics, Friedrich Schiller University Jena, 07743 Jena, Germany
2 Leibniz Institute of Photonic Technology (Leibniz-IPHT), 07745 Jena, Germany
3 Fraunhofer Institute for Applied Optics and Precision Engineering IOF, 07745 Jena, Germany
4 Department of Applied Sciences and Mechatronics, Munich University of Applied Sciences, 80335 Munich, Germany
* Correspondence: david.schmelz@uni-jena.de

Abstract: Antireflective (AR) nanostructures offer an effective, broadband alternative to conventional AR coatings that could be used even under extreme conditions. In this publication, a possible fabrication process based on colloidal polystyrene (PS) nanosphere lithography for the fabrication of such AR structures on arbitrarily shaped fused silica substrates is presented and evaluated. Special emphasis is placed on the involved manufacturing steps in order to be able to produce tailored and effective structures. An improved Langmuir-Blodgett self-assembly lithography technique enabled the deposition of 200 nm PS spheres on curved surfaces, independent of shape or material-specific characteristics such as hydrophobicity. The AR structures were fabricated on planar fused silica wafers and aspherical planoconvex lenses. Broadband AR structures with losses (reflection + transmissive scattering) of <1% per surface in the spectral range of 750–2000 nm were produced. At the best performance level, losses were less than 0.5%, which corresponds to an improvement factor of 6.7 compared to unstructured reference substrates.

Keywords: antireflection; moth-eye nanostructures; colloidal lithography; polystyrene nanospheres; Langmuir-Blodgett; reactive ion etching; fused silica lens

Citation: Schmelz, D.; Jia, G.; Käsebier, T.; Plentz, J.; Zeitner, U.D. Antireflection Structures for VIS and NIR on Arbitrarily Shaped Fused Silica Substrates with Colloidal Polystyrene Nanosphere Lithography. *Micromachines* **2023**, *14*, 1204. https://doi.org/10.3390/mi14061204

Academic Editors: He Yang, Yizhong Huang and Xinyang Su

Received: 30 April 2023
Revised: 1 June 2023
Accepted: 4 June 2023
Published: 7 June 2023

1. Introduction

In the fields of optics and photonics, increasing diversification and specialization with regard to the design and materials of optical systems is a current trend. The development of material- and form-specific solutions benefits from this. A typical case here is the antireflection (AR) effect of particular optical components. For special shapes such as free-form [1–4] or structured [5–7] surfaces, extended spectral ranges, or application in environments with extreme conditions [8,9], conventional AR coatings are often not readily applicable. AR structures, also called moth-eye structures, are a possible alternative for this purpose [10,11]. These are conical subwavelength structures that cause a continuous increase in the effective refractive index due to the continuous increase in the fill factor at the transition from air to the optical device. Reflections based on index jumps are thereby prevented or reduced. The advantages of these structures are their large spectral bandwidth and the material homogeneity. Moreover, they can be combined with coatings for more sophisticated optical solutions [12–14].

AR structures can be differentiated by their arrangement: Deterministic structures [15–18] offer good regularity and control of the structures. However, they are very expensive due to the lithography or direct writing effort required and are difficult to apply on non-planar surfaces. Stochastic structures [1,5,19–22], on the other hand, are inexpensive and more flexible in terms of surface shape. However, they are difficult to control due to the statistical variances regarding their dimensions and due to the process parameter-sensitive fabrication. Their irregular structure produces optical scattering [23] and inevitably requires compromises in terms of optical performance. This prevents specifically customized solutions for specific optical devices. In-between are structures produced with lithographic self-assembly methods [3,23–25]. Due to their regularity, they allow for a good adaptation to the optical requirements. On the other hand, they can be applied in a relatively cost-effective and flexible way on arbitrarily shaped substrates.

In this paper, the application of an improved Langmuir-Blodgett (LB) self-assembly lithography technique is presented. In this process, polystyrene (PS) nanospheres with a selected diameter are deposited on a substrate in a self-arranging monolayer. This creates a hexagonal array that can be used as a pattern for further lithographic processes. The pattern is first transferred into an underlaying Cr layer that subsequently serves as mask for etching into the fused silica substrate. The produced AR structures are afterwards optically evaluated with regard to their AR effect.

Colloidal lithography techniques are often used in combination with metal-assisted chemical etching (MACE) for the fabrication of silicon structures. Typically, PS nanospheres serve as negative masking for the deposition of a noble metal that is used for the etching process [26–29]. MACE is mainly applied for patterning silicon and cannot be transferred to glass materials. In this paper, PS nanospheres serve as a positive mask [9,23,30–32], and reactive ion etching (RIE) techniques are used for the creation of the structures. This approach is much more broadly applicable on different materials and shapes. Using this method for the generation of AR structures on fused silica substrates [9,13,23,33] comes with certain challenges regarding the fabrication of suitable structures. The insertion of an additional Cr layer [15] between the PS spheres and the substrate increases the etch selectivity and thus allows for better process adaptation. Consequently, sophisticated structures with a high aspect ratio, smooth sidewalls, and an adapted structure profile can be fabricated that then achieve better AR performance. This paper aims to present a guideline for the fabrication of such AR structures on fused silica substrates. It focuses on certain fabrication aspects that should be considered in order to obtain structures that are most suitable for reaching a high transmission level.

2. Fabrication

Figure 1 illustrates the steps of fabrication of the AR nanostructures. The process starts from a fused silica substrate with a 30 nm thick Cr layer deposited by ion beam deposition (IBD). The used substrates were IR-grade fused silica wafers of 1 mm thickness and 50.8 mm (2″) diameter (see Figure 2), and planoconvex aspherical lenses of 2 mm thickness, 25.0 mm diameter, and 46.07 mm radius of curvature in Corning 7980 UV-grade fused silica glass. Figure 2 shows a single-side patterned 2-inch wafer in comparison to a double-side-polished (dsp) reference wafer. There are scratches from handling during the investigations on the left and bottom edges of the sample as well as some inhomogeneities.

Regarding the fabrication, the focus is on four manufacturing steps:

1. **Deposition of PS nanospheres:** A compact PS monolayer is formed by a self-organized arrangement of PS spheres on the water surface and subsequent compression with surfactant addition. A slow drainage of the water enables their deposition on a substrate located at the bottom of the water basin.
2. **Shrinkage of PS nanospheres:** An O_2 plasma RIE step ensures the shrinkage of the individual PS nanospheres and thus the adaptation of the masking to the subsequent structuring process.
3. **Transfer into Cr mask:** The PS masking is transferred into the underlaying Cr layer using a chlorine-based RIE process. A key factor here is the generation of rounded edges.

4. **Etching of AR structures:** The Cr mask is etched into the substrate material by a stretched proportional transfer with inductively coupled plasma reactive ion etching (ICP-RIE). This forms vertically tapered AR structures in the substrate.

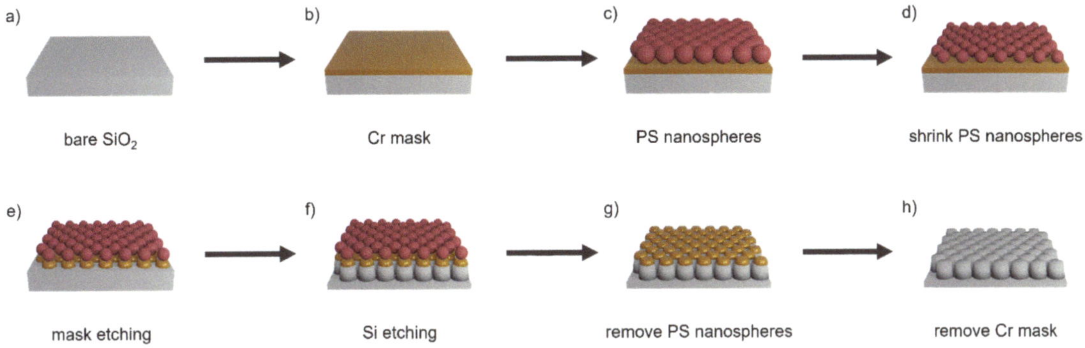

Figure 1. Schematic illustration of the fabrication process of AR nanostructures.

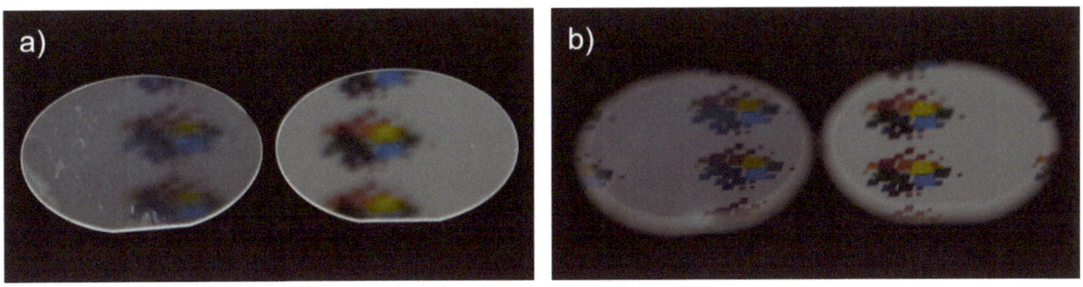

Figure 2. Photographs of two fused silica 2-inch wafers for comparison. A single-side patterned sample with fabricated AR nanostructures (left) is placed next to a double-side polished (dsp) reference wafer (right). The focus of the camera is set once (**a**) on the samples and once (**b**) on the reflected objects.

The method for preparing the PS monolayer is an improved version of the Langmuir-Blodgett (LB) method [26]. Instead of moving mechanical barriers, like in the standard LB process, the compression of the PS spheres is induced by surfactant functionalization. 100 μL of the original suspension is diluted with 1100 μL of ethanol. Subsequently, 5 μL of hexylamine is added for the surface functionalization of the PS nanospheres. The mixture is then sonicated for 30 min. The sample is immersed beneath the water of a Petri dish, and 100 μL of the prepared suspension is slowly injected into the water surface by a syringe. The PS nanospheres spread over the water surface and form a loose monolayer. Then, ca. 50 μL of a 10 wt% sodium dodecyl sulfate (SDS) water solution was injected with another syringe at the edge of the Petri dish. The SDS molecules form a monolayer and dynamically self-assemble themselves during the deposition, so that the pressure at the edge of the PS nanospheres is maintained. The loose PS sphere monolayer is compressed by the SDS and aligned in a hexagonal pattern by the effect of Van der Waals forces. A slow sinking of the water level enables their deposition on the underlaying sample (see Figure 3).

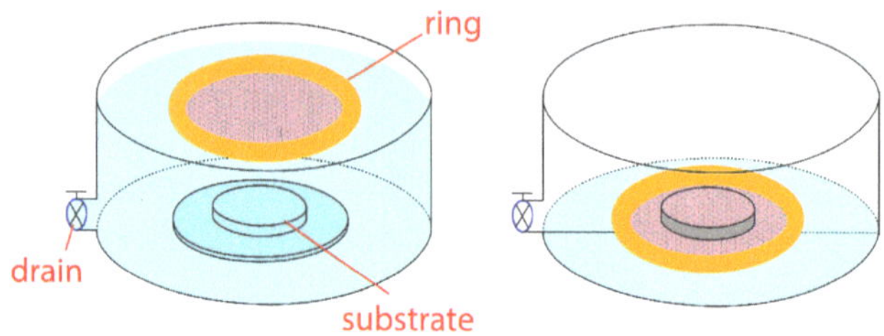

Figure 3. Principle of the monolayer deposition process. Nanospheres are compressed by adding surfactants and deposited on the substrate located at the bottom by draining the water with the aid of a guard ring.

The developed process is significantly simpler, faster, and more fault-tolerant than the standard LB method. In such a way, a conformal deposition of the PS or other nanomaterials such as graphene can be realized on complex 3D shapes and is independent of the surface conditions of the substrate, such as hydrophilicity [34,35]. Moreover, the process offers a good potential for upscaling to larger areas.

The fabrication process was initially developed for structures with a 600 nm period [26], where an aqueous PS nanosphere suspension from Microparticles GmbH was diluted in ethanol and functionalized with hexylamine in an optimized ratio of PS:ethanol: hexylamine = 100:100:5 (in *v:v*). For applications in the VIS and NIR spectral ranges, the 200 nm period and the associated change to nanospheres with 200 nm diameter were modified, so that the ratio of the individual components needed to be optimized. In this work, the ratio of PS:ethanol:hexylamine = 100:1100:5 (in *v:v*) was the optimal condition.

The generated monolayer on the sample serves as a template for transfer into an underlaying 30 nm thick Cr layer. The period of the hexagonal pattern is defined by the diameter of the nanospheres. After the deposition, an O_2 plasma etching step is performed in a RIE plasma etcher SI 591 from SENTECH Instruments GmbH. This shrinks the spheres into a desired size while they remain in their respective positions, and the period stays fixed (see Figure 4). This creates a sufficiently large etch aperture through which the subsequent etching process can attack and form separated structures. If this step is not performed, undesirable contiguous structures result, as shown in Figure 4d. It is important to optimize the amount of shrinkage, so that, on the one hand, the structures are well-separated from each other, and, on the other hand, sufficient etch masking material remains. This can be controlled by selecting the appropriate etching time. For the O_2-RIE process, a standard recipe for the etching of organic photoresists with an RF power of 50 W and an O_2 gas flow of 50 sccm was chosen. The shrinkage of the organic PS nanospheres is time-dependent. Figure 5 shows the measurements of the diameter of the shrunk PS spheres inspected with scanning electron microscopy (SEM). It shows a linear dependence between etching time and diameter. Variations in terms of shrinkage were observed for different sample sizes, which must be considered when setting the etching time.

For the transfer of the mask into the Cr layer, the choice of the etching technology is decisive. The applied chlorine RIE process (2 min, Cl_2 50 sccm, O_2 10 sccm, RF power 100 W) with its isotropic etching component shapes the Cr mask so that it has a slightly descending, rounded sidewall on the outside. This forms the basic shape for the structure profile that can subsequently be transferred into the underlaying fused silica substrate by a stretched proportional transfer. If an ion beam etching (IBE) process with only a physical etching component were used for the transfer instead, the resulting redeposits would produce mask profiles with a slightly outwardly increasing shape. These mask profiles would not be suitable for the following ICP-RIE process. This issue is illustrated in Figure 6, showing PS spheres with 600 nm period on a 30 nm thick Cr layer deposited on a silicon substrate, as silicon is clearly easier to image in an electron microscope than fused silica is. Of the two samples, one was etched with RIE and the other with IBE. The differences in the resulting shape of the Cr masks on the two samples are clearly visible.

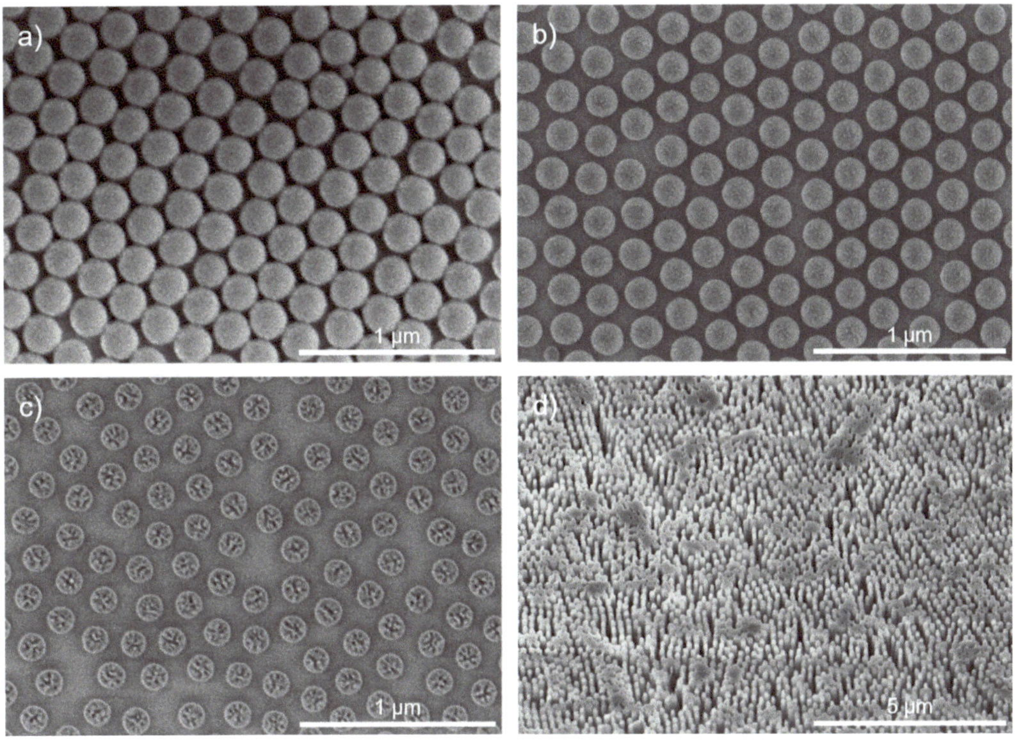

Figure 4. (**a**) PS nanospheres deposited on the surface of a Cr-coated SiO_2 substrate without O_2 plasma treatment. (**b**) PS nanospheres after O_2 plasma treatment with a suitable amount of shrinkage. (**c**) Excessive shrinkage of nanospheres after overlong etching treatment. (**d**) Non-separated structures that emerge when the necessary O_2 plasma step to shrink the PS spheres is omitted. The SEM images were taken (**a–c**) in top view and (**d**) at a 30° tilt angle.

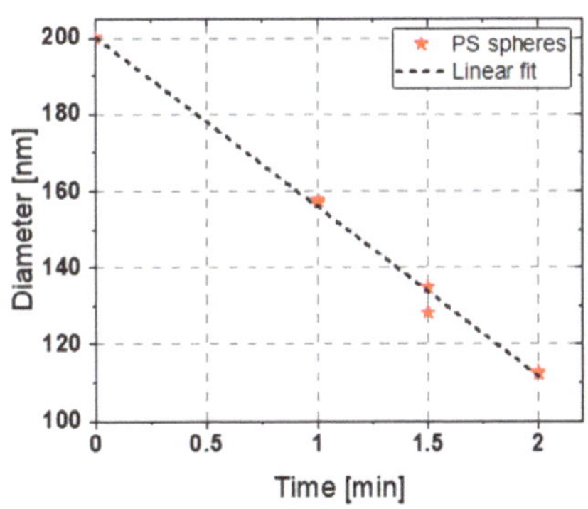

Time of O_2 plasma etch [min]	Diameter of PS spheres [nm]
1:00	158
1:00	157
1:30	128
1:30	135
2:00	112
2:00	113

Figure 5. PS shrinkage after O_2 plasma etching step with respect to the process time on different 2-inch wafer samples.

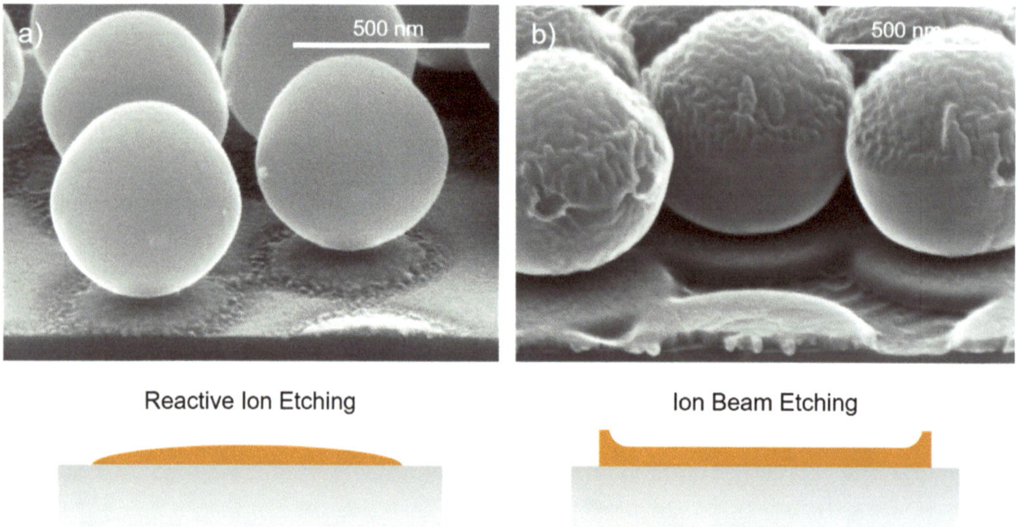

Figure 6. Comparison between (**a**) RIE and (**b**) IBE etching for opening the Cr mask layer. The isotropic component of the RIE etching shapes a rounded mask profile, while the IBE etching creates redeposits resulting in an elevation of the edge region. PS nanospheres with a diameter of 600 nm were used. The process was applied to Si substrates for better visualization in the electron microscope.

The etching of the Cr mask is followed by an ICP-RIE etching step with CHF_3 as the etchant gas at a gas flow of 12 sccm. The process is performed in an ICP-RIE plasma etcher SI 500C etcher by SENTECH Instruments GmbH at relatively high temperatures of 75 °C and a moderate ICP power of 180 W. The process was bias-controlled with a bias voltage of −180 V in the RIE chamber and 20 min duration. It can be considered a stretched proportional transfer of the Cr mask into the SiO_2 substrate. Conical structures were formed in the fused silica 2-inch wafers, as shown in Figure 7a,b. After 20 min, the structures show positive sidewall slopes. In the bottom region, they are connected to each other. Further etching progress is stopped due to the RIE lag respectively aspect ratio dependent etching (ARDE) [36]. The resulting structures have a height of about 600 to 700 nm. With respect to the 200 nm period, this corresponds to an aspect ratio of 3 to 3.5.

The fabrication process is relatively sensitive towards the substrate material and shape. Transferring the process towards the planoconvex SiO_2 lenses requires some adjustments of the process times in the O_2-RIE and ICP-RIE steps 2 and 4. The O_2-RIE process time was extended from 1 min to 1.5 min, as the etch rate of the shrinkage was lower. The ICP-RIE process time was shortened from 20 min to 15 min. This also resulted in slight changes of the structural profile. The structures on the lens are shown in Figure 7c,d. Instead of a conical profile with a constant sidewall angle, a rather bottle-like profile was formed that was close to the Klopfenstein profile [37]. Compared to other structural profiles optimized for the optimal coupling of light, the Klopfenstein profile is very effective in light coupling at comparatively low structural heights [38]. Hence, the structures are very suitable for effective antireflection.

After the etching of the SiO_2 moth-eye structures, the remaining residuals of the PS spheres and the Cr layer are removed with O_2-RIE and Cl-RIE. The fabrication process is then completed. The samples were inspected at different positions. They showed similar structural profiles over the entire sample.

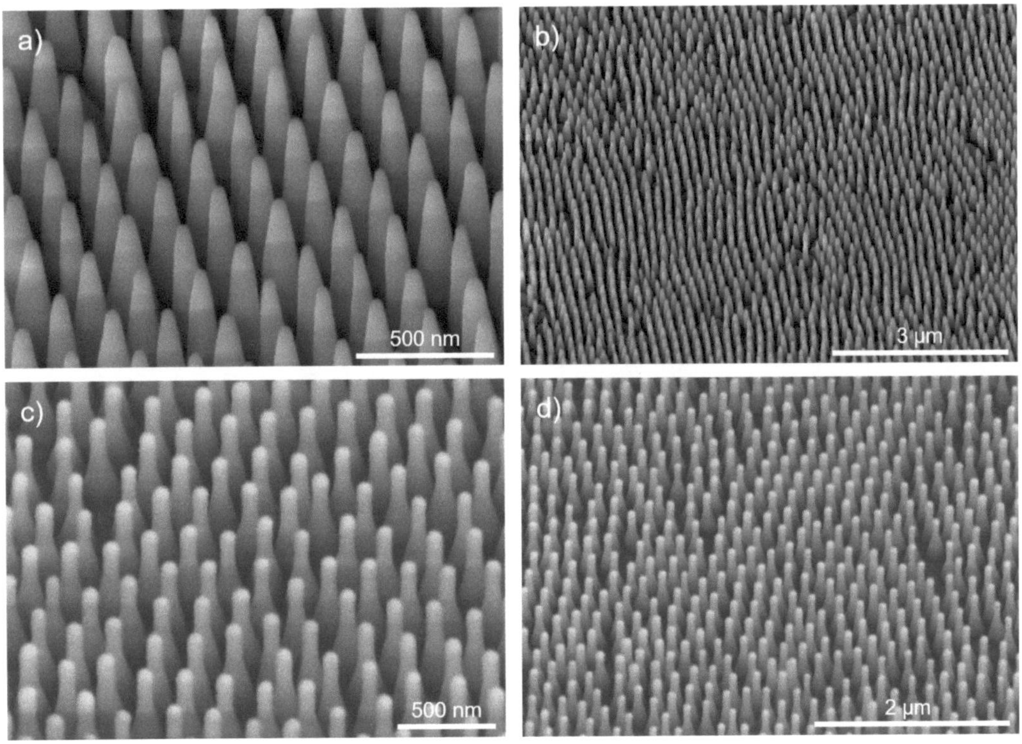

Figure 7. SEM images of fabricated SiO_2 structures after ICP-RIE etching on (**a**,**b**) a plane wafer substrate and (**c**,**d**) on a planoconvex lens at 30° oblique view.

3. Characterization

For the evaluation of the AR properties, the realized structures were characterized using a Lambda 950 spectrometer (Perkin Elmer) with an internal 150 mm diameter integration sphere. The structured samples were measured with respect to their specular transmission over a spectrum of 320–2000 nm and compared to unstructured reference samples. The measurement spot on the sample had a size of about 8×12 mm^2. Besides the specular transmittance, the integration sphere enables the measurement of the total transmittance (consisting of specular and scattered transmissive light). This allows for the differentiation between losses via reflection and transmissive scattering. The former gives an estimate of the AR effect. The latter, on the other hand, is caused by defects in the structure (see Figure 7b). These defects mainly emerge during the deposition process of the PS spheres. In the case of a perfect defect-free assembly of the spheres, the AR structures would be in the subwavelength range. The defects, however, disrupt this subwavelength effect by adding lower spatial frequencies to the pattern, which causes scattering.

Figure 8 shows the measurement results of the structured 2-inch wafer sample (displayed in Figure 7a,b). It plots the specular transmission T_{spec} and total transmission T_{total} of the AR structured sample compared to a double-side polished (dsp) reference SiO_2 sample. The reference sample measurements agree with the theoretical Fresnel reflectance values. From transmittance T of the reference sample, the transmittance of a single polished surface is calculated that is, in the case of a one-side structured sample, equivalent to an ideal AR effect. For an easier assessment of the AR structures, the transmittance of only the single-patterned surface is also determined and is plotted in Figure 8b.

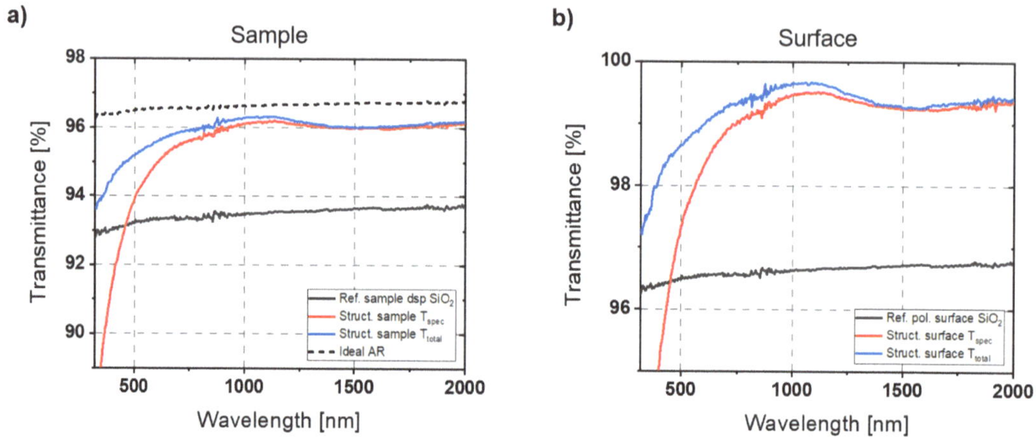

Figure 8. Transmission measurement of one-side structured 2-inch wafer sample and dsp reference sample. (**a**) Specular transmittance T_{spec} is compared to the total transmittance T_{total} (consisting of specular and scattered transmissive light) of the sample. (**b**) For the easier evaluation of the AR effect, the transmittance of only the structured surface is calculated and compared to the polished reference surface.

The fabricated structures have a rounded rather than pointed profile towards the tip. This is beneficial for longer wavelengths where the structural height is lower in relation to the wavelength [23]. The reflectance curves show a broadband AR effect over the entire measurement spectrum. The specular transmission is significantly increased in the range of 500–2000 nm. Between 750 and 2000 nm, the specular transmittance per surface was above 99%. At its maximum at around 1100 nm, losses are less than 0.5%, which corresponds to an improvement factor of 6.7 compared to the polished reference SiO_2 surface.

From 2000 nm towards shorter wavelengths, the transmittance slightly increases and reaches its maximum as the ratio between height and wavelength also increases. Ji et al. [23] describe this part of the spectrum as the range where the shape of the structures is most decisive for the AR effect. Left of the maximum, towards lower wavelengths, the arrangement of the AR structures is the most important factor. For wavelengths above 1000 nm, the transmissive scattering losses are rather neglectable. Towards shorter wavelengths, the gap between T_{spec} and T_{total} grows, indicating an increasing influence of the pattern defects that cause scattering. The decrease in specular reflectance already starts at relatively long wavelengths compared to the structure width of 200 nm. This can be explained by the randomization of the structure due to the presence of defects. Stochastically distributed AR structures exhibit scattering effects even for wavelengths that are significantly longer than the average structure width [23,39]. However, total reflectance T_{total} decreases as well, which indicates an increasing reflectance. This can be explained by the presence of larger areas with low inclination due to defects and the absence of steep structures in these areas. The shorter the wavelengths, the better even small areas are "resolved" by the incident light [40].

The measured values could be reproduced in several measurement series at different positions of the sample, which corresponds to the observations of a homogeneous structural morphology in the SEM images.

In addition to characterizing the planar substrates, which are well-suited for evaluating optical properties, there were also attempts to measure the structured lens sample. The difficulty here is that the refractive behavior of the lens strongly influences the optical beam path in the spectrometer. To reduce this influence, an aperture of 6 mm diameter was used. The aperture blocks the light that is further away from the optical axis, thus not illuminating the areas where strong refraction by the lens occurs. The aperture and the lens sample were placed at a moderate distance in front of the integration sphere. The problem here was that only the central region of the structured lens could be examined. In that area, structuring effects were already macroscopically visible, which influenced the results. Reducing the size of the measurement field results in noisier and less accurate measurement signals. The measured results are shown in Figure 9a. The transmission dip at around 1.4 μm wavelength is caused by OH^- absorption since the used material is UV-grade fused silica in contrast to the IR-grade wafer substrate. Stronger noise is also clearly visible. Apart from that, the values of the reference samples are in a similar range. While the AR effect in the upper wavelength range is still comparable to that of the wafer sample, a drop of the transmittance due to the defects on the lens sample can be seen already at higher wavelengths. Nevertheless, the AR effect based on the AR nanostructures could be principally demonstrated here as well.

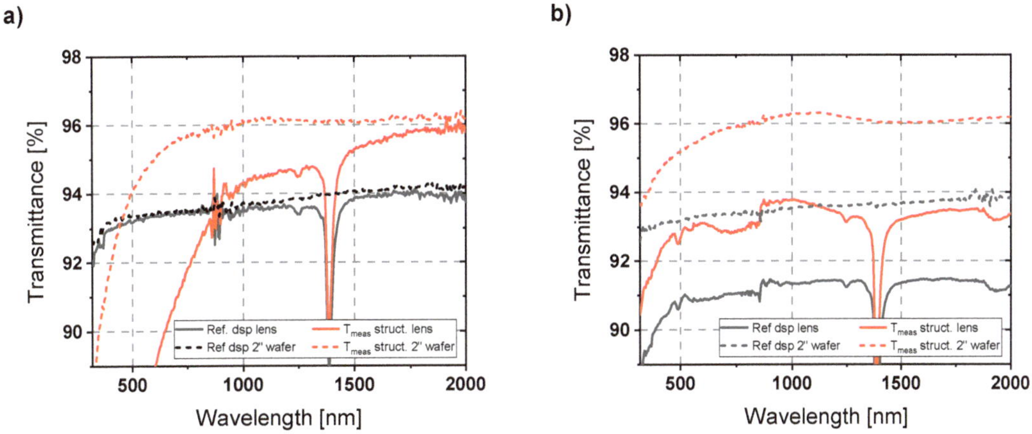

Figure 9. Transmission measurements of the AR structured lens sample compared to an unstructured dsp reference lens. (**a**) The lens was aligned in the center of the optical axis at a moderate distance from the integrating sphere of the spectrometer. An aperture with 6 mm diameter was placed just in front of the lens to cut out stronger refracted light. (**b**) The lens was attached directly in front of the integrating sphere, allowing for the transmissive scattered light to be measured. Control over the alignment of the lens is more difficult in this case, causing deflections in the optical path.

This becomes even clearer in the second attempt that was performed to investigate the AR properties of the patterned lens sample. Here, the 6 mm aperture was omitted. The lens sample was attached directly in front of the integrating sphere using adhesive tape. In this attempt, it is difficult to control the orientation of the lens, which causes deflections in the optical path. However, the measurement spot is larger, which allows for a more general estimation of the AR effect of the lens sample. Furthermore, by placing the lens directly in front of the integration sphere, the transmissive scattered light can also be detected. The measured results are shown in Figure 9b. They show an obvious AR effect over the entire measurement spectrum. This confirms the assumption that, towards shorter wavelengths, losses are mainly caused by transmissive scattering instead of increasing reflection. In both attempts to evaluate the patterned lens, a strong improvement in transmission due to a significant AR effect was observed.

4. Conclusions

In this paper, a readily suitable method for the fabrication of antireflective (AR) moth-eye nanostructures on fused silica was presented. For this purpose, a lithographic method based on self-assembling colloidal polystyrene (PS) nanospheres was applied. An improved Langmuir-Blodgett (LB) process allowed for the nanospheres to be deposited on curved and free-formed substrates. This was demonstrated by the fabrication of AR structures on aspherical lenses of fused silica. In principle, this is applicable to arbitrary materials. Hence, other materials and spectral ranges are worthwhile subjects for future research. An established deposition process gives the opportunities of uncomplicated and prompt processing and modification of optical and photonic components, as no mask layout and no lithographic exposure is required. Apart from classical optical components such as lenses, it may be applied to photonic sensors, modulators, or integrated circuits based on silicon platforms. Conceivable examples are its application on monolithically integrated microlenses [41] or locally backside-thinned devices [42]. On the other hand, there are limitations when the surface area of the substrate to be coated is significantly larger than the area projected onto the water surface.

The insertion of an intermediate Cr layer of a few tens of nanometers significantly expands the design latitude of the desired structures. After transferring the PS pattern, the Cr layer served as masking for the etching of the AR structures. Its high selectivity towards glass in reactive ion etching processes enables the fabrication of structures with high aspect ratios for various materials. For even higher aspect ratios or materials with low etch rates, it might be suitable to apply a multiplexed instead of a static etching process.

Regarding the AR effect of the fabricated nanostructures, a significant suppression of reflection could be demonstrated. However, for the application of transmissive AR structures on glasses, the reduction of transmissive scattering is also essential. The scattering is mainly caused by patterning effects that particularly occur during the deposition of PS nanospheres. Further development of the presented improved LB process towards a defect-free deposition is, therefore, the key factor for the application at shorter wavelengths. Particularly important for this is a homogeneous nanosphere distribution. The optimized composition and high purity of the chemicals prevent the formation of surfactant residues that interfere with perfect compression from the outside. There are ideas and preliminary tests to convert the sequential directed self-assembly process into a continuous process and, ideally, use a roll-to-roll process. In this way, the process parameters can be adjusted very well, and the result can be monitored in situ. In addition to moving towards a defect-free deposition, the deposition on larger substrates might also be of interest for future research, as it enables the integration in common 100 mm and 150 mm fabrication lines.

Author Contributions: Writing—original draft: D.S. and G.J.; Writing—review & editing: D.S., J.P. and U.D.Z.; Methodology: D.S., G.J. and T.K.; Investigation: D.S., G.J. and T.K.; Data curation: D.S.; Supervision: J.P. and U.D.Z.; Funding acquisition: J.P. and U.D.Z. All authors have read and agreed to the published version of the manuscript.

Funding: We acknowledge the funding by the Thuringian Ministry of Economy, Science and Digital Society within the InfraLith project (2021 FGI 0020) co-financed by the European Union within the European Regional Development Fund (ERDF), the funding by the Federal Ministry of Education and Research within the project GraphEnGas (01LY2005B) co-financed by the European Union within NextGenerationEU, and the funding by the Federal Ministry of Education and Research (*BMBF*) (03WKCX1C). We also acknowledge the support by the German Research Foundation Projekt-Nr. 5s12648189 and the Open Access Publication Fund of the Thueringer Universitaets- und Landesbibliothek Jena.

Data Availability Statement: The data that support the findings of this study are available upon reasonable request from the authors.

Conflicts of Interest: The authors declare no conflict of interest.

References

1. Schulze, M.; Lehr, D.; Helgert, M.; Kley, E.-B.; Tünnermann, A. Transmission enhanced optical lenses with self-organized antireflective subwavelength structures for the UV range. *Opt. Lett.* **2011**, *36*, 3924–3926. [CrossRef] [PubMed]
2. Park, S.-C.; Kim, N.; Ji, S.; Lim, H. Fabrication and characterization of moth-eye mimicking nanostructured convex lens. *Microelectron. Eng.* **2016**, *158*, 35–40. [CrossRef]

3. Kohoutek, T.; Parchine, M.; Bardosova, M.; Pemble, M.E. Controlled self-assembly of Langmuir-Blodgett colloidal crystal films of monodispersed silica particles on non-planar substrates. *Colloids Surf. A Physicochem. Eng. Asp.* **2020**, *593*, 124625. [CrossRef]
4. Su, W.-X.; Wu, C.-Y.; Lee, Y.-C. Anti-reflection nano-structures fabricated on curved surface of glass lens based on metal contact printing lithography. *Microelectron. Eng.* **2019**, *214*, 15–20. [CrossRef]
5. Ye, X.; Jiang, X.; Huang, J.; Geng, F.; Sun, L.; Zu, X.; Wu, W.; Zheng, W. Formation of broadband antireflective and superhydrophilic subwavelength structures on fused silica using one-step self-masking reactive ion etching. *Sci. Rep.* **2015**, *5*, 13023. [CrossRef]
6. Gadamsetti, P.; Poutous, M.K. Fresnel reflection suppression from deterministic illumination diffusers using antireflection random nanostructures. *Opt. Eng.* **2022**, *61*, 063106. [CrossRef]
7. Massiot, I.; Trompoukis, C.; Lodewijks, K.; Depauw, V.; Dmitriev, A. Highly conformal fabrication of nanopatterns on non-planar surfaces. *Nanoscale* **2016**, *8*, 11461–11466. [CrossRef]
8. Schulze, M.; Damm, M.; Helgert, M.; Kley, E.-B.; Nolte, S.; Tünnermann, A. Durability of stochastic antireflective structures-analyses on damage thresholds and adsorbate elimination. *Opt. Express* **2012**, *20*, 18348–18355. [CrossRef]
9. Ye, X.; Jiang, X.-D.; Huang, J.; Sun, L.-X.; Geng, F.; Yi, Z.; Zu, X.-T.; Wu, W.-D.; Zheng, W. Subwavelength structures for high power laser antireflection application on fused silica by one-step reactive ion etching. *Opt. Lasers Eng.* **2016**, *78*, 48–54. [CrossRef]
10. Chattopadhyay, S.; Huang, Y.; Jen, Y.; Ganguly, A.; Chen, K.; Chen, L. Anti-reflecting and photonic nanostructures. *Mater. Sci. Eng. R Rep.* **2010**, *69*, 1–35. [CrossRef]
11. Li, Y.; Zhang, J.; Yang, B. Antireflective surfaces based on biomimetic nanopillared arrays. *Nano Today* **2010**, *5*, 117–127. [CrossRef]
12. Yang, Q.; A Zhang, X.; Bagal, A.; Guo, W.; Chang, C.-H. Antireflection effects at nanostructured material interfaces and the suppression of thin-film interference. *Nanotechnology* **2013**, *24*, 235202. [CrossRef] [PubMed]
13. Li, Z.; Song, C.; Xiang, X.; Yang, H.; Wang, X.; Gao, J. Hybrid nanostructured antireflection coating by self-assembled nanosphere lithography. *Coatings* **2019**, *9*, 453. [CrossRef]
14. Schmelz, D.; Gerold, K.; Käsebier, T.; Sergeev, N.; Szeghalmi, A.; Zeitner, U.D. Optical properties of black silicon structures ALD-coated with Al_2O_3. *Nanotechnology* **2022**, *34*, 015704. [CrossRef]
15. Toyota, H.; Takahara, K.; Okano, M.; Yotsuya, T.; Kikuta, H. Fabrication of microcone array for antireflection structured surface using metal dotted pattern. *Jpn. J. Appl. Phys.* **2001**, *40*, L747. [CrossRef]
16. Kanamori, Y.; Kikuta, H.; Hane, K. Broadband antireflection gratings for glass substrates fabricated by fast atom beam etching. *Jpn. J. Appl. Phys.* **2000**, *39*, L735. [CrossRef]
17. Park, K.-C.; Choi, H.J.; Chang, C.-H.; Cohen, R.E.; McKinley, G.H.; Barbastathis, G. Nanotextured silica surfaces with robust superhydrophobicity and omnidirectional broadband supertransmissivity. *ACS Nano* **2012**, *6*, 3789–3799. [CrossRef]
18. Papadopoulos, A.; Skoulas, E.; Mimidis, A.; Perrakis, G.; Kenanakis, G.; Tsibidis, G.D.; Stratakis, E. Biomimetic omnidirectional antireflective glass via direct ultrafast laser nanostructuring. *Adv. Mater.* **2019**, *31*, 1901123. [CrossRef]
19. Isakov, K.; Kauppinen, C.; Franssila, S.; Lipsanen, H. Superhydrophobic antireflection coating on glass using grass-like alumina and fluoropolymer. *ACS Appl. Mater. Interfaces* **2020**, *12*, 49957–49962. [CrossRef]
20. Chen, R.-Y.; Lai, C.-J.; Chen, Y.-J.; Wu, M.-X.; Yang, H. Omnidirectional/unidirectional antireflection-switchable structures inspired by dragonfly wings. *J. Colloid Interface Sci.* **2022**, *610*, 246–257. [CrossRef]
21. Ye, X.; Huang, J.; Geng, F.; Sun, L.; Liu, H.; Jiang, X.; Wu, W.; Zu, X.; Zheng, W. Broadband antireflection subwavelength structures on fused silica using lower temperatures normal atmosphere thermal dewetted Au nanopatterns. *IEEE Photonics J.* **2015**, *8*, 1–10. [CrossRef]
22. Ye, X.; Shao, T.; Sun, L.; Wu, J.; Wang, F.; He, J.; Jiang, X.; Wu, W.-D.; Zheng, W. Plasma-induced, self-masking, one-step approach to an ultrabroadband antireflective and superhydrophilic subwavelength nanostructured fused silica surface. *ACS Appl. Mater. Interfaces* **2018**, *10*, 13851–13859. [CrossRef] [PubMed]
23. Ji, S.; Song, K.; Nguyen, T.B.; Kim, N.; Lim, H. Optimal moth eye nanostructure array on transparent glass towards broadband antireflection. *ACS Appl. Mater. Interfaces* **2013**, *5*, 10731–10737. [CrossRef] [PubMed]
24. Ji, S.; Park, J.; Lim, H. Improved antireflection properties of moth eye mimicking nanopillars on transparent glass: Flat antireflection and color tuning. *Nanoscale* **2012**, *4*, 4603–4610. [CrossRef]
25. Lohmüller, T.; Helgert, M.; Sundermann, M.; Brunner, R.; Spatz, J.P. Biomimetic interfaces for high-performance optics in the deep-UV light range. *Nano Lett.* **2008**, *8*, 1429–1433. [CrossRef]
26. Jia, G.; Westphalen, J.; Drexler, J.; Plentz, J.; Dellith, J.; Dellith, A.; Andrä, G.; Falk, F. Ordered silicon nanowire arrays prepared by an improved nanospheres self-assembly in combination with Ag-assisted wet chemical etching. *Photonics Nanostruct.-Fundam. Appl.* **2016**, *19*, 64–70. [CrossRef]
27. Wendisch, F.J.; Rey, M.; Vogel, N.; Bourret, G.R. Large-scale synthesis of highly uniform silicon nanowire arrays using metal-assisted chemical etching. *Chem. Mater.* **2020**, *32*, 9425–9434. [CrossRef]
28. Van Minh, N.; Van Hieu, D.; Van Tuan, P.; Dung, N.D.; Hoang, C.M. Characteristics of silicon nano-pillars fabricated by nano-sphere lithography and metal assisted chemical etching. *Mater. Sci. Semicond. Process.* **2022**, *142*, 106483. [CrossRef]
29. Mikhael, B.; Elise, B.; Xavier, M.; Sebastian, S.; Johann, M.; Laetitia, P. New silicon architectures by gold-assisted chemical etching. *ACS Appl. Mater. Interfaces* **2011**, *3*, 3866–3873. [CrossRef]
30. Zhang, X.; Zhang, J.; Ren, Z.; Li, X.; Zhang, X.; Zhu, D.; Wang, T.; Tian, T.; Yang, B. Morphology and wettability control of silicon cone arrays using colloidal lithography. *Langmuir* **2009**, *25*, 7375–7382. [CrossRef]

31. Motamedi, M.; Jia, G.; Yao, Y.; Shanks, K.; Yousefi, P.; Hewakuruppu, Y.L.; Rafeie, M.; Lindner, F.; Patterson, R.; Christiansen, S.; et al. Nanopatterned indium tin oxide as a selective coating for solar thermal applications. *Renew. Energy* **2023**, *201*, 386–396. [CrossRef]
32. Jia, G.; Plentz, J.; Höger, I.; Dellith, J.; Dellith, A.; Falk, F. Core–shell diodes for particle detectors. *J. Phys. D Appl. Phys.* **2016**, *49*, 065106. [CrossRef]
33. Li, Y.; Zhang, J.; Zhu, S.; Dong, H.; Jia, F.; Wang, Z.; Tang, Y.; Zhang, L.; Zhang, S.; Yang, B. Bioinspired silica surfaces with near-infrared improved transmittance and superhydrophobicity by colloidal lithography. *Langmuir* **2010**, *26*, 9842–9847. [CrossRef] [PubMed]
34. Jia, G.; Plentz, J.; Presselt, M.; Dellith, J.; Dellith, A.; Patze, S.; Tölle, F.J.; Mülhaupt, R.; Andrä, G.; Falk, F.; et al. A Double Self-Assembly Process for Versatile Reduced-Graphene-Oxide Layer Deposition and Conformal Coating on 3D Structures. *Adv. Mater. Interfaces* **2017**, *4*, 1700758. [CrossRef]
35. Jia, G.; Plentz, J.; Dellith, J.; Dellith, A.; Wahyuono, R.A.; Andrä, G. Large area graphene deposition on hydrophobic surfaces, flexible textiles, glass fibers and 3D structures. *Coatings* **2019**, *9*, 183. [CrossRef]
36. Rangelow, I.W. Critical tasks in high aspect ratio silicon dry etching for microelectromechanical systems. *J. Vac. Sci. Technol. A Vac. Surf. Film.* **2003**, *21*, 1550–1562. [CrossRef]
37. Klopfenstein, R.W. A transmission line taper of improved design. *Proc. IRE* **1956**, *44*, 31–35. [CrossRef]
38. Grann, E.B.; Moharam, M.G.; Pommet, D.A. Optimal design for antireflective tapered two-dimensional subwavelength grating structures. *JOSA A* **1995**, *12*, 333–339. [CrossRef]
39. Steglich, M.; Käsebier, T.; Schrempel, F.; Kley, E.-B.; Tünnermann, A. Self-organized, effective medium Black Silicon for infrared antireflection. *Infrared Phys. Technol.* **2015**, *69*, 218–221. [CrossRef]
40. Steglich, M.; Käsebier, T.; Zilk, M.; Pertsch, T.; Kley, E.-B.; Tünnermann, A. The structural and optical properties of black silicon by inductively coupled plasma reactive ion etching. *J. Appl. Phys.* **2014**, *116*, 173503. [CrossRef]
41. Mangal, N.; Snyder, B.; Van Campenhout, J.; Van Steenberge, G.; Missinne, J. Monolithic integration of microlenses on the backside of a silicon photonics chip for expanded beam coupling. *Opt. Express* **2021**, *29*, 7601–7615. [CrossRef] [PubMed]
42. Schmelz, D.; Steglich, M.; Dietrich, K.; Käsebier, T.; Zeitner, U.D. Black-silicon-structured back-illuminated Ge-on-Si photodiode arrays. *SPIE* **2019**, *11031*, 1103109. [CrossRef]

micromachines

MDPI

Article

Design of 2 μm Low-Loss Hollow-Core Anti-Resonant Fibers

Tianran Sun [1,2], Xinyang Su [1,2,*], Fanchao Meng [1,2], Zaining Wang [1], Jiale Song [1], Chenglong Zhang [1], Tianjia Xu [3], Yunhong Zhang [1,2], Huaiwei Zhang [1,2], Mengdi Cui [1,2] and Yi Zheng [1,2,*]

1 School of Physical Science and Engineering, Beijing Jiaotong University, Beijing 100044, China; 18118043@bjtu.edu.cn (T.S.); mfanchao@126.com (F.M.); 20271163@bjtu.edu.cn (Z.W.); 20271044@bjtu.edu.cn (J.S.); 21271278@bjtu.edu.cn (C.Z.); 19118047@bjtu.edu.cn (Y.Z.); 18126233@bjtu.edu.cn (H.Z.); 21118043@bjtu.edu.cn (M.C.)

2 Key Laboratory of Luminescence and Optical Information, Ministry of Education, Beijing Jiaotong University, Beijing 100044, China

3 School of Optics and Photonics, Beijing Institute of Technology, Beijing 100081, China; 3220220556@bit.edu.cn

* Correspondence: xysu@bjtu.edu.cn (X.S.); yizheng@bjtu.edu.cn (Y.Z.)

Abstract: We systematically studied several of the most traditional hollow-core anti-resonant fiber (HC-ARF) structures, with the aim of achieving low confinement loss, single-mode performance, and high insensitivity to bending in the 2 μm band. Moreover, the propagation loss of fundamental mode (FM), higher-order mode (HOMs), and the higher-order mode extinction ratio (HOMER) under different geometric parameters were studied. Analysis showed that the confinement loss of the six-tube nodeless hollow-core anti-resonant fiber at 2 μm was 0.042 dB/km, and its higher-order mode extinction ratio was higher than 9000. At the same time, a confinement loss of 0.040 dB/km at 2 μm was achieved in the five-tube nodeless hollow-core anti-resonant fiber, and its higher-order mode extinction ratio was higher than 2700.

Keywords: anti-resonant fiber (ARF); higher-order mode extinction ratio (HOMER); confinement loss (CL)

Citation: Sun, T.; Su, X.; Meng, F.; Wang, Z.; Song, J.; Zhang, C.; Xu, T.; Zhang, Y.; Zhang, H.; Cui, M.; et al. Design of 2 μm Low-Loss Hollow-Core Anti-Resonant Fibers. *Micromachines* **2023**, *14*, 1198. https://doi.org/10.3390/mi14061198

Academic Editor: Stefan Wabnitz

Received: 29 April 2023
Revised: 1 June 2023
Accepted: 2 June 2023
Published: 5 June 2023

1. Introduction

Recent years have seen the development of hollow-core fibers. Due to the extraordinary light-guiding properties of hollow-core fibers, they have attracted extensive research and manufacture by research groups around the world [1–8]. Most of the energy (>99.99%) can be concentrated in the air part when the light travels through the hollow-core fibers, and only a small part of the light overlaps with the surrounding glass structure. These fibers can take advantage of the ultra-low Rayleigh scattering and nonlinearity coefficients of the air (one order of magnitude lower than any glass), allowing lasers to propagate at ultralow losses and nonlinearity. In addition, compared with all-solid-state core fibers, light travels 40% faster in hollow-core fibers than in traditional silica fibers, which can achieve nearly 99.7% of the speed of light in a vacuum, and the optical signal transmission speed per kilometer is accelerated by 1.54 μs, with lower time delay and a higher laser damage threshold [4,5,9–11]. In principle, compared with solid-state fibers, hollow-core fibers are less sensitive to environmental disturbances such as mechanical vibrations, magnetic fields, and ionizing radiation [9].

In general, two kinds of hollow-core optical fibers have been proposed according to the waveguide principle: The first is hollow-core photonic bandgap (HC-PBG) fiber, with multiple layers of periodic air pores arranged in the cladding. Light is confined inside the hollow core through the PBG effect [1]. Light in the bandgap cannot propagate in the periodic air pores of the fiber cladding but can only be bound to the core. In 2004, Manyan et al. [12] reported a kind of HC-PBG fiber with a loss of 1.7 dB/km at a wavelength of 1.62 μm, with a transmission bandwidth of 70 nm. In 2005, Roberts et al. [13] reduced the loss of the HC-PBG fiber in this band to 1.2 dB/km, and they pointed out that the loss was close to the limit due to the limitation of surface scattering loss (SSL). In addition to their limited transmission bandwidth, HC-PBG fibers exhibit considerable power overlap and high group delay dispersion (especially near bandgap edges) [14].

Another type of hollow-core fiber is usually referred to as hollow-core anti-resonant fiber (HC-ARF). In 2002, N.M. Litchinitser et al. [15] pioneered the use of the anti-resonant reflecting optical waveguide (ARROW) guiding mechanism, using glass walls to form a Fabry–Perot resonant cavity structure in the cladding. By controlling the incident wavelength and the thickness of the glass wall, the resonance and anti-resonance conditions can be controlled. When the anti-resonance condition is satisfied, the transmission of the resonant cavity is the least and the reflectivity is the most, and the light is restricted in the fiber core by reflection, thereby forming an optical waveguide.

For HC-ARF, because the contact surface between the core and the cladding quartz wall is small, and the core diameter size is large, the light field intensity is weak at the intersection, and the SSL can be reduced by an order of magnitude [9,14]. To control the modal content and attenuation coefficient, HC-ARF nested with several types of anti-resonant tubes has been proposed, studied, and manufactured, including circular anti-resonant tubes [4,10,16–19], "ice cream cone"-shaped anti-resonant tubes [19,20], elliptical anti-resonant tubes [21], nested anti-resonant tubes [5,9,11,22], and more complex shaped anti-resonant tubes [23]. In addition, HC-ARF has become one of the research hotspots in the field of optical fibers due to its advantages, such as ultrawide transmission bandwidth, low confinement loss, and relatively simple preparation. In 2022, Xin Zhang et al. [22] reported the preparation and characterization of a nested five-tube HC-ARF with an attenuation rate of 0.85 dB/km at 2 μm. The bandwidth was 200 nm with a confinement loss below 2 dB/km, and the HC-ARF showed excellent modal purity.

In this study, in order to obtain HC-ARF with better transmission performance in the 2 μm band, and to reduce the difficulty of drawing and obtaining a more easily implemented optical fiber design, several of the most traditional and mature HC-ARF structures were systematically studied to achieve the lowest transmission loss and the best single-mode performance, as well as a design that was insensitive to bending in the 2 μm band. Moreover, to obtain complete information about the modal content of the fiber, the confinement loss of the fundamental mode (FM) and higher-order modes (HOMs) under different geometric parameters, as well as the higher-order mode extinction ratio (HOMER), was studied. Our numerical study provides theoretical guidance for the correct selection of the number and size of the anti-resonant cladding tubes to achieve low transmission loss and high single-mode performance of the 2 μm laser.

2. Fiber Geometry

Figure 1 shows the geometry of the six-tube nested HC-ARF that we analyzed, where D_c is the core diameter, i.e., the diameter of the largest inner circle in the middle of the anti-resonant tube. The core radius of the HC-ARF has a crucial influence on the confinement loss. The loss is inversely proportional to the core diameter, but the way to obtain low loss by increasing D_c is not without limitations, and the higher-order mode loss in the fiber will also decrease with the increase in D_c, which is harmful to the single-mode transmission of the optical fiber. d is the diameter of the inner anti-resonant tube, g is the gap distance between the anti-resonant tubes, and the contact points between the anti-resonant tubes are called nodes. Although the cladding of this fiber structure with nodes is more blocked and there is no gap, it has been shown that these nodes will produce Fano resonance when the light travels through the fiber, which increases the loss of the fiber and adversely affects the transmission of the fiber [24]. A nodeless design can provide better loss attributes. However, excessively large g will lead to light leakage. D is the diameter of the anti-resonant tube, and its numerical calculation formula is as follows:

$$D = \frac{g - D_c \times \sin\left(\frac{\pi}{N}\right)}{\sin\left(\frac{\pi}{N}\right) - 1}, \quad (1)$$

where N is the number of anti-resonant cladding tubes, and t is the wall thickness of the anti-resonant tube. Its numerical calculation formula is as follows:

$$t = \frac{\lambda_m \times (m - 0.5)}{2 \times \sqrt{n^2 - 1}}, \quad (2)$$

where λ_m is the wavelength, m is the resonance order (non-zero integer), and n is the refractive index of the cladding tube.

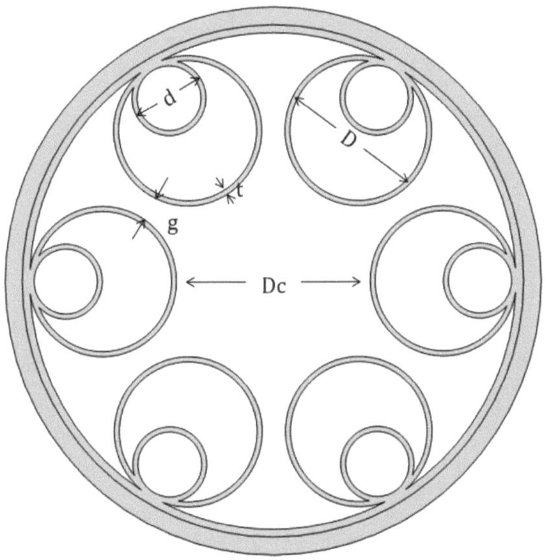

Figure 1. Schematic diagram of the structure of hollow-core nested anti-resonant fiber (HC-NANF or NANF).

3. Numerical Results

For numerical calculations, we used the COMSOL mode solver with the finite element method. To accurately simulate the confinement loss of the fiber, the perfectly matched layers (PMLs) were adopted outside the fiber zone. To achieve accurate results, the mesh size of PML is critical in thin silica walls. Therefore, the fine mesh sizes of $\lambda/4$ and $\lambda/6$ were used in the sections of air and silica, respectively [9]. To ensure the convergence of the numerical results, the code was tested by reproducing the results in [22,25].

3.1. Parameter Optimization of Six-Tube Node-Free Nested HC-ARF

Figure 2 shows the schematic diagram of the classic six-tube HC-AR fiber, which has more confinement loss compared with classic six-tube nested HC-AR fiber, because it has one less layer of the anti-resonant tube. However, its simple construction and easy processing make it convenient for theoretical simulation. With only one layer of anti-resonant tube, the independent parameters that affect the fiber structure are only D_c, g, and t when the number of tubes is six, and the diameter of the anti-resonant tube D can be determined by Equation (1) using the parameters D_c, g, and N.

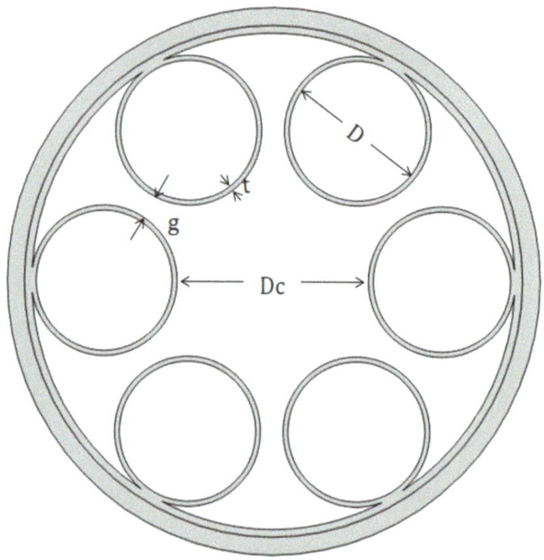

Figure 2. Structure diagram of HC-ARF with circular cladding tubes.

The number of cladding tubes (N) was six in this simulation; the gray area represents quartz, and its refractive index was determined by Sellmeier's formula. The material represented by the white area is air, with a refractive index of 1.0. The wavelength was set to 2 μm. According to known research, the core diameter of optical fibers is generally set to be about 30 times the wavelength [7], so the core diameter D_c was set to be 55 μm. Using Equation (2), it can be inferred that the first-order and the second-order anti-resonance thicknesses are 0.48 μm and 1.46 μm, respectively. In order to determine the best silica strut thickness and simulate the influence of optical fiber thickness on confinement loss, we calculated the confinement loss of the geometric model in Figure 2 with different silica wall thicknesses t and different gap distances g. The results are shown in Figure 3.

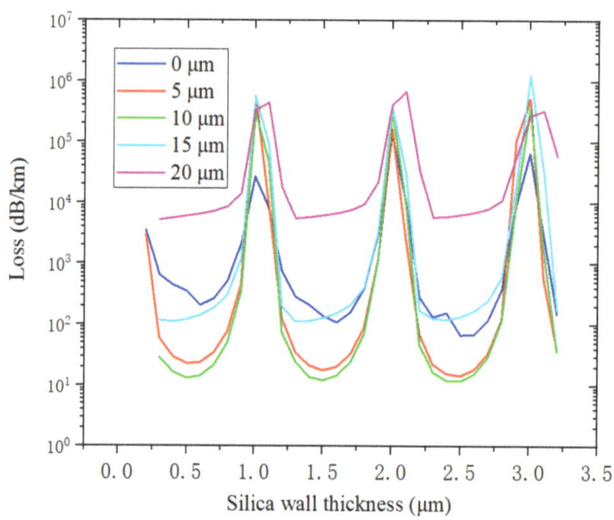

Figure 3. The curve of confinement loss with different silica wall thicknesses t and different gap distances g.

It can be seen from Figure 3 that there are three transmission regions in the wall thickness range of 0.3–3.2 μm. The centers of the first and the second transmission regions are at a wall thickness of 0.5 μm and 1.5 μm, respectively. This is consistent with the ARROW model. Different gap distances have little influence on the transmission region but have a great influence on the loss. Due to the inverse relationship between the resonant wavelength bandwidth and m, the larger the m, the narrower the corresponding bandwidth. When $m = 1$, the wall thickness is too thin. Thus, we chose $t = 1.46$ μm (when $m = 2$) as the thickness of the anti-resonant tube wall for the simulation.

The core radius of an HC-ARF has a crucial influence on the loss, and it can even determine the loss magnitude of the optical fiber. Although the loss is inversely proportional to the diameter of the core, it is limited to attaining low loss by increasing D_c. With the increase in D_c, the loss of higher-order modes will also decrease, which is unfavourable for single-mode propagation, Therefore, D_c is an important index for optimizing HC-ARF. Figure 4 shows the confinement losses and HOMER corresponding to different core diameters and different gap distances.

Figure 4a shows that the confinement loss decreases sharply with the increase in the core diameter, and the white dotted line in Figure 4b marks the lowest loss of different core diameters and shows different gap distances corresponding to different core diameters when the lowest loss occurs. Coincidentally, when the loss is the lowest, the ratio of the diameter of the anti-resonant tube to the diameter of the fiber core D/D_c happens to be about 0.66. As a comparison, when $D_c = 90$ μm, the lowest confinement loss is 0.89 dB/km, while the loss reaches up to 20.85 dB/km at 50 μm. Although the simulation shows that a larger core diameter performs better in terms of confinement loss, the loss of the higher-order mode will also increase with the increase in the core diameter. Figure 4c shows the confinement loss of the LP_{21} mode, which has the lowest loss in propagation compared with other higher-order modes. Figure 4e shows the calculated HOMER, which is defined as the ratio between the lowest confinement loss of the HOM and the confinement loss of FM [9,21]. Current studies suggest that the extinction ratio of hollow-core anti-resonant fiber to higher-order mode should be at least above 100 to have an ideal pseudo-single-mode propagation performance [26]; the white line is the contour line when HOMER is 100. From the figure, it can be seen that there is a wide range on both sides of the gap of the optimal anti-resonant tube, and no matter how large the diameter of the core is, good quasi-single-mode propagation can be achieved. Therefore, the decrease in HOMER is not the factor that limits the core diameter. Figure 4f shows the real part of the effective refractive index of LP_{01}, LP_{21} , and LP_{02} modes.

In addition to the increase in D_c being affected by the decrease in HOMER, the bending loss is also a parameter to be considered when designing a fiber. Bending loss (BL) is caused by the partial mode energy escaping from the cladding when the fiber is bent. The calculation of BL needs to adopt the equivalent straight fiber method in simulation [27], and the refractive index of equivalent straight fiber will vary along the bending direction; this varied refractive index distribution can be described by Equation (3).

$$n' = n \cdot e^{\frac{x}{R_c}} \sim n \cdot \left(1 + \frac{x}{R_c}\right), \quad (3)$$

where R_c represents the bending radius, n is the refractive index distribution of the original optical fiber, and x represents the abscissa position of a point on the equivalent straight fiber (i.e., the distance from the center of the fiber core to the point). After the equivalence, the refractive index property of the material can be redefined in COMSOL. In this simulation, the material refractive index of the fiber model becomes a function of x, and then the effective refractive index is calculated by mode analysis. Because bending loss can be seen as a kind of light leakage, with the imaginary part of the effective refractive index, the corresponding bending loss can be calculated by the formula for calculating the confinement loss. The results are shown in Figure 5.

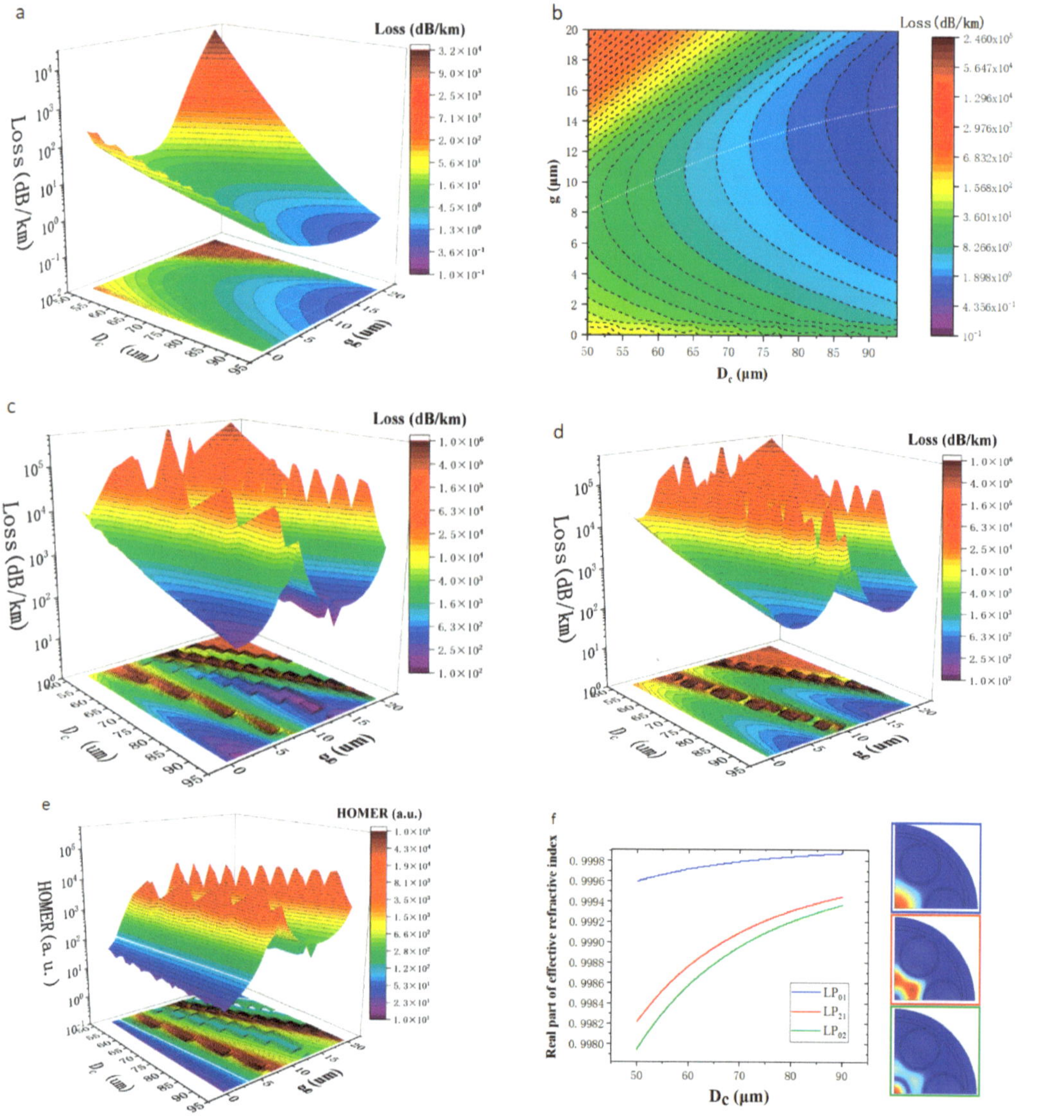

Figure 4. Different core diameters D_c and gap distances g corresponding to different confinement losses and HOMERs: (**a**) Confinement loss of LP_{01} fundamental mode (FM). (**b**) Confinement loss of LP_{01} FM; the white dashed line represents the lowest confinement loss when the core diameter is different. (**c**) Confinement loss of LP_{21} mode. (**d**) Confinement loss of LP_{02} mode. (**e**) HOMER; the white line is the contour line when HOMER is 100. (**f**) The real part of the effective refractive index of LP_{01}, LP_{21} , and LP_{02} modes.

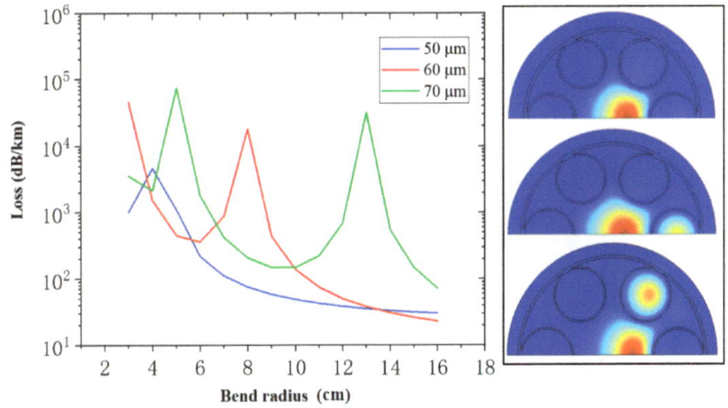

Figure 5. Relationship between bending loss and bending radius at 2 μm wavelength when the core diameters are 50 μm, 60 μm, and 70 μm. All fibers have the same $D/D_c = 0.66$ and $t = 1.46$ μm.

The loss peak observed in the figure is due to the mode coupling between the core and the cladding tube mode, as shown in the calculated modal field profile on the right-hand side. Although the simulation of straight fiber shows that the larger the core diameter, the lower the loss, the simulation of bending loss indicates that a larger core diameter leads to more bending loss. However, bending is inevitable in the normal operation of the optical fiber; thus, the core diameter should not be increased indefinitely. Figure 6 shows the confinement loss in fiber when $R_c = \infty$, $R_c = 13$ cm, and $R_c = 6.5$ cm, while D_c is kept to 50 μm.

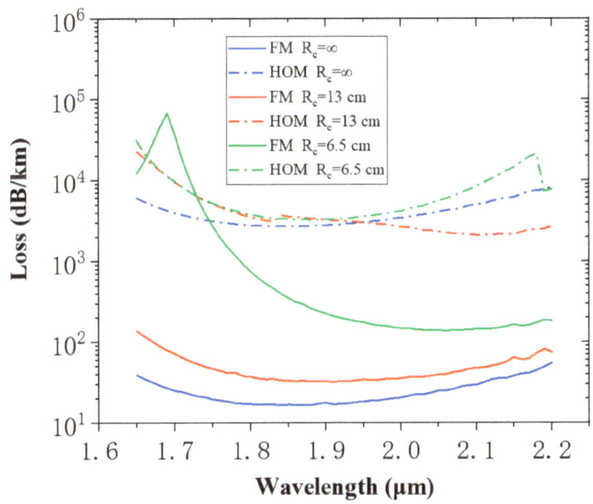

Figure 6. Diagram of the bending loss versus the wavelength in the HC-ARF. All fibers have the same core diameter $D_c = 50$ μm, a uniform silica wall thickness $t = 1.46$ μm, and $D/D_c = 0.66$.

3.2. Parameter Optimization of Six-Tube Node-Free Nested HC-ARF

Hollow-core anti-resonant fibers guide light through anti-resonant reflection conditions, but this kind of cavity-like structure cannot show 100% reflectance—a small part of the light will still transmit through the glass wall, resulting in energy loss. Generally, adding a layer of the anti-resonant tube inside one layer of the tube to form a multilayer nested resonant cavity structure can significantly reduce the confinement loss. With the increase in the number of nested layers, the confinement loss will also be reduced, but the gain in terms of confinement loss is very small, which is accompanied by the complexity of the fiber structure [5]. Therefore, we set the number of nesting layers to one. According to the previous simulation, the confinement loss is correlated with the ratio of the anti-resonant tube's diameter to the core diameter. Therefore, we show a contour plot of the FM and HOMs confinement loss of the six-tube node-free nested HC-ARF in Figure 7, where the wavelength is 2 μm, D_c = 50 μm, and t = 1.46 μm, with the normalized anti-resonant tube diameter (D/D_c) and normalized inner anti-resonant tube diameter (d/D) as variables. From these contour plots, the regions with low confinement loss and single-mode transmission can be effectively determined.

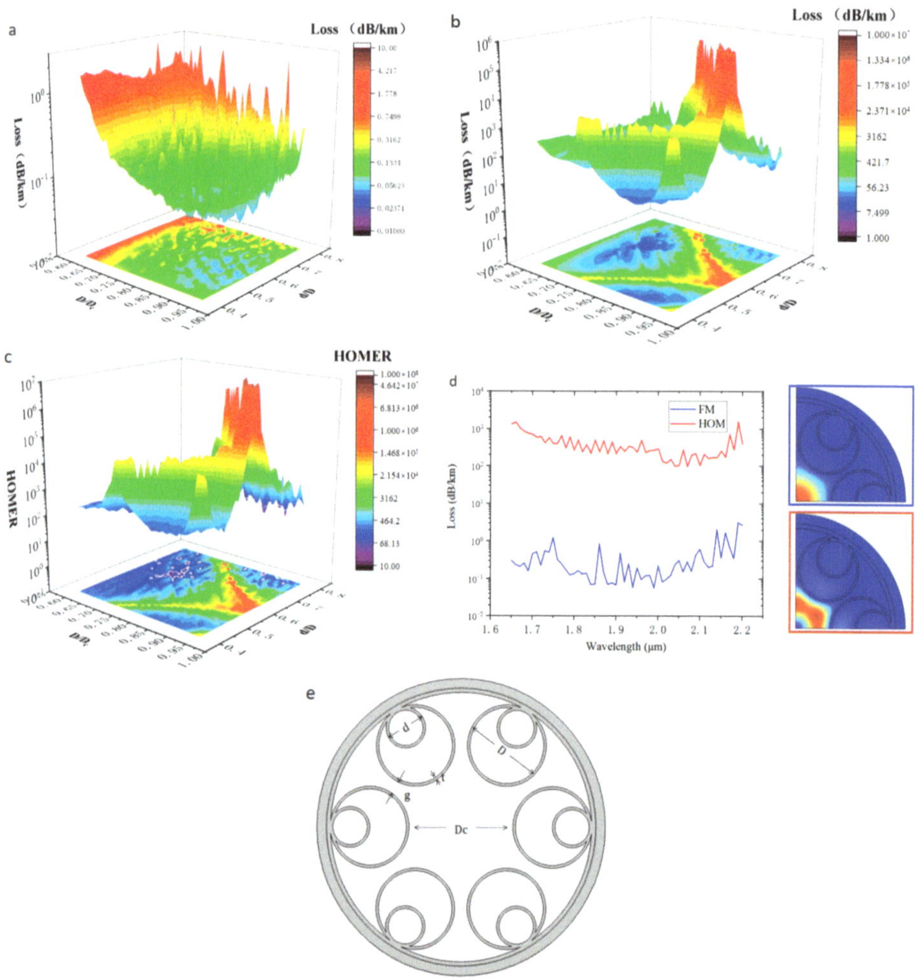

Figure 7. Diagram of confinement loss and extinction ratio of higher-order modes corresponding to different D/D_c and different d/D: (**a**) Confinement loss diagram of LP_{01} FM corresponding to different D/D_c and different d/D. (**b**) Confinement loss diagram of HOM LP_{21} corresponding to different D/D_c and different d/D. (**c**) HOMER, where the white line is the contour line when HOMER is 100. (**d**) Calculated confinement loss spectrum of D/D_c = 0.89, d/D = 0.6. (**e**) The geometry of a node-free nested HC-ARF with 6 anti-resonant tubes.

The simulation results show that the six-tube node-free nested HC-ARF with a wavelength of 2 μm has the lowest confinement loss at D/D_c = 0.89, d/D = 0.6, and the confinement loss is 0.042 dB/km. HOMER exceeds 100 in most parts; for D/D_c = 0.89, d/D = 0.6, HOMER is more than 9000, and the HC-ARF has excellent single-mode transmission performance.

3.3. Parameter Optimization of Five-Tube Node-Free Nested HC-ARF

According to the theory of Habib et al. [25], the five-tube nested HC-ARF has lower confinement loss and a wider transmission window than the six-tube and seven-tube nested HC-ARF. Figure 8 shows the geometric structure diagram of the nested HC-ARF with five anti-resonant tubes.

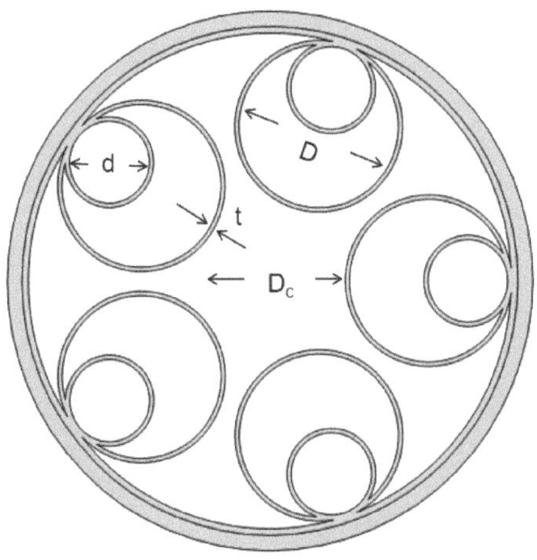

Figure 8. The geometric structure of a node-free nested HC-ARF with 5 anti-resonant tubes.

We show contour plots of the FM and HOMs confinement losses of the five-tube node-free nested HC-ARF in Figure 9, where the wavelength is 2 μm, D_c = 50 μm, t = 1.46 μm, and the normalized anti-resonant tube diameter (D/D_c) and the normalized nested bushing diameter (d/D) are taken as variables. From these contour plots, the regions with low confinement losses and effective single-mode transmission can be determined.

The five-tube nested HC-ARF has lower confinement loss than the six-tube nested HC-ARF. However, with the decrease in the number of anti-resonant tubes, the anti-resonant tube diameter and the area of the air in the tube will also increase. In few-tube designs, the ultra-large area of air in the anti-resonant tube will result in the same or even a higher effective refractive index compared with that of the core, resulting in higher FM confinement loss [9] and much higher bending loss—that is, mode coupling between the core and the cladding tube mode is more likely to occur. A smaller pseudo-single-mode output region is shown in this simulation, as presented in Figure 9c, where the white line is a contour line with the HOMER of 100, such as the HOMER diagram in Figure 7c obtained by the nested HC-ARF with six tubes. The region of HOMER greater than 100 is much smaller. At the lowest point of FM confinement loss when D/D_c = 1.07, d/D = 0.61, the confinement loss of the wavelength at 2 μm is 0.035 dB/km, but HOMER is only 86.5, which does not meet the condition of the pseudo-single-mode transmission. However, when D/D_c = 1.12, d/D = 0.53, the confinement loss of the wavelength at 2 μm is 0.040 dB/km, and HOMER is >2700. Therefore, a five-tube structure can obtain better confinement loss compared with that of a six-tube structure, but the optimal parameters of the fiber used need to be more accurate and sensitive—that is, the design with minimum confinement loss cannot be easy to adopt. Figure 10 shows the confinement loss of five-tube (D/D_c = 1.12, d/D = 0.53) and six-tube (D/D_c = 0.89, d/D = 0.6) structures at different bending radii.

When the radius is less than 12 cm, the confinement loss of the five-tube structure is greater than that of the six-tube structure, indicating that the larger air area in the anti-resonant tube has a negative effect on the confinement loss during bending. When the radius is greater than 12 cm, the confinement loss of a five-tube structure is less than that of a six-tube structure, with the five-tube structure having lower confinement loss than that of the six-tube structure in the straight fiber.

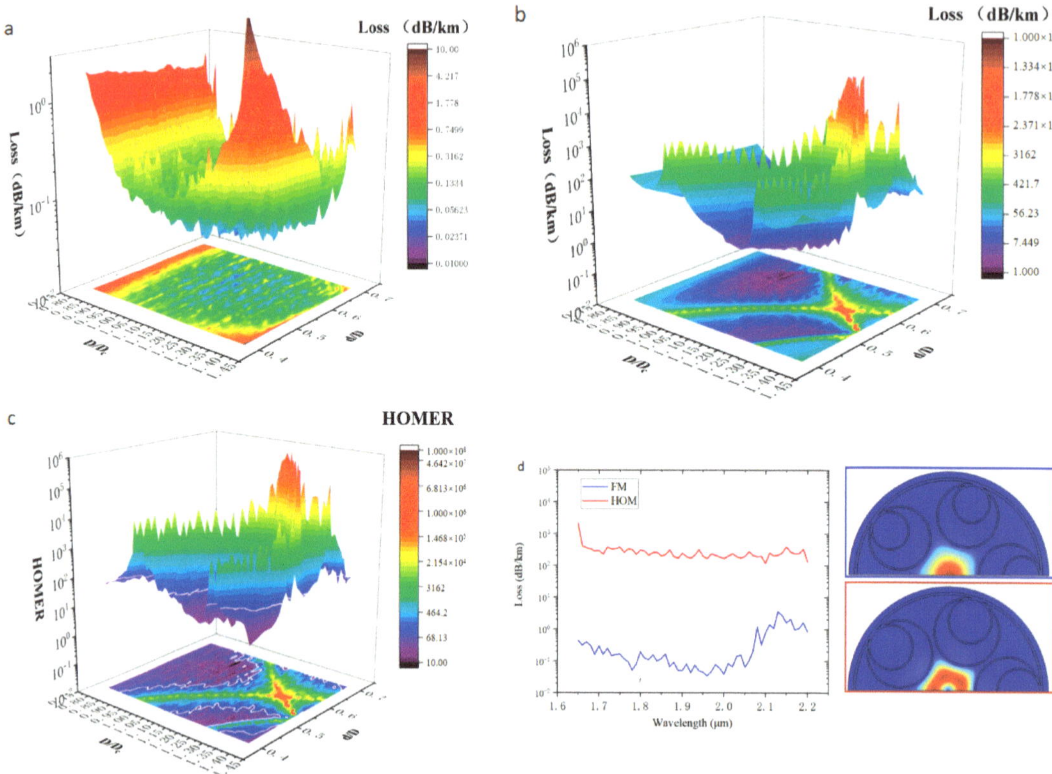

Figure 9. Diagram of confinement loss and extinction ratio of higher-order modes corresponding to different D/D_c and different d/D: (**a**) Confinement loss diagram of LP_{01} FM corresponding to different D/D_c and different d/D. (**b**) Confinement loss diagram of higher-order mode LP_{21} corresponding to different D/D_c and different d/D. (**c**) HOMER, where the white line is the contour line when HOMER is 100. (**d**) Calculated confinement loss spectrum of D/D_c = 1.12, d/D = 0.53.

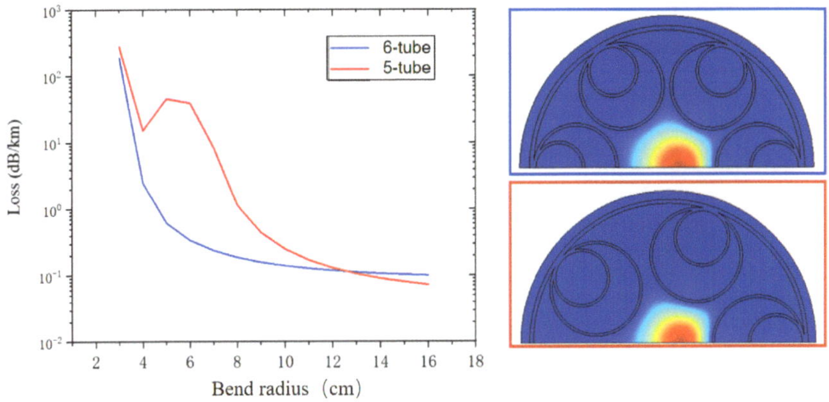

Figure 10. Confinement loss at the wavelength of 2 μm with different bending radii.

4. Conclusions

In this paper, to obtain HC-ARF with better transmission performance in the 2 μm band, and to reduce the difficulty of drawing and obtaining a more easily implemented optical fiber design, we determined the design of the mature and traditional five-tube and six-tube node-free nested HC-ARF, with the features of confinement loss < 0.05 dB/km, low bending confinement loss, and single-mode propagation. To obtain the complete parameter information of the optical fiber modal content, we analyzed and optimized the confinement loss and single-mode characteristics by carefully adjusting the normalized anti-resonant tube diameter (D/D_c) and the normalized inner tube diameter (d/D). We proved that the details of the cladding structure and the number of anti-resonant cladding tubes significantly affected the confinement loss and single-mode operation. We found that the six-tube node-free nested HC-ARF had the lowest confinement loss at $D/D_c = 0.89$ and $d/D = 0.6$, with a confinement loss of 0.042 dB/km, HOMER > 9000, and excellent single-mode transmission performance. Moreover, we found that the five-tube node-free nested HC-ARF had a wider transmission bandwidth and lower confinement loss compared with the six-tube system, while the region of HOMER greater than 100 was much smaller, and the bending loss was also larger. Therefore, although a five-tube HC-ARF can obtain better confinement loss compared with that of a six-tube system, the optimal parameters that the HC-ARF used need to be more accurate and sensitive—that is, the design with minimum confinement loss cannot be easy to adopt. Our numerical study provides theoretical guidance for correctly selecting the number and size of anti-resonant cladding tubes to achieve low confinement loss and high single-mode propagation performance in the 2 μm region.

Author Contributions: Conceptualization, T.S., X.S. and Y.Z. (Yi Zheng); investigation, X.S., F.M., Y.Z. (Yunhong Zhang) and H.Z.; writing—original draft preparation, T.S.; writing—review and editing, X.S., F.M., Z.W., J.S., C.Z., Y.Z. (Yunhong Zhang), H.Z., T.X. and M.C.; visualization, T.S. All authors have read and agreed to the published version of the manuscript.

Funding: This research was funded by National Natural Science Foundation of China (61735005, 61935010, 12275017); Natural Science Foundation of Beijing Municipality (4182054, 4212052); Fundamental Research Funds for the Central Universities (2021RC206, 2022YJS128); High-end Foreign Expert Introduction Program (G2022104010L).

Data Availability Statement: Data are available upon request from the corresponding author.

Conflicts of Interest: The authors declare no conflict of interest.

References

1. Cregan, R.F.; Mangan, B.J.; Knight, J.C.; Birks, T.A.; Russell, P.S.J.; Roberts, P.J.; Allan, D.C. Single-mode photonic band gap guidance of light in air. *Science* **1999**, *285*, 1537–1539. [CrossRef] [PubMed]
2. Knight, J.C.; Broeng, J.; Birks, T.A.; Russell, P.S.J. Photonic band gap guidance in optical fibers. *Science* **1998**, *282*, 1476–1478. [CrossRef]
3. Benabid, F.; Knight, J.C.; Antonopoulos, G.; Russell, P.S.J. Stimulated Raman scattering in hydrogen-filled hollow-core photonic crystal fiber. *Science* **2002**, *298*, 399–402. [CrossRef]
4. Pryamikov, A.D.; Biriukov, A.S.; Kosolapov, A.F.; Plotnichenko, V.G.; Semjonov, S.L.; Dianov, E.M. Demonstration of a waveguide regime for a silica hollow-core microstructured optical fiber with a negative curvature of the core boundary in the spectral region > 35 μm. *Opt. Express* **2011**, *19*, 1441–1448. [CrossRef] [PubMed]
5. Belardi, W.; Knight, J.C. Hollow antiresonant fibers with reduced attenuation. *Opt. Lett.* **2014**, *39*, 1853–1856. [CrossRef] [PubMed]
6. Wei, C.; Joseph Weiblen, R.; Menyuk, C.R.; Hu, J. Negative curvature fibers. *Adv. Opt. Photonics* **2017**, *9*, 504–561. [CrossRef]
7. Yu, F.; Knight, J.C. Negative curvature hollow-core optical fiber. *IEEE J. Sel. Top. Quantum Electron* **2016**, *22*, 146–155. [CrossRef]
8. Hasan, M.I.; Akhmediev, N.; Chang, W. Positive and negative curvatures nested in an antiresonant hollow–core fiber. *Opt. Lett.* **2017**, *42*, 703–706. [CrossRef]
9. Poletti, F. Nested antiresonant nodeless hollow core fiber. *Opt. Express* **2014**, *22*, 23807–23828. [CrossRef]
10. Kolyadin, A.N.; Kosolapov, A.F.; Pryamikov, A.D.; Biriukov, A.S.; Plotnichenko, V.G.; Dianov, E.M. Light transmission in negative curvature hollow core fiber in extremely high material loss region. *Opt. Express* **2013**, *21*, 9514–9519. [CrossRef]
11. Habib, M.S.; Bang, O.; Bache, M. Low-loss hollow-core silica fibers with adjacent nested anti-resonant tubes. *Opt. Express* **2015**, *23*, 17394–17406. [CrossRef] [PubMed]
12. Mangan, B.J.; Fair, L.; Langford, A.; Roberts, P.J.; Williams, D.P.; Couny, F.; Lawman, M.; Mason, M.; Coupland, S.; Flea, R.; et al. Low loss (1.7 dB/km) hollow core photonic bandgap fiber. In Proceedings of the Conference on Optical Fiber Communication, Technical Digest Series, Los Angeles, CA, USA, 22 February 2004; Volume 2.
13. Roberts, P.J.; Couny, F.; Sabert, H.; Mangan, B.J.; Williams, D.P.; Farr, L.; Mason, M.W.; Tomlinson, A.; Birks, T.A.; Knight, J.C.; et al. Ultimate low loss of hollow-core photonic crystal fibres. *Opt. Express* **2005**, *13*, 236–244. [CrossRef] [PubMed]
14. Benabid, F.; Roberts, P.J. Linear and nonlinear optical properties of hollow core photonic crystal fiber. *J. Mod. Opt.* **2011**, *58*, 87–124. [CrossRef]
15. Litchinitser, N.M.; Abeeluck, A.K.; Headley, C.; Eggleton, B.J. Antiresonant reflecting photonic crystal optical waveguides. *Opt. Lett.* **2002**, *27*, 1592–1594. [CrossRef] [PubMed]
16. Debord, B.; Amsanpally, A.; Chafer, M.; Baz, A.; Maurel, M.; Blondy, J.M.; Hugonnot, E.; Scol, F.; Vincetti, L.; Gérôme, F.; et al. Ultralow transmission loss in inhibited-coupling guiding hollow fibers. *Optica* **2017**, *4*, 209–217. [CrossRef]
17. Belardi, W.; Knight, J.C. Hollow antiresonant fibers with low bending loss. *Opt. Express* **2014**, *22*, 10091–10096. [CrossRef] [PubMed]
18. Kosolapov, A.F.; Pryamikov, A.D.; Biriukov, A.S.; Shiryaev, V.S.; Astapovich, M.S.; Snopatin, G.E.; Plotnichenko, V.G.; Churbanov, M.F.; Dianov, E.M. Demonstration of CO_2-laser power delivery through chalcogenide-glass fiber with negative-curvature hollow core. *Opt. Express* **2011**, *19*, 25723–25728. [CrossRef]
19. Van Newkirk, A.; Antonio-Lopez, J.E.; Anderson, J.; Alvarez-Aguirre, R.; Eznaveh, Z.S.; Lopez-Galmiche, G.; Amezcua-Correa, R.; Schülzgen, A. Modal analysis of antiresonant hollow core fibers using S^2 imaging. *Opt. Lett.* **2016**, *41*, 3277–3280. [CrossRef]
20. Yu, F.; Wadsworth, W.J.; Knight, J.C. Low loss silica hollow core fibers for 3–4 μm spectral region. *Opt. Express* **2012**, *20*, 11153–11158. [CrossRef]
21. Habib, M.S.; Bang, O.; Bache, M. Low-loss single-mode hollow-core fiber with anisotropic anti-resonant elements. *Opt. Express* **2016**, *24*, 8429–8436. [CrossRef]
22. Zhang, X.; Song, W.; Dong, Z.; Yao, J.; Wan, S.; Hou, Y.; Wang, P. Low loss nested hollow-core anti-resonant fiber at 2 μm spectral range. *Opt. Lett.* **2022**, *47*, 589–592. [CrossRef] [PubMed]
23. Huang, X.; Qi, W.; Ho, D.; Yong, K.-T.; Luan, F.; Yoo, S. Hollow core anti-resonant fiber with split cladding. *Opt. Express* **2016**, *24*, 7670–7678. [CrossRef] [PubMed]
24. Vincetti, L.; Setti, V. Extra loss due to Fano resonances in inhibited coupling fibers based on a lattice of tubes. *Opt. Express* **2012**, *20*, 14350–14361. [CrossRef] [PubMed]
25. Habib, M.S.; Antonio-Lopez, J.E.; Markos, C.; Schülzgen, A.; Amezcua-Correa, R. Single-mode, low loss hollow-core anti-resonant fiber designs. *Opt. Express* **2019**, *27*, 3824–3836. [CrossRef]
26. Uebel, P.; Günendi, M.C.; Frosz, M.H.; Ahmed, G.; Edavalath, N.N.; Ménard, J.-M.; Russell, P.S.J. Broadband robustly single-mode hollow-core PCF by resonant filtering of higher-order modes. *Opt. Lett.* **2016**, *41*, 1961–1964. [CrossRef]
27. Heiblum, M.; Harris, J.H. Analysis of Curved Optical Waveguides by conformal transformation. *IEEE J. Quantum Electron.* **1974**, *11*, 75–83. [CrossRef]

micromachines

Article

Dynamic Flow Control over Optical Properties of Liquid Crystal–Quantum Dot Hybrids in Microfluidic Devices

Artem Bezrukov * and Yury Galyametdinov

Department of Physical and Colloid Chemistry, Kazan National Research Technological University, 420015 Kazan, Russia; yugal2002@mail.ru

* Correspondence: a.bezrukov.knrtu@yandex.ru

Abstract: In this paper, we report developing approaches to tuning the optical behavior of microfluidic devices by infusing smart hybrids of liquid crystal and quantum dots into microchannel confinement. We characterize the optical responses of liquid crystal–quantum dot composites to polarized and UV light in single-phase microflows. In the range of flow velocities up to 10 mm/s, the flow modes of microfluidic devices were found to correlate with the orientation of liquid crystals, dispersion of quantum dots in homogeneous microflows and the resulting luminescence response of these dynamic systems to UV excitation. We developed a Matlab algorithm and script to quantify this correlation by performing an automated analysis of microscopy images. Such systems may have application potential as optically responsive sensing microdevices with integrated smart nanostructural components, parts of lab-on-a-chip logic circuits, or diagnostic tools for biomedical instruments.

Keywords: microfluidics; lab on chip; liquid crystals; quantum dots; confinement; optical response

Citation: Bezrukov, A.; Galyametdinov, Y. Dynamic Flow Control over Optical Properties of Liquid Crystal–Quantum Dot Hybrids in Microfluidic Devices. *Micromachines* **2023**, *14*, 990. https://doi.org/10.3390/mi14050990

Academic Editor: Aaron Hawkins

Received: 3 April 2023
Revised: 27 April 2023
Accepted: 29 April 2023
Published: 30 April 2023

1. Introduction

Microfluidic devices and laboratories-on-a-chip broaden horizons for the synthesis and application of photonic particles [1–3]. Microfluidic circuits offer additional options for the tailored synthesis of photonic micro- and nanoparticles [4–6] and their applications in biosensing [7,8], medical diagnostics [9,10], and fabricating labels with anti-counterfeiting capabilities [11–13].

Integrating microfluidic devices with optically active functional nanomaterials such as quantum dots (QD) makes a substantial contribution to advancing applications of fluidic micromachines in nanotechnology [14,15], chemical and bioanalysis [16–18], the synthesis of target drug delivery systems [19–21] or nanomedicine [22]. Current research activities also focus on fine-tuning the microscale synthesis of quantum dots [23–26] and conjugating them with soft matter [15,21,27].

Among organized media studied and applied in microfluidic confinement, liquid crystals (LC) are among the most suitable and promising materials for photonics [28–30]. Compared to quantum dots at a microscale, integrating liquid crystals with microfluidic chips offers an alternative approach to designing microscale temperature and flow sensors [29,31–33] or optically active functional microcapsules [34–36]. Confinement offers a simultaneous and programmable control over the optical behavior of LC systems by a number of factors such as flow [37–39], temperature [29,33], or light [40].

Doping liquid crystal materials with quantum dots allows to combine their advantages as optically active materials [41–43] and offers a synergistic effect to applications of the resulting composites such as displays [44] or security labeling [45]. As opposed to studying and applying such composites in macroscopic conditions, microfluidic research in composites of LC with quantum dots or similar functional nanomaterials is an emerging area represented by mostly pioneering publications, which focus, for example, on composite microdroplets synthesized by microfluidic techniques [45,46]. In this respect, a detailed analysis on the impact of various factors such as flow conditions on optical properties of

LC-QD composites in microfluidic channels may bring new insights into fine-tuning their optical responses at the microscale and enhancing the photonic application capabilities of fluidic micromachines.

This work continues our research in nematic LC systems [47] and LC-QD composites [43,48] at the microscale and macroscale. In our previous works, we focused on analyzing the impact of flow conditions in microchannels on the orientation behavior of liquid crystals in single-phase and multiphase microflows and characterized the sensitivity of such systems to variable flows in polarized light.

On the other hand, luminescence is an intrinsic property of quantum dots, which is the cornerstone of their sensing capabilities. This work, therefore, aims specifically at adding the luminescence behavior of quantum dots to the portfolio of possible optical responses of LC-QD composites to variable microfluidic flows. We studied the optical behavior of LC-QD hybrid materials in variable flow conditions in both polarized and UV light in single-phase microflows in straight rectangular microchannels. We analyze a correlation between orientation states of the LC matrix in a microchannel, optical responses of the LC material under polarizers, and the luminescence behavior of quantum dots. The novel contribution of this work is also to offer automated polarized and UV microscopy image analysis capabilities. We offer a Matlab algorithm for image analysis that obtains a quantitative correlation between the optical state of the LC-QD composite and the flow velocity.

2. Materials and Methods

2.1. Materials

Microfluidic devices were fabricated from polydimethylsiloxane (PDMS) Sylgard™ 184 silicone elastomer. PDMS was purchased from Dow Corning (Midland, MI, USA) and used as received. It came as a two-part elastomer kit (the pre-polymer and curing agent). SU-8 3050 photoresist (Microchem Corp., Westborough, MA, USA) and was used to produce a mold for microfluidic chips.

For the liquid crystal phase, the nematic liquid crystal N-(4-methoxybenzylidene)-4-butylaniline (MBBA) was used. It was purchased from Reachem, Moscow, Russia, and used as received. It exhibits liquid crystal properties at room temperature.

In this work, pre-synthesized core–shell quantum dots were used. In quantum dots, the CdSe core was coated by two shells: the internal CdS shell and the external ZnS shell. The details of synthesis of quantum dots are provided in Supplementary Materials.

The PDMS surface is initially hydrophobic, and its contact angle was high (>100°) according to our contact angle measurement experiments. Before performing microfluidic experiments, microchannel walls were pre-processed by infusing surfactant solutions (sodium dodecyl sulphate—SDS) for 10 min to increase the wettability of internal PDMS walls and favor the orientation of liquid crystal molecules with respect to microchannel walls.

SDS was purchased from BDH Limited, Poole, UK, and used as received. SDS was sold as a powder. Bulk samples of pre-micellar 5×10^{-3} mol/L SDS were produced by dissolving dry surfactant in water. At this concentration, SDS is below its critical micellization concentration, which is 8.4×10^{-3} mol/L [49]. The addition of surfactant provides a smooth and reproducible formation of uniform droplets in microfluidic confinement [50]. Bidistilled water was used for the aqueous phase. Before performing microfluidic experiments, the solvent was passed through 0.45 μm Millipore polytetrafluoroethylene (PTFE) filters by Merck, Darmstadt, Germany.

2.2. Methods

2.2.1. Synthesis of LC-QD Composite and Characterization of Components

Composites of liquid crystal and quantum dots were prepared by doping the LC phase with the dispersion of quantum dots in hexane. The composites were agitated by a magnetic stirrer at 50 °C for 5 h to remove hexane. The concentration of quantum dots in the composite was 0.25 wt %.

The photoluminescence emission spectra of quantum dots and the composite were recorded by a Varian Cary Eclipse spectrofluorimeter (Agilent, Santa Clara, CA, USA). The hydrodynamic diameter of quantum dots was measured by dynamic light scattering (DLS) using a Malvern Zetasizer Nano ZS system (Malvern Panalytical, Malvern, UK).

2.2.2. Recording and Processing Microscopy Images of Microflows

The preliminary characterization of LC and LC-QD flows in microfluidic confinement was performed by digital optical microscopy using a Levenhuk D320 optical microscope (Levenhuk, Tampa, FL, USA). Microscopy images were captured at 100× magnification using a ToupCam E3ISPM08300KPB camera (Touptek, Hangzhou, China).

The orientation behavior of the LC material and LC-QD composites in microfluidic flows were studied by polarized optical microscopy (POM) using an Olympus BX51 microscope (Olympus, Tokyo, Japan) equipped with a high-precision Linkam heating system. Microscopy images were captured at 100× and 500× magnification using a ToupCam E3ISPM08300KPC camera (Touptek, Hangzhou, China).

The luminescent properties of LC-QD composites in microchannels were studied by an Olympus BX43 fluorescent microscope (Olympus, Tokyo, Japan). Microscopy images were captured at 100× magnification using a ToupCam E3ISPM05000KPA camera (Touptek, Hangzhou, China).

Polarized and fluorescent microscopy images were processed by Matlab 2021b software. For processing, microscopy images were taken at identical microscope settings. The luminance components of polarized microscopy images and the red color components of fluorescent microscopy images were extracted and processed by the pre-developed Matlab script.

2.2.3. Fabricating Microfluidic Devices and Preparing Experimental Setup

Microfluidic devices were fabricated using standard photolithography techniques [51]. The chips with rectangular microchannels were produced by this technology. The length, width, and height of all the microchannels were 15 mm, 200 μm, and 100 μm, respectively. SU-8 photoresist and a transparent photomask with a negative image of a microchip were used to produce a 100 μm thick mold of microfluidic chips on top of a 3-inch silicon wafer. PDMS pre-polymer was mixed with a curing agent, poured over the mold, and allowed to cure for 4 h in a 60 °C oven. Once cured, PDMS was peeled off the mold and bonded to a flat PDMS slab via 1 min plasma treatment by Harrick Plasma Cleaner PDC-23G, Ithaca, NY, USA. The PDMS device was then heated in an oven at 180 °C for 1 h to complete the bonding of the two polymer layers.

The LC and aqueous phases were infused into microfluidic devices using Shenchen ISPLab01 syringe pumps (Baoding Shenchen Precision Pump Co. Ltd., Baoding, China), which provide a minimal flow rate of 0.001 μL/min. In this work, the flow velocities of the LC material, LC-QD composite, and the aqueous phase were varied in the range up to 10 mm/s. To provide the same hydraulic paths for fluids to all the inlets, PTFE tubes of identical lengths (10 cm) and internal diameters that fit the same needle tips inserted into microchip outputs (20 G-type needles, 0.9 mm diameter) were used. These tubes were connected to identical 1 mL syringes installed into syringe pumps.

3. Results

3.1. *Preliminary Characterization of LC and LC-QD Composites in Microfluidic Confinement*

Microfluidic devices and the confined environment of microchannels often require specific conditions for infusing and processing both organic and aqueous fluids. The viscosity, temperature behavior and optical characteristics of liquid crystal and quantum dot systems are important factors for designing applicable microfluidic devices and ensuring the compatibility of media with microfluidic experiments.

At the first stage of this work, we performed a preliminary analysis of components of the LC-QD composite and its compatibility with microchannel confinement. Figure 1 shows a schematic representation of LC molecules, quantum dots, and their composites and the results of their characterization by applicable spectroscopy and microscopy methods.

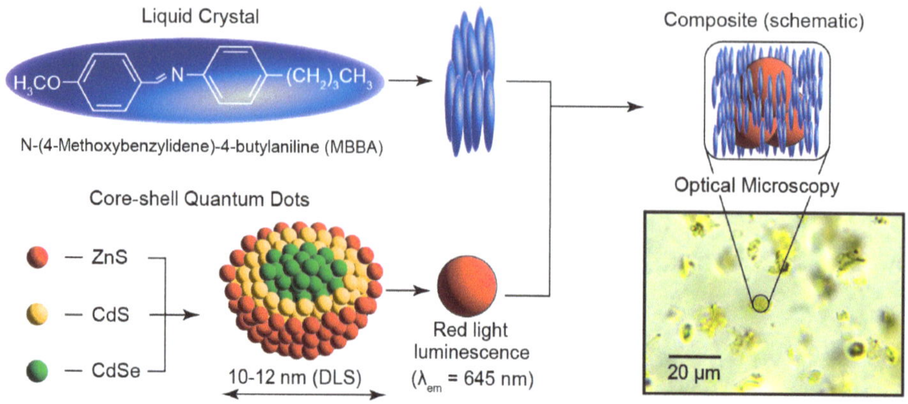

Figure 1. Components, schematic microstructure and optical microscopy images of MBBA composites with CdSe-CdS-ZnS core–shell quantum dots.

According to the synthesis procedure, the core–shell quantum dots used in this work consist of the internal CdSe core that is covered with CdS and ZnS shells. According to dynamic light scattering experiments, the average hydrodynamic diameter of quantum dots dispersed in hexane is about 10–12 nm. As compared with smaller size QD particles, the emission maximum of such quantum dots is supposed to be in a longer-wave visible range [43]. It agrees with spectrofluorimetry studies, which report that the emission peak of these quantum dots is in the red light range both individually and in composite ($\lambda_{em} \approx 650$ nm). This wavelength is within the transmittance range of PDMS [52], so we can expect a smooth detection of QD emission in microflows.

MBBA is an intensively studied liquid crystal material, which attracts a sustainable research interest both individually and as a matrix for various composites [53,54]. It exhibits nematic mesophase at room temperature (its clearing point is about 38 °C [54]). At room temperature, MBBA is represented by a yellow turbid liquid, which could be easily infused into standard PDMS microchips, according to our preliminary microfluidic tests. Such microchips can operate at room temperature and require no internal or external heating system to obtain a nematic MBBA mesophase.

The optical microscopy photo in Figure 1 shows an image of the LC-QD composite at 500× magnification. We can see that the structure of the composite is non-homogeneous and includes clusters of quantum dots in the size range of approximately 5–30 μm. The size of such QD clusters is much smaller than the standard width of microfluidic channels (~100–200 μm), so the composites can be smoothly infused into conventional microfluidic devices.

Therefore, we designed and fabricated microfluidic devices for MBBA-QD composites with the channel widths of 200 μm and heights of ≈100 μm, so they could smoothly incorporate such clusters of quantum dots. Figure 2 demonstrates the designs of the respective microfluidic devices and test microscopy photos of single-phase and two-phase microflows.

For single-phase experiments (Figure 2a), we fabricated single-channel microfluidic chips with a channel width of 200 μm and length of 15 mm. Such microchips allowed us to generate a smooth flow of the LC phase (Figure 2b) or a composite in the studied flow velocity range (up to 10 mm/s) and above.

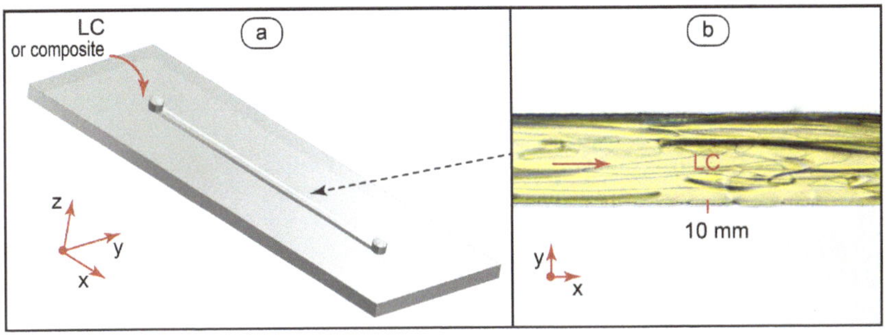

Figure 2. Design of microfluidic chips and flow testing: (**a**) microfluidic device with a straight microchannel for generating single-phase LC and LC-QD microflows; (**b**) bright field optical microscopy image of a single-phase flow of the pure MBBA liquid crystal at 0.5 mm/s.

Thus, MBBA-QD composites are represented by microscale QD clusters in an LC matrix. Such disperse systems are compatible with microfluidic confinement and can be processed by standard microfluidic devices. The optical properties of the LC phase and QD emission can be studied by applicable microscopy methods. The next stage of this work focused on characterizing orientation behavior of the LC phase by polarized optical microscopy and luminescence of quantum dots by fluorescence microscopy.

3.2. Optical Behavior of LC-QD Microflows in Polarized and UV Light

3.2.1. Orientation Behavior of Confined MBBA and MBBA-QD Composite

Nematic liquid crystal flows demonstrate a flow-dependent orientation behavior in microfluidic confinement [37]. Depending on the orientation of LC molecules with respect to the view direction, a variety of optical responses are generated by microchannels with a mobile mesophase in polarized light: from a homeotropic orientation and a planar orientation of a continuous LC phase to characteristic cross-shaped textures of LC microdroplets [28,39].

A useful approach to quantify orientation of the LC director such as its tilt angle is comparing their interference colors with a Michel–Levy chart [55–58]. The authors in [56] evaluated the in-plane orientation of the liquid crystal by rotating the LC cell between crossed polarizers and observing changes in interference colors. In [59], the rotation of polarizers to 45° was used to confirm the orientation of LC specifically in microchannel confinement.

To characterize the orientation behavior of the LC matrix in a LC-QD composite in single-phase flows, we performed polarized optical microscopy studies of LC microflows at flow velocities up to 10 mm/s. The results are summarized in Figure 3.

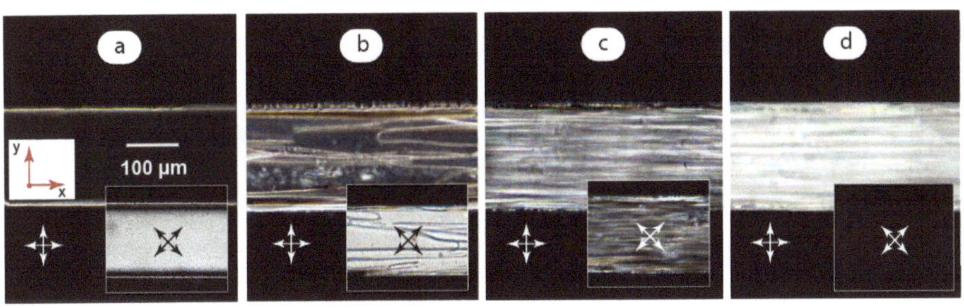

Figure 3. Polarized microscopy images of LC microflows: (**a**) U → 0; (**b**) U = 0.5 mm/s; (**c**) U = 1 mm/s; (**d**) U = 10 mm/s. Red arrows demonstrate coordinate axes with respect to Figure 2. Crossed arrows indicate positions of polarizers.

Figure 3a was obtained by infusing LC to the microchannel first and then turning off the syringe pump to obtain an immobile LC phase. At flows below approximately 0.1 mm/s, the microchannel environment is represented by a nearly uniform dark field at crossed polarizers except for ~10 μm bright stripes at microchannel boundaries. Such an optical response was reported in [37,39]. We also observed it in our previous studies of similar nematic LC flows [47]. Such an image is supposed to represent a homeotropic orientation of the LC phase with the domains aligned perpendicular to microchannel walls and the view direction.

The POM image with polarizers rotated by 45° (Figure 3a inset) shows, however, a bright field image in such conditions. No visible differences in color are observed along the microchannel, which could indicate no differences in LC alignment to be distinguished by the Michel–Levy chart. Such an alignment was reported to be more typical for a uniform director orientation in a position tilted to the flow axis [37] or aligned along it [59].

It should be noted that the alignment state shown in Figure 3a was stable only in weak (<0.1 mm/s) flow conditions. The immobile LC phase demonstrated sequential transitions from uniform bright to dark colors of different brightness at crossed polarizers both in 0° and 45° positions within 30 min observation after the flow was fully stopped. The coincidence of dark fields at both positions of crossed polarizers did take place, but it was very unstable.

Such a behavior can be associated with the fact that achieving a stable homeotropic alignment requires processing microchannels with aligning reagents [37]. Although microchannels were pre-processed with surfactant (1 mmol SDS) before LC experiments, added surfactant turned out to be insufficient to provide a stable homeotropic orientation. In their turn, weak flows may perform as a stabilizing orientation factor for LC molecules that favors their alignment with respect to the microchannel axis. It should be emphasized that the director orientation stayed uniform along the microchannel in the observed zero flow or weak flow conditions: no separate zones of different colors or light intensities were detected along the microchannel.

At flow velocities higher than 0.2 mm/s, non-uniform stripes appear in the LC flow under polarized light (Figure 3b). Polarizers rotated to 45° show (Figure 3b inset) show a residual bright field with the color close to that demonstrated in the Figure 3a inset. It indicates that LC molecules still tend to align uniformly with respect to the flow axis in faster flow conditions.

A further increase of the flow velocity up to 1 mm/s makes this non-uniform structure predominant (Figure 3c). The rotation of polarizers (Figure 3c inset) provides a similar irregular pattern with a variety of light intensities and colors. If we apply the Michel–Levy color chart approach to Figure 3c, we can conclude that it demonstrates a rather irregular alignment of LC domains with arbitrary angles with respect to the microchannel axis, although a certain ordering effect of flow on the stripes is observed.

At higher flow velocities, LC domains start to align along the flow direction. At the flow velocity above ~5 mm/s, the orientation of LC domains is predominantly planar (Figure 3d) and shows a characteristic bright field in polarized light with crossed polarizers at 0° and a darker field at 45°.

At the next stage of this work, we studied the impact of added quantum dots on the behavior of the LC matrix under polarized light. Figure 4 summarizes the results obtained in the same flow conditions as those demonstrated in Figure 3.

The impact of quantum dots on the optical state of the composite at $U \rightarrow 0$ is shown in Figure 4a. We can see a mostly uniform alignment of the LC matrix in slow flow or zero flow conditions. Quantum dot aggregates, however, also perform as aligning additives for the LC molecules and show bright spots around QD clusters in polarized light. The rotation of polarizers changes the colors of the microchannel medium in a way similar to that shown in Figure 3a for a pure LC matrix.

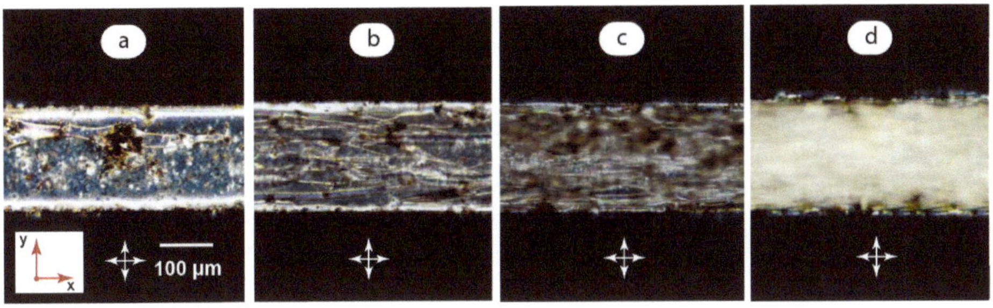

Figure 4. Polarized microscopy images of LC-QD microflows: (**a**) U → 0; (**b**) U = 0.5 mm/s; (**c**) U = 1 mm/s; (**d**) U = 10 mm/s. Red arrows demonstrate coordinate axes with respect to Figure 2. Crossed arrows indicate the positions of polarizers.

QD aggregates were found to be unstable in microchannel flows. An increase of flow velocities above 0.5 mm/s results in the decomposition of QD aggregates to smaller clusters (Figure 4b,c). Increasing shear stresses in microscale flows favors the homogenization of the LC-QD composite that agrees with our previous studies of a similar hybrid system [48]. At the same time, QD aggregates favor the formation of instabilities in the LC matrix: a chaotic pattern in Figure 4b is more pronounced as compared to that of a pure LC matrix in the same flow conditions (Figure 3b). Transition to an irregular pattern occurs in the 0.5–1 mm/s range, which is approximately 0.3 mm/s lower than in a pure LC matrix: the composites in Figure 4b,c show much more similarities than a pure LC matric at the same flow velocities in Figure 3b,c.

Finally, a uniform bright field is achieved in fast flows approaching 10 mm/s (Figure 4d) similar to that demonstrated by a pure LC matrix in Figure 3d upon its transition to a predominant orientation along the flow.

A suggested confined orientation of LC molecules in the presence of QD aggregates is shown in Figure 5. In slow flow conditions (Figure 5a), the orientation is predominantly uniform. In slow flows, LC molecules are supposed to align in a certain direction with respect to the microchannel axis. QD aggregates coexist with the LC phase and also perform as additional directors for its molecules. It should be noted that an alignment of LC molecules in zero flow conditions was found to be unstable but still uniform.

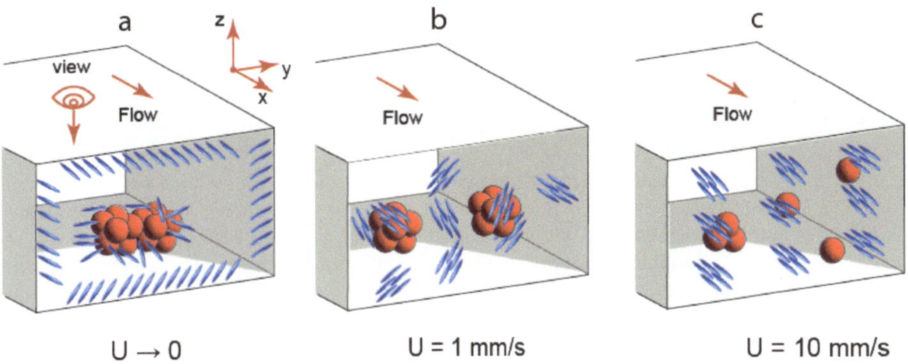

Figure 5. Suggested orientation of LC molecules and state of QD aggregates in LC-QD composites at various flow velocities: (**a**) U = 0.1 mm/s; (**b**) U = 1 mm/s; (**c**) U = 10 mm/s.

In microflows at 1 mm/s (Figure 5b), a uniform orientation of LC molecules is replaced with an irregular pattern of LC domains. The size of QD aggregates decreases in faster flows.

Finally, LC domains tend to align along the flow axis (Figure 5c). However, such an alignment is not perfect, and residual inhomogeneities can be still detected in Figures 3d and 4d.

The Reynolds number is a key parameter that characterizes flow transitions in single-phase microfluidic systems, so it is interesting to evaluate it for the conditions shown in Figures 3 and 4. For aqueous microflows, for example, this parameter is generally small and corresponds to a laminar flow mode, which was observed in our previous experiments [60]. The Reynolds number depends on the fluid viscosity, and liquid crystals demonstrate non-uniform flow-dependent viscosities. As opposed to aqueous flows, the Reynolds number is expected to provide only an approximate characterization of LC flows at various flow velocities and non-uniform viscosities resulting from changes in the alignment of LC molecules in variable flows. The data in [61] allow us to evaluate MBBA viscosities according to Leslie–Ericksen theory for various Miesowicz geometries of LC molecules aligned parallel or orthogonal to the velocity vector. According to Figure 3, MBBA may exhibit intermediate orientation states as compared to these geometries in our microfluidic experiments. We can consider, therefore, the viscosities found for these geometries as the boundaries of a possible MBBA viscosity range in our experiments and evaluate the resulting Reynolds number. All the viscosity coefficients reported in [61] provided $Re < 1$ for the microchannel width of 200 μm and velocities up to 10 mm/s. The LC flow transitions shown in Figure 3 occur, therefore, at low Reynolds numbers.

It should be noted that microfluidic devices remained stable and operational at increasing flow velocities. Although flow velocities of several millimeters per second are quite high for microfluidic experiments performed in microscale width channels, we observed no deformation of such a microchannel at LC flow rates up to 20 μm/s. A possible effect of flow pressure can be expected for long channels of serpentine microchips with much larger hydraulic resistance. The major flow-induced effects of LC alignment and QD luminescence were observed, however, at flow velocities below 1 mm/s. Thus, such operation modes can be potentially applicable to a broader range of microchip designs.

The aggregates of quantum dots were large at zero flow conditions. Their aspect ratio (aggregate diameter to microchannel height) was greater than 0.3 in zero flow conditions or flows slower than 0.2 mm/s. The aggregates, however, did not exert a negative impact on microchannel environment. They turned out to be easily deformable inside a microchannel and were therefore supposed to put virtually no pressure on microchannel walls. The aggregates were also sensitive to flow: their size reduced to approximately 20–30 μm upon an applied flow of 0.3–0.5 mm/s and less in faster flows.

Thus, MBBA-QD composites demonstrate a flow-dependent orientation behavior of the LC matrix and aggregative the behavior of quantum dots. It resulted in a variety of optical responses to polarized light. At the next stage of this work, we analyzed an impact of flow-dependent properties of the composite on the luminescence properties of quantum dots in UV light.

3.2.2. Luminescence Behavior of MBBA-QD Composite Microflows

Fluorescence microscopy is a key method to unleash the sensing potential of lab-on-a-chip, especially in cutting-edge biomedical applications [17,62,63] and organ-on-a-chip platforms [64]. Therefore, analyzing the luminescence behavior of LC-QD hybrids in microfluidic confinement may reveal new approaches to sensing applications of such composites in lab-on-a-chip devices. At this stage of the work, we studied the impact of flow velocity on the luminescence behavior of the composite. Figure 6 summarizes the results of the respective fluorescence microscopy experiments.

Excitation by UV light initiates a red-light emission of quantum dots in a confined immobile composite (Figure 6a). Some of the QD clusters are absorbed on microchannel walls and represent bright red light spots on the microscopy image. The largest clusters are about 50–60 μm in diameter. The height of the microchannel is 100 μm, so all the QD clusters can be potentially fully immersed in the LC matrix. The luminescence from such clusters is easily detected in Figure 6a as turbid red spots throughout the microchannel.

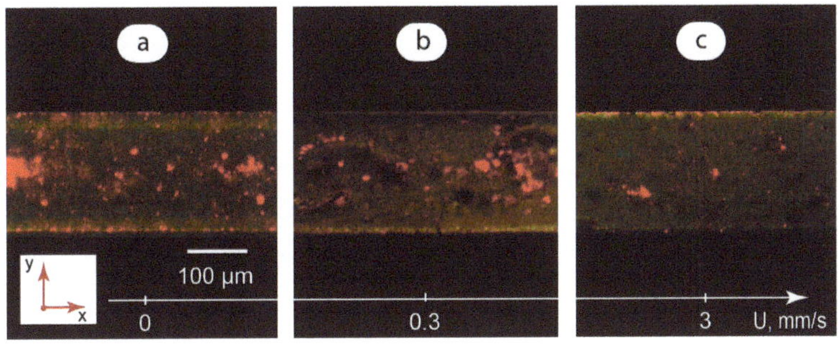

Figure 6. Fluorescence microscopy images of single-phase LC-QD composite microflows: (**a**) U → 0; (**b**) U = 0.3 mm/s; (**c**) U = 3 mm/s.

In the flow of the composite at 0.3 mm/s (Figure 6b), large clusters of quantum dots still exist. They exhibit, however, a much weaker luminescence than in immobile composite. The clusters are mostly represented by dark spots with limited red color areas. The major contribution to luminescence is provided by smaller QD clusters.

Finally, Figure 6c shows a residual red light emission from small QD aggregates at 3 mm/s. The majority of QD aggregates are represented by turbid dark spots inside the LC matrix.

The luminescence behavior of the composite demonstrated in Figure 6 correlates with the orientation states of the LC matrix (Figure 3). The most intensive luminescence was observed for a uniformly aligned LC matrix in zero or slow flow conditions. The transition from a uniformly aligned state to an irregular pattern of domains reduced QD emission.

A possible reason for such an optical behavior of quantum dots is the Rayleigh scattering of electromagnetic radiation and anisotropy of the LC matrix. In slow flows of microfluidic LC-QD composites, the molecules of the LC matrix are supposed to be aligned uniformly with respect to microchannel walls. Such an orientation may initiate a less intensive scattering of UV light on its path to the QD cluster inside a microchannel as compared to faster flows with an irregular pattern of LC domains.

In addition, the scattering intensity is inversely proportional to λ^4, according to the Rayleigh scattering law, where λ is the wavelength of light. Excitation UV light (λ = 365 nm), therefore, is scattered more intensively than emission light from QD clusters (λ = 645 nm) with the factor of $(645/365)^4 \approx 10$. In LC-QD flows, UV light intensity can be insufficient to excite a substantial emission from QD clusters inside a microchannel. Such clusters are represented by dark spots in Figure 6b,c. The major contribution to luminescence is made by small QD clusters absorbed on microchannel walls or segments of large QD clusters near microchannel walls, where a thickness of the LC matrix and the resulting scattering are minimal.

Finally, the decomposition of aggregates and reduction in their adsorption to microchannel walls favors their uniform distribution inside the LC matrix and reduces a contact with microchannel walls in fast flows that should also contribute to minimizing the emission from the composite (Figure 6c).

3.2.3. Quantifying the Impact of Flow Conditions on Optical Responses of Confined LC-QD Composite

The automated analysis of microscopy data with image processing software offers a convenient tool to characterize the optical behavior of microfluidic flows. To reveal possible quantitative correlations between microflow conditions and optical responses of the LC-QD composite in polarized and UV light, we performed a microscopy image analysis with a pre-developed Matlab script. The details of the image processing algorithm and the respective Matlab script are provided in the Supplementary Materials.

To perform this analysis, a series of LC and LC-QD composite microscopy images were taken in polarized and UV light in addition to the images shown in Figures 3, 4 and 6.

All the images in each series were taken with identical microscope settings. These images are shown in Figure S1 in Supplementary Materials.

For polarized microscopy images, their average luminance $\overline{Y}$ of the YCbCr color space [65] was calculated by the Matlab script. For fluorescence microscopy images, their average red color brightness $\overline{R}$ from the RGB data was calculated by the Matlab script. Figure 7 demonstrates the results.

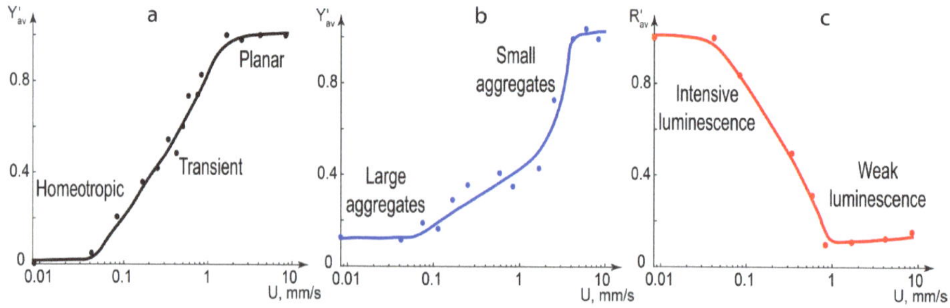

Figure 7. Optical characteristics of microfluidic channels with infused LC or LC-QD composites obtained from polarized and fluorescence microscopy images processed with Matlab: (**a**) reduced average luminance of the pure LC phase in polarized light; (**b**) reduced average luminance of the LC-QD composite in polarized light; (**c**) reduced average red color brightness of the LC-QD composite in UV light.

The following reference values were used for image processing. The average luminance $\overline{Y}_0$ of the pure immobile LC phase was considered to be zero. The average luminance of the pure LC phase at 10 mm/s $\overline{Y}_{max}$ was considered to be maximum. The average red color brightness of the pure LC phase at 10 mm/s $\overline{R}_0$ was considered to be zero. The average red color brightness of the immobile LC-QD composite at 10 mm/s $\overline{R}_{max}$ was considered to be maximum.

The average reduced luminance (Figure 7a,b) was calculated by the following equation:

$$Y' = \frac{\overline{Y} - \overline{Y}_0}{\overline{Y}_{max} - \overline{Y}_0} \quad (1)$$

The average reduced red color brightness (Figure 6c) was calculated by the following equation:

$$R' = \frac{\overline{R} - \overline{R}_0}{\overline{R}_{max} - \overline{R}_0} \quad (2)$$

Both Y' and R' varied, therefore, in the [0;1] range in all the images.

Figure 7a shows the optical states of pure LC phase microflows. The zero luminance zone at the 0° position of crossed polarizers corresponds to a uniform orientation of LC molecules in microflows below 0.1 mm/s. It transforms into the maximum luminance at velocities above 2–3 mm/s through a transient zone where the Y' changes nearly linearly with logarithmic flow velocity upon a growing contribution of irregular LC patterns.

In Figure 7b, the orientation of LC molecules around QD aggregates in composites increases the initial luminance at flow velocities below 0.1 mm/s. Large and dark QD aggregates decrease the growth of luminance in the 0.1–1 mm/s range as compared with pure LC phase flows. The decomposition of aggregates in faster flows quickly increases the luminance up to its maximum value.

Figure 7c shows that the red color brightness is maximum in flows slower than 0.1 mm/s, which corresponds to undisturbed red light emission from quantum dots in the uniformly aligned LC matrix. In faster flows, the red color intensity decrease is close to linear with logarithmic flow velocity. In flows faster than 1 mm/s, irregular LC patterns are supposed to be responsible for intensive scattering of the excitation UV light. The brightness reaches its minimum corresponding to a residual luminescence from QD clusters absorbed on the microchannel walls.

Thus, processing polarized and fluorescence microscopy images allowed us to quantify the optical response of the LC-QD composite to variable microchannel flow velocity and correlate it with orientation state of the QD matrix, flow behavior of QD aggregates and their luminescence properties.

4. Discussion

Composites of MBBA with core–shell CdSe quantum dots were found to generate a variety of optical responses in mobile microchannel confinement both in polarized and UV light. Such an optical behavior of the composite may be attributed to changes in the orientation of domains in the LC matrix. It can be, therefore, conveniently controlled by a single microfluidic parameter (flow velocity up to 10 mm/s).

A quantitative correlation between the flow velocity and average brightness of LC-QD hybrids in polarized and UV light was detected in the flow range up to 2–3 mm/s, which is a typical operation mode of microfluidic devices. Such hybrids can be, therefore, used as luminescence flow detectors or quantitative flow sensors in this range.

At low (below 0.05 mm/s) and high (over 1–2 mm/s) flow velocities, there are two opposite optical states by both the LC matrix in polarized light and QD luminescence in UV. In polarized light, the LC matrix shows a dark field microchannel environment in slow flows and a bright field image in fast flows. Quantum dots demonstrate bright luminescence in UV light in zero or slow flows, which decreases to almost zero in fast flow conditions. Both these effects can be suitable for microfluidic logic applications and setting "0" or "1" logic states and designing flow-controlled optofluidic logic gates. Such logic states can be detected and processed by a simple automated microscopy image analysis with a respective algorithm and used for automated sensing application in lab-on-a-chip tools.

Despite their application potential, the revealed optical effects, however, represent fundamental research results. Studying effects of additional factors on optical properties of confined LC-QD hybrids may contribute to clarifying their potential as nanostructural components of optofluidic devices. The future research activities will focus on studying the impact of temperature and chemical agents (such as surfactants at various concentrations) on optical properties of the studied LC-QD hybrids. Special attention will be given to analyzing the flow and temperature behavior of multiphase systems with LC-QD hybrids confined in microdroplets suitable for further applications as functional photonic microparticles. In this respect, hydrodynamic trapping is a promising approach that allows us to perform experiments with colloid particles and anisotropic systems in microfluidic confinement with a precise control of these parameters [66,67]. Available microfluidic chip designs will allow us to create Stokes traps for LC-QD hybrids and control their behavior with flowrates and temperature as well as additional parameters such as magnetic field.

5. Conclusions

MBBA liquid crystal composites with CdSe core–shell quantum dots were found to reversibly generate opposite optical responses in both polarized and UV light in a typical microflow velocity range up to 10 mm/s. In flows slower than 0.05 mm/s, the composite is dark under crossed polarizers at the 0° position and shows intensive red light emission in UV light. These effects change to a bright field in polarized light and almost zero emission in UV light in flows faster than 1–2 mm/s. The change of brightness upon transition between these two states occurs almost linearly with the logarithmic flow velocity. Such a behavior is considered to correlate with a transition of the LC matrix from a uniform orientation at zero or slow flows to an irregular pattern of domains, which then gradually align along the flow axis.

Such optofluidic microdevices offer new options for applications in photonics as luminescent flow sensors or lab-on-a-chip binary logic circuits as well as optically active tools for medical diagnostics.

Supplementary Materials: The following supporting information can be downloaded at: https://www.mdpi.com/article/10.3390/mi14050990/s1, Figure S1. Processed microfluidic images of MBBA flows in microfluidic channel at various flow velocities [68,69].

Author Contributions: Conceptualization, A.B. and Y.G.; methodology, A.B.; software, A.B.; validation, Y.G.; formal analysis, A.B. and Y.G.; investigation, A.B.; resources, Y.G.; data curation, Y.G.; writing—original draft preparation, A.B.; writing, A.B., review and editing, Y.G.; visualization, A.B.; supervision, Y.G. All authors have read and agreed to the published version of the manuscript.

Funding: The work was performed using the financial support of the Ministry of Science and Higher Education of the Russian Federation within the framework of the state task for delivery of state services (performing works) of 9.12.2022 # 075-01508-23-00. Topic of research: "Development of Scientific Foundations for Synthesis of Novel Multifunctional Materials with a Broad Scope of Application".

Data Availability Statement: Not applicable.

Acknowledgments: The study was performed using the equipment of the Center of Collective Usage "Nanomaterials and Nanotechnologies" of the Kazan National Research Technological University.

Conflicts of Interest: The authors declare no conflict of interest.

References

1. Ramirez, J.C.; Grajales García, D.; Maldonado, J.; Fernández-Gavela, A. Current Trends in Photonic Biosensors: Advances towards Multiplexed Integration. *Chemosensors* **2022**, *10*, 398. [CrossRef]
2. Dawson, H.; Elias, J.; Etienne, P.; Calas-Etienne, S. The Rise of the OM-LoC: Opto-Microfluidic Enabled Lab-on-Chip. *Micromachines* **2021**, *12*, 1467. [CrossRef] [PubMed]
3. Cai, Z.; Li, Z.; Ravaine, S.; He, M.; Song, Y.; Yin, Y.; Zheng, H.; Teng, J.; Zhang, A. From colloidal particles to photonic crystals: Advances in self-assembly and their emerging applications. *Chem. Soc. Rev.* **2021**, *50*, 5898–5951. [CrossRef]
4. Xavier, J.; Vincent, S.; Meder, F.; Vollmer, F. Advances in optoplasmonic sensors–combining optical nano/microcavities and photonic crystals with plasmonic nanostructures and nanoparticles. *Nanophotonics* **2018**, *7*, 1–38. [CrossRef]
5. Kim, Y.G.; Park, S.; Kim, S.-H. Designing photonic microparticles with droplet microfluidics. *Chem. Commun.* **2022**, *58*, 10303–10328. [CrossRef] [PubMed]
6. Dou, Y.; Wang, B.; Jin, M.; Yu, Y.; Zhou, G.; Shui, L. A review on self-assembly in microfluidic devices. *J. Micromechanics Microengineering* **2017**, *27*, 113002. [CrossRef]
7. Rifat, A.A.; Ahmed, R.; Yetisen, A.K.; Butt, H.; Sabouri, A.; Mahdiraji, G.A.; Yun, S.H.; Adikan, F.M. Photonic crystal fiber based plasmonic sensors. *Sens. Actuators B Chem.* **2017**, *243*, 311–325. [CrossRef]
8. Liu, Y.; Zhang, X. Microfluidics-based plasmonic biosensing system based on patterned plasmonic nanostructure arrays. *Micromachines* **2021**, *12*, 826. [CrossRef]
9. Maia, F.R.; Reis, R.L.; Oliveira, J.M. Finding the perfect match between nanoparticles and microfluidics to respond to cancer challenges. *Nanomed. Nanotechnol. Biol. Med.* **2020**, *24*, 102139. [CrossRef]
10. Guimarães, C.F.; Cruz-Moreira, D.; Caballero, D.; Pirraco, R.P.; Gasperini, L.; Kundu, S.C.; Reis, R.L. Shining a Light on Cancer—Photonics in Microfluidic Tumor Modeling and Biosensing. *Adv. Healthc. Mater.* **2022**, 2201442. [CrossRef]
11. Hou, J.; Li, M.; Song, Y. Patterned colloidal photonic crystals. *Angew. Chem. Int. Ed.* **2018**, *57*, 2544–2553. [CrossRef] [PubMed]
12. Ren, W.; Lin, G.; Clarke, C.; Zhou, J.; Jin, D. Optical nanomaterials and enabling technologies for high-security-level anticounterfeiting. *Adv. Mater.* **2020**, *32*, 1901430. [CrossRef] [PubMed]
13. Visaveliya, N.R.; Köhler, J.M. Softness Meets with Brightness: Dye-Doped Multifunctional Fluorescent Polymer Particles via Microfluidics for Labeling. *Adv. Opt. Mater.* **2021**, *9*, 2002219. [CrossRef]
14. Valencia, P.M.; Basto, P.A.; Zhang, L.; Rhee, M.; Langer, R.; Farokhzad, O.C.; Karnik, R. Single-step assembly of homogenous lipid− polymeric and lipid− quantum dot nanoparticles enabled by microfluidic rapid mixing. *ACS Nano* **2010**, *4*, 1671–1679. [CrossRef] [PubMed]
15. Wojnicki, M.; Hessel, V. Quantum materials made in microfluidics-critical review and perspective. *Chem. Eng. J.* **2022**, *438*, 135616. [CrossRef]
16. Fattahi, Z.; Hasanzadeh, M. Nanotechnology-assisted microfluidic systems platform for chemical and bioanalysis. *TrAC Trends Anal. Chem.* **2022**, 116637. [CrossRef]
17. Kashaninejad, N.; Yaghoobi, M.; Pourhassan-Moghaddam, M.; Bazaz, S.R.; Jin, D.; Warkiani, M.E. Biological diagnosis based on microfluidics and nanotechnology. In *Nanotechnology and Microfluidics*; Wiley: Hoboken, NJ, USA, 2020; pp. 211–238.

18. Şahin, S.; Ünlü, C.; Trabzon, L. Affinity biosensors developed with quantum dots in microfluidic systems. *Emergent Mater.* **2021**, *4*, 187–209. [CrossRef]
19. Yong, K.-T.; Wang, Y.; Roy, I.; Rui, H.; Swihart, M.T.; Law, W.-C.; Kwak, S.K.; Ye, L.; Liu, J.; Mahajan, S.D. Preparation of quantum dot/drug nanoparticle formulations for traceable targeted delivery and therapy. *Theranostics* **2012**, *2*, 681–694. [CrossRef]
20. Onoshima, D.; Yukawa, H.; Baba, Y. Multifunctional quantum dots-based cancer diagnostics and stem cell therapeutics for regenerative medicine. *Adv. Drug Deliv. Rev.* **2015**, *95*, 2–14. [CrossRef]
21. Li, Y.; Männel, M.J.; Hauck, N.; Patel, H.P.; Auernhammer, G.K.; Chae, S.; Fery, A.; Li, J.; Thiele, J. Embedment of quantum dots and biomolecules in a dipeptide hydrogel formed in situ using microfluidics. *Angew. Chem. Int. Ed.* **2021**, *60*, 6724–6732. [CrossRef]
22. Bian, F.; Sun, L.; Cai, L.; Wang, Y.; Zhao, Y. Quantum dots from microfluidics for nanomedical application. *Wiley Interdiscip. Rev. Nanomed. Nanobiotechnol.* **2019**, *11*, e1567. [CrossRef]
23. Hu, Y.; Wang, J.; Wang, H.; Wang, Q.; Zhu, J.; Yang, Y. Microfluidic fabrication and thermoreversible response of core/shell photonic crystalline microspheres based on deformable nanogels. *Langmuir* **2012**, *28*, 17186–17192. [CrossRef] [PubMed]
24. Baek, J.; Shen, Y.; Lignos, I.; Bawendi, M.G.; Jensen, K.F. Multistage microfluidic platform for the continuous synthesis of III–V core/shell quantum dots. *Angew. Chem. Int. Ed.* **2018**, *57*, 10915–10918. [CrossRef] [PubMed]
25. Li, S.; Hsiao, J.C.; Howes, P.D.; de Mello, A.J. Microfluidic tools for the synthesis of bespoke quantum dots. In *Nanotechnology and Microfluidics*; Wiley: Hoboken, NJ, USA, 2020; pp. 109–148.
26. Kim, E.B.; Tomczak, K.M.; Chandrasiri, H.B.; Pálmai, M.; Ghaznavi, A.; Gritsenko, D.; Xu, J.; Snee, P.T. Nucleation control of quantum dot synthesis in a microfluidic continuous flow reactor. *Front. Nanotechnol.* **2023**, *4*, 104. [CrossRef]
27. Du, X.Y.; Wang, C.F.; Wu, G.; Chen, S. The rapid and large-scale production of carbon quantum dots and their integration with polymers. *Angew. Chem. Int. Ed.* **2021**, *60*, 8585–8595. [CrossRef]
28. Chen, H.Q.; Wang, X.Y.; Bisoyi, H.K.; Chen, L.J.; Li, Q. Liquid Crystals in Curved Confined Geometries: Microfluidics Bring New Capabilities for Photonic Applications and beyond. *Langmuir* **2021**, *37*, 3789–3807. [CrossRef]
29. Pan, Y.; Xie, S.; Wang, H.; Huang, L.; Shen, S.; Deng, Y.; Ma, Q.; Liu, Z.; Zhang, M.; Jin, M. Microfluidic Construction of Responsive Photonic Microcapsules of Cholesteric Liquid Crystal for Colorimetric Temperature Microsensors. *Adv. Opt. Mater.* **2022**, *11*, 2202141. [CrossRef]
30. Kim, J.G.; Park, S.Y. Photonic Spring-Like Shell Templated from Cholesteric Liquid Crystal Prepared by Microfluidics. *Adv. Opt. Mater.* **2017**, *5*, 1700243. [CrossRef]
31. Copar, S.; Ravnik, M.; Žumer, S. Introduction to colloidal and microfluidic nematic microstructures. *Crystals* **2021**, *11*, 956. [CrossRef]
32. Takenaka, Y.; Škarabot, M.; Muševič, I. Nematic Liquid-Crystal Necklace Structure Made by Microfluidics System. *Langmuir* **2020**, *36*, 3234–3241. [CrossRef] [PubMed]
33. Segura, R.; Cierpka, C.; Rossi, M.; Joseph, S.; Bunjes, H.; Kähler, C.J. Non-encapsulated thermo-liquid crystals for digital particle tracking thermography/velocimetry in microfluidics. *Microfluid. Nanofluidics* **2013**, *14*, 445–456. [CrossRef]
34. Priest, C.; Quinn, A.; Postma, A.; Zelikin, A.N.; Ralston, J.; Caruso, F. Microfluidic polymer multilayer adsorption on liquid crystal droplets for microcapsule synthesis. *Lab A Chip* **2008**, *8*, 2182–2187. [CrossRef]
35. Schwartz, M.; Lenzini, G.; Geng, Y.; Rønne, P.B.; Ryan, P.Y.; Lagerwall, J.P. Cholesteric liquid crystal shells as enabling material for information-rich design and architecture. *Adv. Mater.* **2018**, *30*, 1707382. [CrossRef] [PubMed]
36. Hamlington, B.D.; Steinhaus, B.; Feng, J.J.; Link, D.; Shelley, M.J.; Shen, A.Q. Liquid crystal droplet production in a microfluidic device. *Liq. Cryst.* **2007**, *34*, 861–870. [CrossRef]
37. Čopar, S.; Kos, Ž.; Emeršič, T.; Tkalec, U. Microfluidic control over topological states in channel-confined nematic flows. *Nat. Commun.* **2020**, *11*, 59. [CrossRef]
38. Sengupta, A.; Tkalec, U.; Bahr, C. Nematic textures in microfluidic environment. *Soft Matter* **2011**, *7*, 6542–6549. [CrossRef]
39. Sengupta, A.; Tkalec, U.; Ravnik, M.; Yeomans, J.M.; Bahr, C.; Herminghaus, S. Liquid crystal microfluidics for tunable flow shaping. *Phys. Rev. Lett.* **2013**, *110*, 048303. [CrossRef] [PubMed]
40. Gleeson, H.F.; Wood, T.A.; Dickinson, M. Laser manipulation in liquid crystals: An approach to microfluidics and micromachines. *Philos. Trans. R. Soc. A Math. Phys. Eng. Sci.* **2006**, *364*, 2789–2805. [CrossRef]
41. Basu, R.; Iannacchione, G.S. Evidence for directed self-assembly of quantum dots in a nematic liquid crystal. *Phys. Rev. E* **2009**, *80*, 010701. [CrossRef]
42. Mirzaei, J.; Reznikov, M.; Hegmann, T. Quantum dots as liquid crystal dopants. *J. Mater. Chem.* **2012**, *22*, 22350–22365. [CrossRef]
43. Kurilov, A.D.; Chausov, D.N.; Osipova, V.V.; Kucherov, R.N.; Belyaev, V.V.; Galyametdinov, Y.G. Highly luminescent nanocomposites of nematic liquid crystal and hybrid quantum dots CdSe/CdS with ZnS shell. *J. Mol. Liq.* **2021**, *339*, 116747. [CrossRef]
44. Huang, B.-L.; Guo, T.-L.; Xu, S.; Ye, Y.; Chen, E.-G.; Lin, Z.-X. Color converting film with quantum-dots for the liquid crystal displays based on inkjet printing. *IEEE Photonics J.* **2019**, *11*, 1–9. [CrossRef]
45. Gollapelli, B.; Suguru Pathinti, R.; Vallamkondu, J. Carbon Quantum Dots doped Cholesteric Liquid Crystal Films and Microdroplets for Anti-Counterfeiting. *ACS Appl. Nano Mater.* **2022**, *5*, 11912–11922. [CrossRef]
46. Gollapelli, B.; Ganji, S.R.R.; Tatipamula, A.K.; Vallamkondu, J. Bio-derived chlorophyll dye doped cholesteric liquid crystal films and microdroplets for advanced anti-counterfeiting security labels. *J. Mol. Liq.* **2022**, *363*, 119952. [CrossRef]

47. Bezrukov, A.; Galyametdinov, Y. Orientation Behavior of Nematic Liquid Crystals at Flow-Wall Interfaces in Microfluidic Channels. *Coatings* **2023**, *13*, 169. [CrossRef]
48. Bezrukov, A.; Osipova, V.; Galyametdinov, Y.G. Orientational behavior of a nematic liquid crystal and its composite with quantum dots in a microfluidic channel. *Russ. Chem. Bull.* **2022**, *71*, 2092–2097. [CrossRef]
49. Holmberg, K.; Jönsson, B.; Kronberg, B.; Lindman, B. *Surfactants and Polymers in Aqueous Solution*, 2nd ed.; Wiley: Hoboken, NJ, USA, 2003; 545p.
50. Anna, S.L. Droplets and Bubbles in Microfluidic Devices. *Annu. Rev. Fluid Mech.* **2016**, *48*, 285–309. [CrossRef]
51. McDonald, J.C.; Duffy, D.C.; Anderson, J.R.; Chiu, D.T.; Wu, H.; Schueller, O.J.A.; Whitesides, G.M. Fabrication of microfluidic systems in poly(dimethylsiloxane). *Electrophoresis* **2000**, *21*, 27–40. [CrossRef]
52. Wu, M.-H.; Paul, K.; Whitesides, G. Patterning Flood Illumination with Microlens Arrays. *Appl. Opt.* **2002**, *41*, 2575–2585. [CrossRef]
53. Ibragimov, T.D. Dielectric and conductivity properties of liquid crystal MBBA doped with SWCNTs. *Fuller. Nanotub. Carbon Nanostructures* **2022**, *30*, 848–852. [CrossRef]
54. Antharjanam, P.S.; Prasad, E. Nematic to smectic texture transformation in MBBA by in situ synthesis of silver nanoparticles. *N. J. Chem.* **2010**, *34*, 420–425. [CrossRef]
55. Price, A.D.; Schwartz, D.K. DNA Hybridization-Induced Reorientation of Liquid Crystal Anchoring at the Nematic Liquid Crystal/Aqueous Interface. *J. Am. Chem. Soc.* **2008**, *130*, 8188–8194. [CrossRef]
56. Shah, R.R.; Abbott, N.L. Orientational Transitions of Liquid Crystals Driven by Binding of Organoamines to Carboxylic Acids Presented at Surfaces with Nanometer-Scale Topography. *Langmuir* **2003**, *19*, 275–284. [CrossRef]
57. Zou, Y.; Namkung, J.; Lin, Y.; Ke, D.; Lindquist, R. Interference colors of nematic liquid crystal films at different applied voltages and surface anchoring conditions. *Opt. Express* **2011**, *19*, 3297–3303. [CrossRef]
58. Brake, J.M.; Mezera, A.D.; Abbott, N.L. Active Control of the Anchoring of 4′-Pentyl-4-cyanobiphenyl (5CB) at an Aqueous−Liquid Crystal Interface By Using a Redox-Active Ferrocenyl Surfactant. *Langmuir* **2003**, *19*, 8629–8637. [CrossRef]
59. Sengupta, A.; Herminghaus, S.; Bahr, C. Liquid crystal microfluidics: Surface, elastic and viscous interactions at microscales. *Liq. Cryst. Rev.* **2014**, *2*, 73–110. [CrossRef]
60. Bezrukov, A.; Galyametdinov, Y. On-Chip Control over Polyelectrolyte-Surfactant Complexation in Nonequilibrium Microfluidic Confinement. *Polymers* **2022**, *14*, 4109. [CrossRef]
61. Kneppe, H.; Schneider, F. Determination of the Viscosity Coefficients of the Liquid Crystal MBBA. *Mol. Cryst. Liq. Cryst.* **2011**, *65*, 23–37. [CrossRef]
62. Jamiruddin, M.R.; Meghla, B.A.; Islam, D.Z.; Tisha, T.A.; Khandker, S.S.; Khondoker, M.U.; Haq, M.A.; Adnan, N.; Haque, M. Microfluidics technology in SARS-CoV-2 diagnosis and beyond: A systematic review. *Life* **2022**, *12*, 649. [CrossRef]
63. Gao, D.; Ma, Z.; Jiang, Y. Recent advances in microfluidic devices for foodborne pathogens detection. *TrAC Trends Anal. Chem.* **2022**, *157*, 116788. [CrossRef]
64. Buchanan, B.C.; Yoon, J.-Y. Microscopic imaging methods for organ-on-a-chip platforms. *Micromachines* **2022**, *13*, 328. [CrossRef]
65. Gupta, N.; Shukla, D. A Reconstruction of Gray Scale Image into RGB Color Space Image Using YCbCr Color Spacing and Luminance Mapping in Matlab. *i-Manag. J. Pattern Recognit.* **2017**, *4*, 17.
66. Kumar, D.; Richter, C.M.; Schroeder, C.M. Conformational dynamics and phase behavior of lipid vesicles in a precisely controlled extensional flow. *Soft Matter* **2020**, *16*, 337–347. [CrossRef] [PubMed]
67. Kumar, D.; Shenoy, A.; Deutsch, J.; Schroeder, C.M. Automation and flow control for particle manipulation. *Curr. Opin. Chem. Eng.* **2020**, *29*, 1–8. [CrossRef]
68. Qidwai, U.; Chen, C.H. *Digital Image Processing: An Algorithmic Approach with MATLAB*; CRC Press: Boca Raton, FL, USA, 2009; p. 294.
69. Galyametdinov, Y.G.; Krupin, A.S.; Sagdeev, D.O.; Karyakin, M.E.; Shamilov, R.R.; Knyazev, A.A. Luminescent Composites Based on Liquid Crystalline Europium({III}) Complex and {CdSe}/{CdS}/{ZnS} Quantum Dots. *Liq. Cryst. Appl.* **2022**, *22*, 27–38. [CrossRef]

micromachines

Review

Hardware Error Correction for MZI-Based Matrix Computation

Huihuang Hou [1,2], Pengfei Xu [3], Zhiping Zhou [3,4,*] and Hui Su [1,5,*]

1 Key Laboratory of Optoelectronic Materials Chemistry and Physics, Fujian Institute of Research on the Structure of Matter, Chinese Academy of Sciences, Fuzhou 350002, China; houhuihuang@fjirsm.ac.cn
2 University of Chinese Academy of Sciences, Beijing 100049, China
3 State Key Laboratory of Advanced Optical Communication Systems and Networks, School of Electronics, Peking University, Beijing 100871, China; xupengf@pku.edu.cn
4 Beijing Aijie Optoelectronic Technology Co., Ltd., Beijing 100190, China
5 Fujian Science & Technology Innovation Laboratory for Optoelectronic Information of China, Fuzhou 350108, China
* Correspondence: zjzhou@pku.edu.cn (Z.Z.); huisu@fjirsm.ac.cn (H.S.)

Abstract: With the rapid development of artificial intelligence, the electronic system has fallen short of providing the needed computation speed. It is believed that silicon-based optoelectronic computation may be a solution, where Mach–Zehnder interferometer (MZI)-based matrix computation is the key due to its advantages of simple implementation and easy integration on a silicon wafer, but one of the concerns is the precision of the MZI method in the actual computation. This paper will identify the main hardware error sources of MZI-based matrix computation, summarize the available hardware error correction methods from the perspective of the entire MZI meshes and a single MZI device, and propose a new architecture that will largely improve the precision of MZI-based matrix computation without increasing the size of the MZI's mesh, which may lead to a fast and accurate optoelectronic computing system.

Keywords: hardware error correction; matrix computation; Mach–Zehnder interferometer

Citation: Hou, H.; Xu, P.; Zhou, Z.; Su, H. Hardware Error Correction for MZI-Based Matrix Computation. *Micromachines* **2023**, *14*, 955. https://doi.org/10.3390/mi14050955

Academic Editors: He Yang, Xinyang Su and Yizhong Huang

Received: 20 March 2023
Revised: 23 April 2023
Accepted: 24 April 2023
Published: 27 April 2023

1. Introduction

Benefiting from the era of big data, the rapid growth of the Internet provides sufficient training datasets for functioning artificial intelligence (AI), which has also become a hot spot in the current technological revolution. However, artificial intelligence has conventionally relied on electronic processors, with its computing power being greatly determined by transistor numbers and capabilities; managing the massive amount of data necessitates additional computing resources. In the post-Moore era, the increasing transistor numbers can no longer keep up with the computing power demand of artificial intelligence, and electronic computation is therefore stuck in a bottleneck [1–3]. On the other hand, photons are bosons, which have the advantages of a higher transmission speed and can work well with electrons through the Einstein coefficients, which is conducive to the realization of ultra-high-speed optoelectronic computing. Consequently, more and more researchers are beginning to focus on optoelectronic computing [4–9].

The fabrication techniques for silicon photonics are advancing rapidly, enabling the creation of large-scale and complicated circuits and paving the way for low-loss and low-cost optoelectronic devices [10]. These devices can be produced using complementary metal–oxide–semiconductor (CMOS) fabs [11]. As a result, silicon-based optoelectronic computation has emerged as a prominent area of research [3]. At present, optoelectronic matrix computation mainly includes three implementation methods [12]: multi-plane light conversion (MPLC), the Mach–Zehnder interferometer method (MZI), and wavelength division multiplexing (WDM). The Mach–Zehnder interferometer method is simple and easy to integrate. It is one of the best choices to implement optoelectronic matrix computation [13].

The optoelectronic matrix computation implemented by the Mach–Zehnder interferometer is a kind of analog computation, in which computational precision is the most important and should be treated with special attention. However, the performance of silicon photonic devices is sensitive to the influence of environmental disturbances [14–17], fabrication [15,18], and device aging [15,19], including MZI devices. Particularly, the hardware errors generated by fabrication are the focus of research [20,21], which result in deviations between the matrix implemented by MZI and the actual needed matrix. The hardware errors involve beam splitter errors [20], phase errors [22], and errors caused by optical loss differences [23]. The hardware errors will accumulate with the increase of the scale of the MZI meshes, which limits the implementation of large-scale matrix computation based on MZI meshes. In order to overcome this limitation of hardware error, especially beam splitter error, researchers have proposed different MZI meshes or different MZI devices with higher precision.

In this paper, the Mach–Zehnder interferometer method of achieving optoelectronic matrix computation is introduced in Section 2. The origins of the hardware error of the Mach–Zehnder interferometer are analyzed in Section 3, followed by a synthesis of the studies of MZI hardware error correction in Section 4. Finally, a feasible approach to mitigate the MZI hardware error is proposed in Section 5.

2. Methods

In 1994, Reck et al. [24] proposed one type of triangular mesh based on the Mach–Zehnder interferometer and demonstrated that the unitary matrix transformation of any finite dimension could be implemented by MZI devices. They decomposed the N-dimensional unitary matrix into a series of two-dimensional unitary matrices, which were presented as a triangular mesh formed by the arrangement of MZIs, as shown in Figure 1.

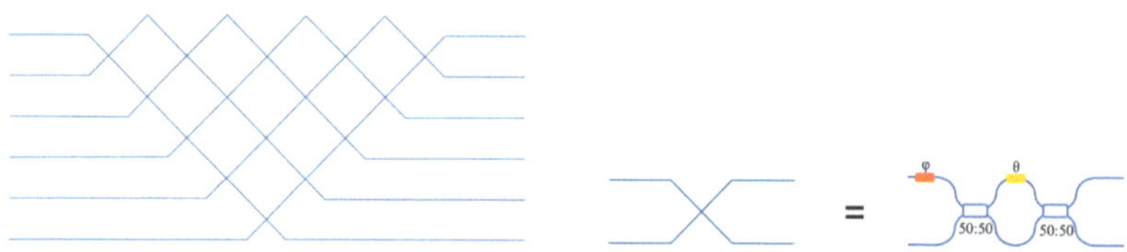

Figure 1. Triangular mesh.

A common Mach–Zehnder interferometer consists of an external phase shifter (φ), an internal phase shifter (θ), and two 50:50 beam splitters (directional couplers (DC)) or multimode interference couplers (MMI)), as shown in Figure 2.

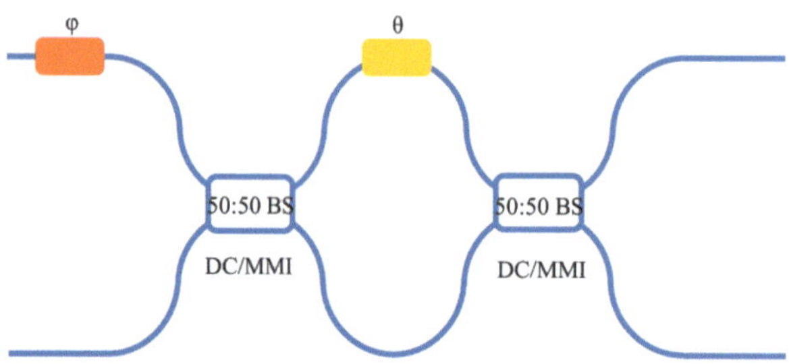

Figure 2. The common MZI.

For an MZI device, transmission matrix S is

$$S = C_4 P_3 C_2 P_1 \tag{1}$$

where P_1 and P_3 are the transmission matrices of the external phase shifter and the internal phase shifter, respectively; C_2 and C_4 are the transmission matrices of the beam splitter on the left side of the MZI and the beam splitter on the right side of the MZI, respectively.

Among them, the left and right beam splitters are considered as ideal 50:50 beam splitters. Hence, MZI transmission matrix S can be written as [25]

$$S = \begin{bmatrix} 1/\sqrt{2} & i/\sqrt{2} \\ i/\sqrt{2} & 1/\sqrt{2} \end{bmatrix} \begin{bmatrix} e^{i\theta} & 0 \\ 0 & 1 \end{bmatrix} \begin{bmatrix} 1/\sqrt{2} & i/\sqrt{2} \\ i/\sqrt{2} & 1/\sqrt{2} \end{bmatrix} \begin{bmatrix} e^{i\varphi} & 0 \\ 0 & 1 \end{bmatrix} \tag{2}$$

$$S = ie^{i\theta/2} \begin{bmatrix} e^{i\varphi} sin\theta/2 & cos\theta/2 \\ e^{i\varphi} cos\theta/2 & -sin\theta/2 \end{bmatrix} \tag{3}$$

where φ and θ are phases produced by the external phase shifter and the internal phase shifter, respectively. Obviously, we can produce different phases through these two phase shifters to implement two-dimensional unitary matrices with different elements. For an N-dimensional unitary matrix, the MZI's transmission matrix can be generalized as $T_{n,m}$ (θ, φ):

$$T_{n,m}(\theta,\varphi) = \begin{bmatrix} 1 & 0 & & \cdots & & \cdots & & 0 & 0 \\ 0 & 1 & & & & & & 0 & 0 \\ & & \ddots & & & & & & \\ \vdots & & & e^{i\varphi} sin\theta/2 & & cos\theta/2 & & & \vdots \\ \vdots & & & e^{i\varphi} cos\theta/2 & & -sin\theta/2 & & & \vdots \\ & & & & & & \ddots & & \\ 0 & 0 & & & & & & 1 & 0 \\ 0 & 0 & & \cdots & & \cdots & & 0 & 1 \end{bmatrix} \tag{4}$$

where n and m represent the transmission matrix of the MZI between the nth and mth input ports of the signal entering the mesh. The dimension of an N-dimensional unitary matrix, $U(N)$, can be reduced by right multiplying $T_{n,m}$ (θ, φ), namely,

$$U(N)T_{N,N-1}T_{N,N-2}\cdots T_{N,1} = \begin{bmatrix} U(N-1) & 0 \\ 0 & 1 \end{bmatrix} \tag{5}$$

The N-dimensional unitary matrix's dimension can be continued to be reduced according to the above dimensionality reduction method, and finally, a diagonal matrix, D, with elements of modulo 1 is obtained.

Let

$$R(N) = T_{N,N-1}T_{N,N-2}\cdots T_{N,1} \tag{6}$$

$$U(N)R(N)R(N-1)R(2) = \begin{bmatrix} 1 & & \cdots & & 0 \\ & \ddots & & & \\ \vdots & & 1 & & \vdots \\ & & & \ddots & \\ 0 & & \cdots & & 1 \end{bmatrix} = D \tag{7}$$

It can be seen from the previous theoretical derivation that the final implementation of the N-dimensional unitary matrix will be a triangular mesh with N-1 MZIs in the N row and N-2 MZIs in the N-1 row. Based on the above theory, in 2017, Shen et al. [26] experimentally demonstrated a cascaded mesh of 56 programmable MZIs with this triangular mesh, which improved the computational speed and power efficiency compared with that of traditional electronic processors.

3. Hardware Error

The N-dimensional unitary matrix implemented by MZI meshes has some deviations from the needed matrix in actual computation. On the one hand, the triangular mesh of the Reck design leads to an inconsistent optical loss of each output, and the larger mesh is, the greater the optical loss difference of each output, and the higher the total losses. On the other hand, more importantly, the beam splitter in MZI cannot implement the ideal 50:50 beam splitter due to the hardware error generated by the fabrication and the non-uniformity of the material, which leads to the deviation of the transmission matrix and affects the final computation.

The non-ideal 50:50 beam splitter transmission matrix can be expressed as follows:

$$\begin{bmatrix} cos\varphi & isin\varphi \\ isin\varphi & cos\varphi \end{bmatrix} \tag{8}$$

where, when $\varphi = \pi/4$, it is the ideal 50:50 beam splitter; φ will generally deviate from $\pi/4$ for the non-ideal 50:50 beam splitter, which leads to the deviation of the MZI transmission matrix. For the transmission matrix of the beam splitter on the left side of the MZI and the beam splitter on the right side of the MZI, φ can be expressed as φ_1 and φ_2, respectively, and then MZI transmission matrix S' can be expressed as

$$S' = \begin{bmatrix} cos\varphi_2 & isin\varphi_2 \\ isin\varphi_2 & cos\varphi_2 \end{bmatrix} \begin{bmatrix} e^{i\theta} & 0 \\ 0 & 1 \end{bmatrix} \begin{bmatrix} cos\varphi_1 & isin\varphi_1 \\ isin\varphi_1 & cos\varphi_1 \end{bmatrix} \begin{bmatrix} e^{i\varphi} & 0 \\ 0 & 1 \end{bmatrix} \tag{9}$$

$$S' = ie^{i\theta/2} \begin{bmatrix} e^{i\varphi} \begin{pmatrix} cos(\varphi_1 - \varphi_2)sin(\theta/2) - \\ icos(\varphi_1 + \varphi_2)cos(\theta/2) \end{pmatrix} & \begin{pmatrix} sin(\varphi_1 + \varphi_2)cos(\theta/2) + \\ isin(\varphi_1 - \varphi_2)sin(\theta/2) \end{pmatrix} \\ e^{i\varphi} \begin{pmatrix} sin(\varphi_1 + \varphi_2)cos(\theta/2) - \\ isin(\varphi_1 - \varphi_2)sin(\theta/2) \end{pmatrix} & -\begin{pmatrix} cos(\varphi_1 - \varphi_2)sin(\theta/2) + \\ icos(\varphi_1 + \varphi_2)cos(\theta/2) \end{pmatrix} \end{bmatrix} \tag{10}$$

Obviously, the MZI's transmission matrix changes because of two non-ideal beam splitters, which is inconsistent with the transmission matrix in the ideal case, which leads to the deviation between the N-dimensional unitary matrix implemented by the MZI meshes and the theoretical one.

4. Error Correction

In terms of the hardware error in the N-dimensional unitary matrix transformation implemented by the MZI meshes introduced in Section 3, researchers have made a lot of effort. This section will introduce the main hardware error correction methods for (1) the entire MZI meshes and (2) a single MZI device.

4.1. The Entire MZI Meshes

4.1.1. Rectangular Mesh

In 2016, Clements et al. [23] proposed a rectangular mesh, which was an improvement of the triangular mesh of Reck, as shown in Figure 3. Compared to the triangular mesh of Reck, the rectangular mesh has higher symmetry and a lower optical loss difference of each output. Additionally, the longest optical depth of the rectangular mesh is about half that of the triangular mesh, and the total optical loss is only half that of the triangular mesh.

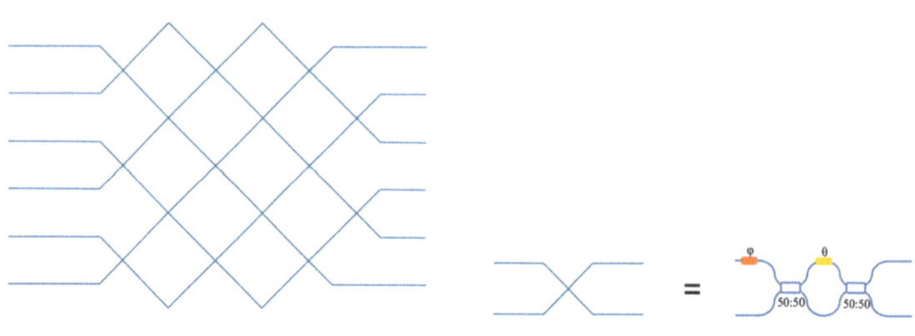

Figure 3. Rectangular mesh.

The improvement of the rectangular mesh is specifically manifested in the process of decomposition of the N-dimensional unitary matrix into a series of two-dimensional unitary matrices. The decomposition process does not reduce the dimensions by right multiplying $T_{n,m}(\theta, \varphi)$, but by both right multiplying $T_{n,m}(\theta, \varphi)$ and left multiplying $T^{-1}{}_{n,m}(\theta, \varphi)$, as shown in Figure 4 and seen in the following:

$$T_{4,5}T_{3,4}T_{2,3}T_{1,2}T_{4,5}T_{3,4}UT^{-1}_{1,2}T^{-1}_{3,4}T^{-1}_{2,3}T^{-1}_{1,2} = D \tag{11}$$

that is,

$$U = T^{-1}_{3,4}T^{-1}_{4,5}T^{-1}_{1,2}T^{-1}_{2,3}T^{-1}_{3,4}T^{-1}_{4,5}DT_{1,2}T_{2,3}T_{3,4}T_{1,2} \tag{12}$$

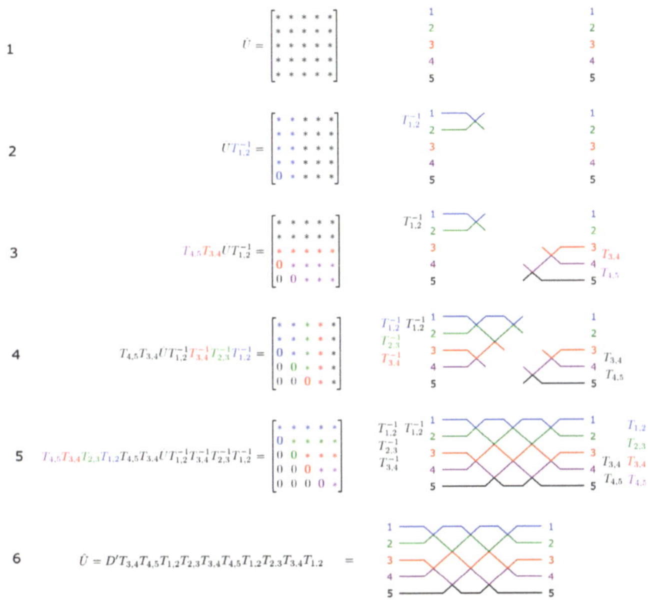

Figure 4. The decomposition method of the rectangular mesh. Reprinted under the terms of the CC-BY license [23]. Copyright 2016, Clements et al., published by Optica.

For a matrix $T_{n,m}(\theta, \varphi)$ and a diagonal matrix D, there are matrix $T^{-1}{}_{n,m}(\theta, \varphi)$ and another diagonal matrix D', with $T^{-1}{}_{n,m}(\theta, \varphi)\ D' = D\ T_{n,m}(\theta, \varphi)$, and we obtain a right-multiplication dimensionality reduction operation similar to that in Equation (7).

$$U = D'T_{3,4}T_{4,5}T_{1,2}T_{2,3}T_{3,4}T_{4,5}T_{1,2}T_{2,3}T_{3,4}T_{1,2} \tag{13}$$

Shokraneh et al. [27] experimentally proved that due to the asymmetric distribution of MZIs in the triangular mesh, optical loss is greater on the triangular mesh than on the rectangular mesh in the computation process. They used a dataset that was perfectly classifiable to assess the classification performance of the two meshes in optical neural networks (ONN). Compared to the triangular mesh, the rectangular mesh is more phase-error-tolerant and loss-tolerant. Thus, the rectangular mesh is commonly adopted as the fundamental unit for constructing ONN in various research studies [28–30]. However, the beam splitter errors will still impact the computed result of the rectangular mesh.

4.1.2. Fourier Structure

In order to solve hardware errors introduced by the non-ideal 50:50 beam splitter, Lopez-Pastor et al. [31] presented one kind of mesh composed of Fourier transforms and phase masks, which can implement the unitary matrix transformation of any finite dimension.

For an N-dimensional unitary matrix, the above Equation (13) can be written as

$$U = D\prod_{i=1}^{N/2}\prod_{k=1}^{N/2-1} T_{2k}\left(\chi_k^{(i)},\eta_k^{(i)}\right)\prod_{j=1}^{N/2} T_{2j-1}\left(\theta_j^{(i)},\varphi_j^{(i)}\right), \tag{14}$$

$\prod_{k=1}^{N/2-1} T_{2k}\left(\chi_k^{(i)},\eta_k^{(i)}\right)$ and $\prod_{j=1}^{N/2} T_{2j-1}\left(\theta_j^{(i)},\varphi_j^{(i)}\right)$ can be decomposed by the diagonal matrix, permutation matrix and circulant matrix, and the decomposition of the unitary matrix can be obtained as follows:

$$U = DG\left[\prod_{i=1}^{N/2} B^{(i)}A^{(i)}\right]G, \tag{15}$$

where $A^{(i)}$ and $B^{(i)}$ are

$$A^{(i)} = \left\{E, G, H, p\left(\Gamma\left(\theta^{(i)}\right)\right), E, G\Gamma\left(\varphi^{(i)}\right)\right\}_{F'} \tag{16}$$

and,

$$B^{(i)} = \left\{E, p(G), H, \Gamma\left(\chi^{(i)}\right), E, p\left(G\Gamma\left(\eta^{(i)}\right)\right)\right\}_{F'} \tag{17}$$

Finally, The N-dimensional unitary matrix, $U(N)$, can be decomposed into a product of phase masks and Fourier transforms. As shown in Figure 5, the gray rounded rectangles represent the Fourier transforms, and the colored rectangles represent the phase-mask diagonal matrices. Only two diagonal matrices per layer (denoted by red and yellow rectangles) depend on the unitary matrix being implemented, while the rest (denoted by blue rectangles) are fixed. In this structure, the MZI of each layer in the mesh is decomposed into the form of Fourier transforms and phase masks, and the Fourier transforms can be implemented by MMI. This mesh avoids the use of a large number of beam splitters, reduces the source of error to a certain extent, and is more conducive to the realization of a large-scale mesh for optoelectronic computation. However, this decomposition method is not as simple and easy to implement as the triangular mesh and rectangular mesh methods are. Additionally, an even-dimension unitary matrix requires 6N DFTs (the discrete Fourier transform) and 6N + 1 controllable phase masks, and thus the size of the mesh will also significantly increase.

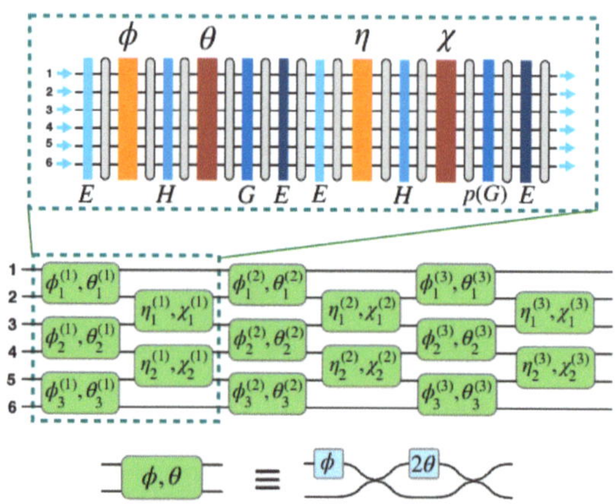

Figure 5. Mesh composed of Fourier transforms and phase masks. Reprinted under the terms of the OSA Open Access Publishing Agreement [31]. Copyright 2021, Lopez-Pastor et al., published by Opt. Express.

4.1.3. Redundant Rectangular Mesh

Pai et al. [32] proposed two mesh improvements. The first is adding redundant tunable layers in the rectangular Mesh, called the redundant rectangular mesh (RRM), which can accelerate the optimization process of the mesh to implement the unitary matrix. The second is adding a redundant mesh composed of low-loss waveguide crossings or MZIs with fixed cross-state phase shifts, called the permuting rectangular mesh (PRM). As shown in Figure 6, Figure 6a is a schematic diagram of the RRM, where the green part represents the redundant tunable layers, and Figure 6b shows the PRM, where the gray part represents the additional low-loss waveguide crossings or MZIs with fixed cross-state phase shifts. This method of adding additional MZIs in the mesh increases the tunable degrees of freedom in the MZI's mesh. For a given unitary matrix, there is a supersaturated implementation schemes. Some unitary matrices that cannot be realized due to MZI imperfections can be realized by new equivalent schemes brought by the extra degrees of freedom in the redundant mesh [33].

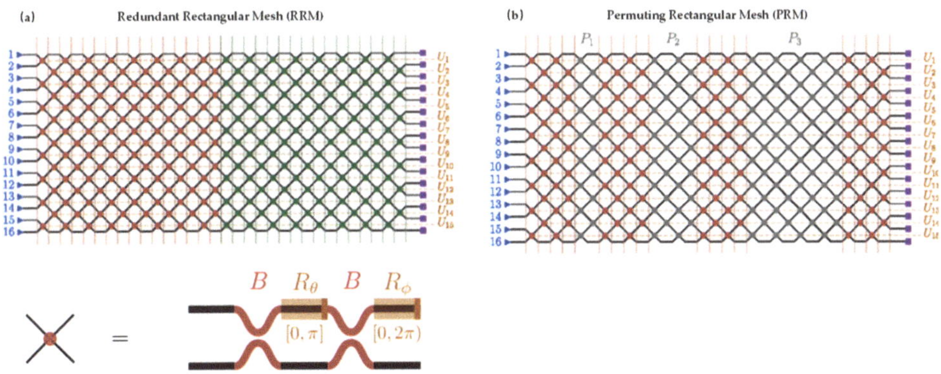

Figure 6. (**a**) Redundant rectangular mesh; (**b**) permuting rectangular mesh. Reprinted with permission from ref. [32]. Copyright 2019, APS.

However, the redundant mesh increases the size of the MZI meshes, which inevitably leads to an increase in the optical loss of the entire mesh. In a small mesh, the increased optical loss is within an acceptable range, and the computational precision of the MZI meshes is better improved.

4.2. A Single MZI Device

In addition to the above improvements in the MZI's mesh, more researchers focus on the improvement of a single MZI device. The improvement of a single MZI device is mainly carried out by adding redundant phase shifters and beam splitters to improve the tunable degrees of freedom of the MZI, so as to correct its hardware errors.

For a MZI with non-ideal beam splitters, where neither of the two beam splitters has an ideal 50:50 beam splitter ratio, the transmission matrix is given by Equation (10). When we input optical signal E_1 from only one port of the MZI, the output optical powers P_{out_1} and P_{out_2} of the two output ports are

$$P_{out\_1} = |E_1|^2\left[cos^2(\varphi_1+\varphi_2)cos^2(\theta/2)+cos^2(\varphi_1-\varphi_2)sin^2(\theta/2)\right], \quad (18)$$

$$P_{out\_2} = |E_1|^2\left[sin^2(\varphi_1+\varphi_2)cos^2(\theta/2)+sin^2(\varphi_1-\varphi_2)sin^2(\theta/2)\right], \quad (19)$$

When $\varphi_1+\varphi_2=\pi/2$, $P_{out\_1}=|E_1|^2cos^2(\varphi_1+\varphi_2)cos^2(\theta/2)$, complete extinction can be achieved. Suzuki et al. [34] proposed a MZI architecture, replacing a beam splitter in the common MZI with a MZI, as shown in Figure 7. They adjusted this MZI as a 50:50 beam splitter so that its split ratio was complementary to that of another beam splitter to meet the above complete extinction conditions. Moreover, they demonstrated this 2 × 2 MZI with the highest extinction ratio (50.4 dB).

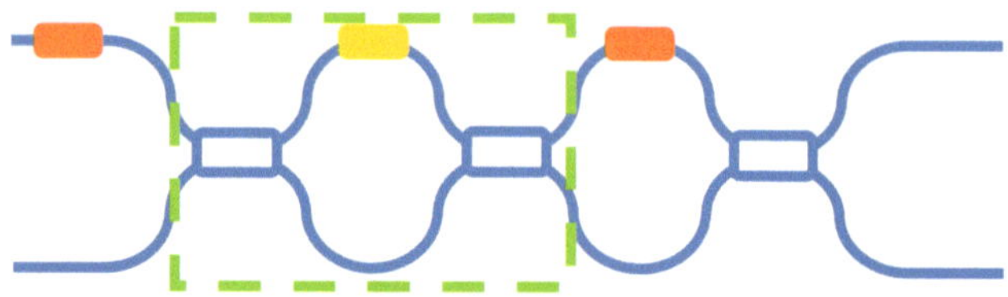

Figure 7. MZI architecture with a variable splitter as the front 3-dB splitter (MZI as a beam splitter in green frame).

Although this architecture improves the extinction ratio of the MZI and corrects its transmission matrix to a certain extent, the correction of the transmission matrix is not complete. When complete extinction is achieved, MZI transmission matrix S'' can be expressed as

$$S'' = ie^{i\theta/2}\begin{bmatrix} 2e^{i\varphi}\cos\varphi_2 sin\varphi_2 sin(\theta/2) & (cos(\theta/2) + i\cos(2\varphi_2)sin(\theta/2)) \\ e^{i\varphi}(cos(\theta/2) - icos(2\varphi_2)sin(\theta/2)) & -2\cos\varphi_2 sin\varphi_2 sin(\theta/2) \end{bmatrix}, \quad (20)$$

It is evident that the transmission matrix of MZI undergoes a slight correction, yet it still deviates from the ideal MZI transmission matrix's S (Equation (3)). Consequently, errors may still arise during matrix computations, especially when dealing with large-scale MZI meshes.

Miller et al. [35] presented a double Mach–Zehnder interferometer (DMZI), as shown in Figure 8. Two MZIs were used as tunable beam splitters to replace the two splitters on the left and right of the common MZI, forming a new MZI architecture.

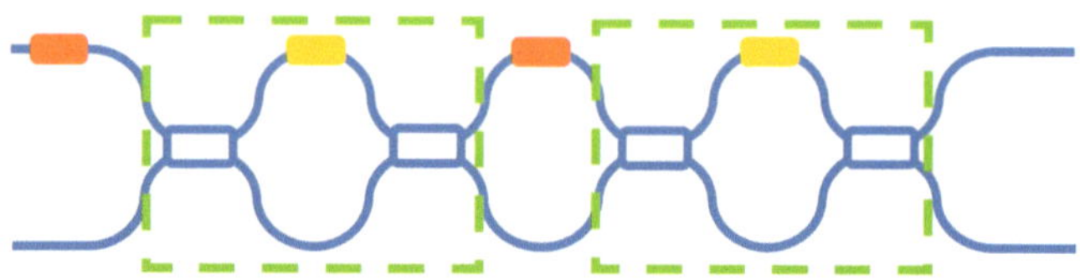

Figure 8. A double MZI architecture (MZI as a beam splitter in green frame).

Now, we will illustrate the error correction process of DMZI. For instance, considering the MZI on the left, we should adjust its beam splitters to a 50:50 split. The transmission matrix of two beam splitters in this MZI can be expressed as P_a and P_b:

$$P_a = \begin{bmatrix} \sqrt{\frac{1}{2} - R_a} & i\sqrt{\frac{1}{2} + R_a} \\ i\sqrt{\frac{1}{2} + R_a} & \sqrt{\frac{1}{2} - R_a} \end{bmatrix}, \quad (21)$$

$$P_b = \begin{bmatrix} \sqrt{\frac{1}{2} - R_b} & i\sqrt{\frac{1}{2} + R_b} \\ i\sqrt{\frac{1}{2} + R_b} & \sqrt{\frac{1}{2} - R_b} \end{bmatrix}, \quad (22)$$

where, R_a and R_b represent errors of the left beam splitter and the right beam splitter, respectively. The transmission matrix, P_L, of the MZI as a beam splitter can be expressed as

$$P_L = \begin{bmatrix} \sqrt{\frac{1}{2} - R_b} & i\sqrt{\frac{1}{2} + R_b} \\ i\sqrt{\frac{1}{2} + R_b} & \sqrt{\frac{1}{2} - R_b} \end{bmatrix} \begin{bmatrix} e^{i\theta} & 0 \\ 0 & 1 \end{bmatrix} \begin{bmatrix} \sqrt{\frac{1}{2} - R_a} & i\sqrt{\frac{1}{2} + R_a} \\ i\sqrt{\frac{1}{2} + R_a} & \sqrt{\frac{1}{2} - R_a} \end{bmatrix} \begin{bmatrix} e^{i\varphi} & 0 \\ 0 & 1 \end{bmatrix}, \quad (23)$$

This can be rewritten as

$$P_L = \begin{bmatrix} e^{i\theta}\sqrt{\alpha_b\alpha_a} - \sqrt{\beta_b\beta_a} & ie^{i\theta}\sqrt{\alpha_b\beta_a} + i\sqrt{\beta_b\alpha_a} \\ ie^{i\theta}\sqrt{\beta_b\alpha_a} + i\sqrt{\alpha_b\beta_a} & \sqrt{\alpha_b\alpha_a} - e^{i\theta}\sqrt{\beta_b\beta} \end{bmatrix}, \quad (24)$$

where, $\alpha_b = \frac{1}{2} - R_b$; $\beta_b = \frac{1}{2} + R_b$; $\alpha_a = \frac{1}{2} - R_a$; $\beta_a = \frac{1}{2} + R_a$. When we input optical signals, E_1, from only one port of MZI, the output optical power, P_{out}, of one of the two output ports is

$$P_{out} = |E_1|^2 \left\{ \frac{1}{2} + 2\left[R_a R_b - \sqrt{\left(\frac{1}{4} - R_a^2\right)\left(\frac{1}{4} - R_b^2\right)} \times cos\theta \right] \right\}, \quad (25)$$

Thus, if this MZI is used as a 50:50 beam splitter, $P_3 = \frac{1}{2}|E_1|^2$; hence, from Equation (25)

$$R_a R_b = \sqrt{\left(\frac{1}{4} - R_a^2\right)\left(\frac{1}{4} - R_b^2\right)} \times cos\theta, \quad (26)$$

considering $cos^2\theta \leq 1$; hence, from Equation (26)

$$R_a^2 R_b^2 \leq \left(\frac{1}{4} - R_a^2\right)\left(\frac{1}{4} - R_b^2\right), \quad (27)$$

which can be derived to give

$$|R_a| \leq \sqrt{\frac{1}{8}} \simeq 0.35\&|R_b| \leq \sqrt{\frac{1}{8}} \simeq 0.35, \quad (28)$$

Therefore, the split ratios of the fabricated power from 85:15 to 15:85 can be compensated by adjusting the split ratio of the two tunable beam splitters back to 50:50. Two redundant beam splitters and two redundant phase shifters can completely correct the transmission matrix of MZI. Wilkes et al. [36] proposed a configuration algorithm for this DMZI, eventually achieving a 60 dB extinction ratio.

However, the size of the MZI also increases significantly. The increase in optical loss brought about by redundant beam splitters and phase shifters is also a problem to be considered in large-scale MZI meshes. Moreover, the transmission matrix of a real MZI as a beam splitter is different from that of the ideal one; when the ideal MZI is used as a 50:50 beam splitter, its transmission matrix R, from Equation (3), is:

$$R = \begin{bmatrix} 1 & 1 \\ 1 & -1 \end{bmatrix}, \quad (29)$$

where, $x = cos(\varphi_1 - \varphi_2); y = cos(\varphi_1 + \varphi_2); s = sin(\theta/2); c = cos(\theta/2)$. The corresponding output optical phase of the four elements in the matrix R can be represented on a complex plane, as shown in Figure 9.

For a real MZI, when it is used as a 50:50 beam splitter, its transmission matrix R', from Equation (10), is

$$R' = \begin{bmatrix} sx - icy & c\sqrt{1-y^2} + is\sqrt{1-x^2} \\ c\sqrt{1-y^2} - is\sqrt{1-x^2} & -sx - icy \end{bmatrix}, \quad (30)$$

where, $x = cos(\varphi_1 - \varphi_2); y = cos(\varphi_1 + \varphi_2); s = sin(\theta/2); c = cos(\theta/2)$. The corresponding output optical phase of the four elements in the matrix R' can be represented on a complex plane, as shown in Figure 10.

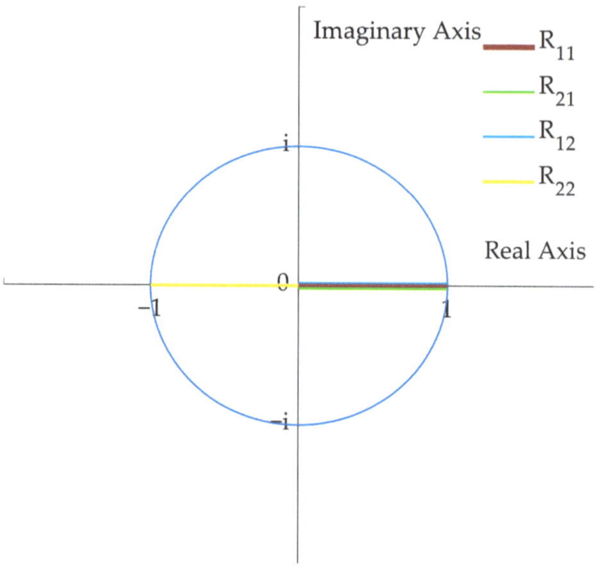

Figure 9. The corresponding output optical phase of the four elements in the matrix R.

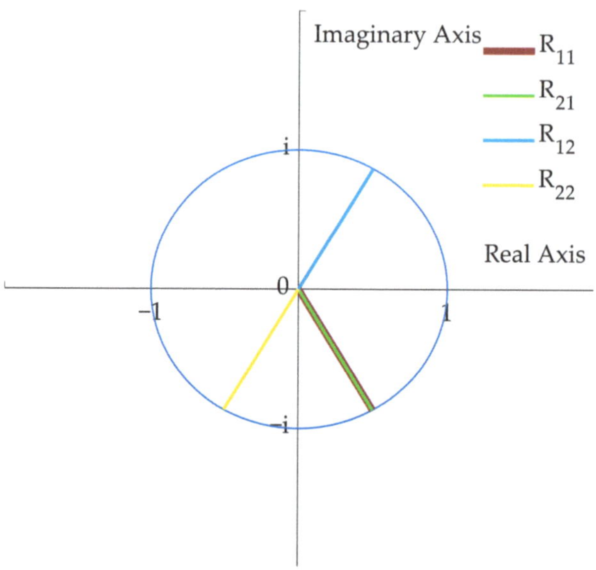

Figure 10. The corresponding output optical phase of the four elements in the matrix R'.

By comparison with Figures 11 and 12, it is evident that for a MZI with a beam splitter that is not an ideal 50:50 beam splitter, when it is used as a 50:50 beam splitter, the phase of the output optical signal will be altered, thus ultimately impacting the interference result in the subsequent optical path of the mesh and thus the final computed result.

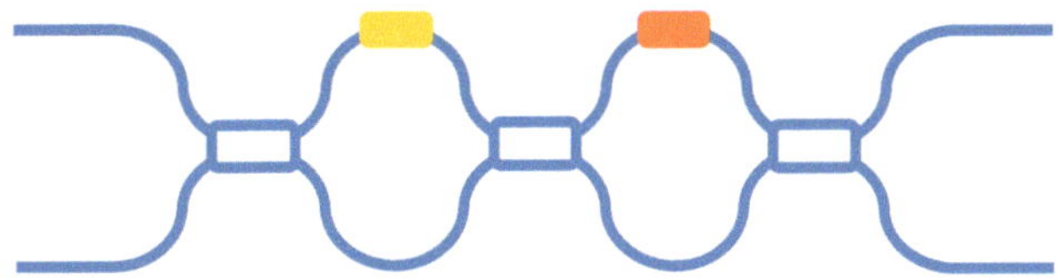

Figure 11. Three-splitter MZI.

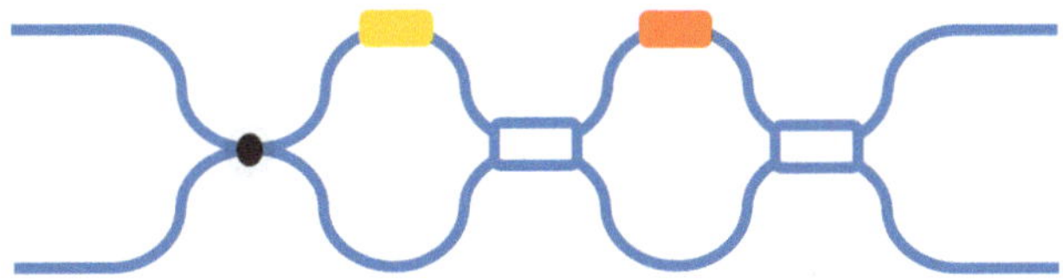

Figure 12. MZI + crossing.

Based on the above works, Hamerly et al. [20] proposed two MZI architectures (Figures 11 and 12). One is of a three-splitter MZI, which can correct generic errors and achieve a full range of split ratios. To realize the full range of the split ratio, it changes the position of "forbidden regions" caused by the error of the beam splitter, which are some unitary matrices that cannot be implemented because of the beam splitter error. The "forbidden regions" are displaced away from the cross state, rather than being eliminated completely. However, similarly to the architecture proposed by Suzuki, it does not completely correct the transmission matrix of the MZI. Furthermore, this MZI architecture does not incorporate an external phase shifter. Therefore, when it is formed into a mesh, it cannot correct the phase error from the previous layeR's MZIs.

The other one is MZI + crossing, which can only correct correlated device errors. Because the errors of the right and left beam splitters are consistent, the added cross waveguide rotates the "forbidden regions" by 180° to achieve complete extinction. Thus, this architecture has bandwidth tolerance. However, due to the errors of the right and left beam splitters being usually inconsistent, this architecture only exhibits good bandwidth tolerance but cannot correct the beam splitter error and eliminate limitations to matrix computation. Compared to the Suzuki design and Miller design, these two MZI architectures are smaller in size and do not add redundant phase shifters, but the hardware error correction is not good enough.

Bandyopadhyay et al. [21] used redundant phase shifters to correct hardware errors and proposed a method of adding phase shifters to two ports of the output end of an MZI to locally correct the hardware error within an individual MZI. In this method, no additional beam splitters are added, and the increase in the size of a single device is small, as shown in Figure 13.

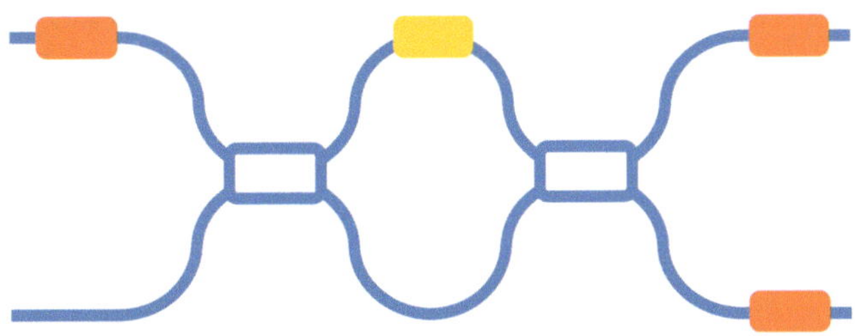

Figure 13. Bandyopadhyay et al.'s MZI architecture with a local error correction design.

For the MZI with non-ideal 50:50 beam splitters, the transmission matrix S', from Equation (10), can be rewritten as:

$$S'(\theta',\varphi') = ie^{i\theta'/2}\begin{bmatrix} e^{i\varphi'}\begin{pmatrix} cos(\varphi_1-\varphi_2)sin(\theta'/2)- \\ icos(\varphi_1+\varphi_2)cos(\theta'/2) \end{pmatrix} & \begin{pmatrix} sin(\varphi_1+\varphi_2)cos(\theta'/2)+ \\ isin(\varphi_1-\varphi_2)sin(\theta'/2) \end{pmatrix} \\ e^{i\varphi'}\begin{pmatrix} sin(\varphi_1+\varphi_2)cos(\theta'/2)- \\ isin(\varphi_1-\varphi_2)sin(\theta'/2) \end{pmatrix} & -\begin{pmatrix} cos(\varphi_1-\varphi_2)sin(\theta'/2)+ \\ icos(\varphi_1+\varphi_2)cos(\theta'/2) \end{pmatrix} \end{bmatrix}, \tag{31}$$

To implement a desired unitary S (Equation (3)), we should find the θ', φ' of each $S'(\theta',\varphi') = S$. The condition to each $S'(\theta',\varphi') = S$ produces the following expression for θ':

$$\theta' = 2arcsin\sqrt{\frac{sin^2(\theta/2) - cos^2(\varphi_1+\varphi_2)}{cos^2(\varphi_1-\varphi_2) - cos^2(\varphi_1+\varphi_2)}}, \tag{32}$$

Therefore, the beam splitter errors restrict θ to the range

$$2\left|\varphi_1+\varphi_2-\frac{\pi}{2}\right| < \theta < \pi - 2|\varphi_1-\varphi_2|, \tag{33}$$

Assuming that θ is in this range, the transmission matrix's $S'(\theta',\varphi')$, from Equation (31), can be rewritten as

$$S'' = ie^{i\theta'/2}\begin{bmatrix} e^{i\varphi'}e^{i\delta_a}sin(\theta/2) & e^{i\delta_b}cos(\theta/2) \\ e^{i\varphi'}e^{i\delta_c}cos(\theta/2) & -e^{i\delta_d}sin(\theta/2) \end{bmatrix}, \tag{34}$$

$$= ie^{i\theta'/2}\begin{bmatrix} e^{i\delta_b} & 0 \\ 0 & e^{i\delta_d} \end{bmatrix}\begin{bmatrix} e^{i\varphi'}e^{i(\delta_a-\delta_b)}sin(\theta/2) & cos(\theta/2) \\ e^{i\varphi'}e^{i(\delta_c-\delta_d)}cos(\theta/2) & -sin(\theta/2) \end{bmatrix}, \tag{35}$$

$$= ie^{i\theta'/2}\begin{bmatrix} e^{i\delta_b} & 0 \\ 0 & e^{i\delta_d} \end{bmatrix}\begin{bmatrix} e^{i(\varphi'+\delta_a-\delta_b)}sin(\theta/2) & cos(\theta/2) \\ e^{i(\varphi'+\delta_a-\delta_b)}cos(\theta/2) & -sin(\theta/2) \end{bmatrix}, \tag{36}$$

where δ_a, δ_b, δ_c and δ_d are phase errors the elements of $S'(\theta',\varphi')$, and for the unitary matrix requiring that $\delta_a+\delta_d = \delta_b+\delta_c$, we can correct those phase errors from Equation (36) to set $S'(\theta',\varphi')$ as equal to S. In accordance with Equation (36), the architecture shown in Figure 13 can be obtained. For different unitary matrices, corresponding error correction procedures must be implemented, and the correction method is complicated. It is important to note that this architecture can only correct the phase error. Although the transmission matrix is corrected, θ is restricted, meaning that some matrices cannot be implemented accurately. When θ is not in this range, there is a deviation between the desired matrix and the actual matrix.

5. Discussion

To overcome hardware errors in MZI meshes during the implementation of matrix computation, the primary strategies include increasing the tunable degrees of freedom of MZI meshes. This involves adding redundant beam splitters and phase shifters. However, incorporating a redundant mesh and improving a single MZI device may result in an increase in the size of the MZI mesh, as shown in Table 1. Therefore, it is crucial to develop a MZI mesh that can perform high-precision computation without increasing the device's size to achieve large-scale matrix computation.

In 2001, a study report [37] was conducted on tunable multimode interference couplers (MMI). By changing the position of the four-fold images of MMI out of the phase shift of the four fields, the split ratio of the two-fold images was controlled, which was then realized in the InP material. In 2008, May-Arrioja et al. [38] used local electrical modulation to control the split ratio, and in addition, changed the phase of the two-fold images [39,40] to realize the arbitrary split ratio of the self-image. A thermally modulated MMI using polymer materials was also introduced [41]. In 2019, Perez et al. [42] reported a thermally modulated dual-drive directional coupler (DD-DC) and experimentally proved that it can realize an arbitrary split ratio. The size of the tunable beam splitters proposed in these studies is relatively large, and some beam splitters are not integrated on a silicon platform.

However, these studies provide us with a novel perspective; by applying thermal or electrical modulation interference to the local optical field of the static beam splitter, we can also create a tunable beam splitter on a silicon platform without needing to increase its size.

Table 1. Characteristics of major MZI schemes.

Architecture	Number of Beam Splitters	Number of Phase Shifts	Size	Hardware Error Correction
The common MZI	2	2	1	none
Suzuki	3	3	1.5	+++
Miller	4	4	2	++++
3-splitter MZI	3	2	1.2	++
MZI + crossing	2	2 + 1 crossing	1.2	+
Bandyopadhyay	2	4	1.2	++

Here, we propose a new MZI architecture, where we replace the 50:50 beam splitter in the common MZI with a tunable DC/MMI that takes into account hardware error correction and mesh size, as shown in Figure 14. This improves the computational precision of the MZI's mesh without increasing its size. By adjusting the split ratios of the tunable DC/MMI, we can eliminate its split ratio deviations from 50:50. For instance, for a tunable MMI, when the phase of optical fields at the position of the four-fold images in the multimode interferometer is altered, it affects the interference results of the optical fields, culminating in changes in the intensity of the two images at the two-fold image position. This enables the adjustment of the beam split ratios. The tunable MMI has the theoretical capability to adjust split ratios from 100:0 to 0:100, enabling the adjustment of any fabricated power split ratios in the physical 2×2 MMI to the ideal 50:50 split. This correction completely eliminates beam splitter errors and enables high-precision MZI-based matrix computation.

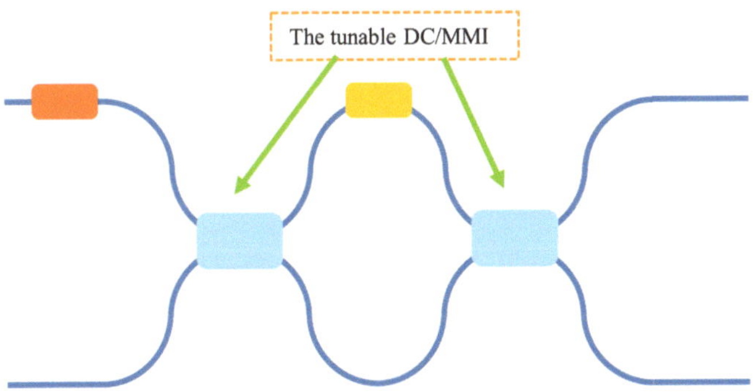

Figure 14. A new MZI architecture.

6. Conclusions

To summarize, this paper introduces the method of using a MZI's mesh to implement any finite dimensional unitary matrix transformation, along with an analysis of the effects of hardware errors during matrix computation based on the MZI. Addressing MZIs' hardware errors is crucial to achieving large-scale matrix computation. To eliminate the hardware error of an MZI, various improvement works have been carried out on the entire MZI mesh and a single MZI device. It is important to note that correcting these errors can lead to an increase in an MZI's mesh size, which is also a significant concern. The trade-off between hardware error correction and mesh size requires more innovative works in artificial intelligence, materials, optics, device manufacturing and other related fields.

In this paper, a new MZI architecture is proposed which replaces the 50:50 beam splitter in the common MZI with the tunable DC/MMI. This MZI is designed to take into account both hardware error correction and mesh size concerns. The tunable beam splitter-based MZI provides a new approach for more accurate large-scale matrix computation based on the MZI's mesh.

Author Contributions: Conceptualization, H.H. and Z.Z.; writing-original draft preparation, H.H.; writing-review and editing, H.H., Z.Z., P.X. and H.S. All authors have read and agreed to the published version of the manuscript.

Funding: This research was funded by the Key Program of the National Natural Science Foundation of China (62035001).

Institutional Review Board Statement: Not applicable.

Informed Consent Statement: Not applicable.

Data Availability Statement: No new data were created or analyzed in this study. Data sharing is not applicable to this article.

Acknowledgments: The authors would like to thank the Beijing Aijie Optoelectronic Technology Co., Ltd. for the technical and partial finacial support.

Conflicts of Interest: The authors declare no conflict of interest.

References

1. Zhou, Z. *Silicon Based Optoelectronics*, 2nd ed.; Science Press: Beijing, China, 2021; pp. 430–447.
2. Zhou, Z.; Xu, P.; Dong, X. Computing on Silicon Photonic Platform. *Chin. J. Lasers* **2020**, *47*, 0600001. [CrossRef]
3. Xu, P.; Zhou, Z. Silicon-based optoelectronics for general-purpose matrix computation: A review. *Adv. Photonics* **2022**, *4*, 044001. [CrossRef]
4. Carolan, J.; Harrold, C.; Sparrow, C.; Martín-López, E.; Russell, N.J.; Silverstone, J.W.; Shadbolt, P.J.; Matsuda, N.; Oguma, M.; Itoh, M.; et al. Universal linear optics. *Science* **2015**, *349*, 711–716. [CrossRef] [PubMed]
5. Bogaerts, W.; Pérez, D.; Capmany, J.; Miller, D.A.B.; Poon, J.; Englund, D.; Morichetti, F.; Melloni, A. Programmable photonic circuits. *Nature* **2020**, *586*, 207–216. [CrossRef]
6. Pérez-López, D.; López, A.; DasMahapatra, P.; Capmany, J. Multipurpose self-configuration of programmable photonic circuits. *Nat. Commun.* **2020**, *11*, 6359. [CrossRef] [PubMed]
7. Dong, M.; Clark, G.; Leenheer, A.J.; Zimmermann, M.; Dominguez, D.; Menssen, A.J.; Heim, D.; Gilbert, G.; Englund, D.; Eichenfield, M. High-speed programmable photonic circuits in a cryogenically compatible, visible–near-infrared 200 mm CMOS architecture. *Nat. Photonics* **2022**, *16*, 59–65. [CrossRef]
8. Enomoto, Y.; Yonezu, K.; Mitsuhashi, Y.; Takase, K.; Takeda, S. Programmable and sequential Gaussian gates in a loop-based single-mode photonic quantum processor. *Sci. Adv.* **2021**, *7*, j6624. [CrossRef]
9. Fu, T.; Zang, Y.; Huang, Y.; Du, Z.; Huang, H.; Hu, C.; Chen, M.; Yang, S.; Chen, H. Photonic machine learning with on-chip diffractive optics. *Nat. Commun.* **2023**, *14*, 70. [CrossRef]
10. Bogaerts, W.; Chrostowski, L. Silicon Photonics Circuit Design: Methods, Tools and Challenges. *Laser Photonics Rev.* **2018**, *12*, 1700237. [CrossRef]
11. Boeuf, F.; Cremer, S.; Temporiti, E.; Fere, M.; Shaw, M.; Vulliet, N.; Orlando, B.; Ristoiu, D.; Farcy, A.; Pinguet, T.; et al. Recent progress in silicon photonics R and D and manufacturing on 300mm wafer platform. In Proceedings of the 2015 Optical Fiber Communications Conference and Exhibition (OFC), Los Angeles, CA, USA, 22–26 March 2015.
12. Zhou, H.; Dong, J.; Cheng, J.; Dong, W.; Huang, C.; Shen, Y.; Zhang, Q.; Gu, M.; Qian, C.; Chen, H.; et al. Photonic matrix multiplication lights up photonic accelerator and beyond. *Light. Sci. Appl.* **2022**, *11*, 30. [CrossRef]
13. Du, Y.; Su, K.; Yuan, X.; Li, T.; Liu, K.; Man, H.; Zou, X. Implementation of optical neural network based on Mach–Zehnder interferometer array. *Iet Optoelectron.* **2023**, *17*, 1–11. [CrossRef]
14. Jacques, M.; Samani, A.; El-Fiky, E.; Patel, D.; Xing, Z.; Plant, D.V. Optimization of thermo-optic phase-shifter design and mitigation of thermal crosstalk on the soi platform. *Opt. Express* **2019**, *27*, 10456. [CrossRef] [PubMed]
15. AlTaha, M.W. *Automatic Tuning Circuits for Mach-Zehnder Interferometer Optical Switches*; University of British Columbia: Vancouver, BC, Canada, 2017. [CrossRef]
16. Zhang, Y.; Ashe, J.J. *Designing a High-Performance Tec Controller*; SPIE: Washington, DC, USA, 2002; Volume 4913, pp. 177–183. [CrossRef]
17. Shastri, B.J.; Tait, A.N.; Ferreira De Lima, T.; Pernice, W.H.P.; Bhaskaran, H.; Wright, C.D.; Prucnal, P.R. Photonics for artificial intelligence and neuromorphic computing. *Nat. Photonics* **2021**, *15*, 102–114. [CrossRef]
18. Xie, Y.; Shi, Y.; Liu, L.; Wang, J.; Priti, R.; Zhang, G.; Liboiron-Ladouceur, O.; Dai, D. Thermally-reconfigurable silicon photonic devices and circuits. *IEEE J. Sel. Top. Quantum Electron.* **2020**, *26*, 1–21. [CrossRef]
19. Chung, H.S.; Chang, S.H.; Lee, J.C.; Lee, J.H.; Kim, K. Field experiment of 112 gb/s dual-carrier dqpsk signal transmission with automatic bias control of optical iq modulator. In Proceedings of the 2013 Optical Fiber Communication Conference and Exposition and the National Fiber Optic Engineers Conference (OFC/NFOEC), Anaheim, CA, USA, 17–21 March 2013.
20. Hamerly, R.; Bandyopadhyay, S.; Englund, D. Asymptotically fault-tolerant programmable photonics. *Nat. Commun.* **2022**, *13*, 6831. [CrossRef] [PubMed]

21. Bandyopadhyay, S.; Hamerly, R.; Englund, D. Hardware error correction for programmable photonics. *Optica* **2021**, *8*, 1247. [CrossRef]
22. Yang, Y.; Ma, Y.; Guan, H.; Liu, Y.; Danziger, S.; Ocheltree, S.; Bergman, K.; Baehr-Jones, T.; Hochberg, M. Phase coherence length in silicon photonic platform. *Opt. Express* **2015**, *23*, 16890. [CrossRef]
23. Clements, W.R.; Humphreys, P.C.; Metcalf, B.J.; Kolthammer, W.S.; Walsmley, I.A. Optimal design for universal multiport interferometers. *Optica* **2016**, *3*, 1460. [CrossRef]
24. Reck, M.; Zeilinger, A.; Bernstein, H.J.; Bertani, P. Experimental realization of any discrete unitary operator. *Phys. Rev. Lett.* **1994**, *73*, 58–61. [CrossRef]
25. Macho Ortiz, A.; Pérez López, D.; Capmany, J. Optical Implementation of 2 × 2 Universal Unitary Matrix Transformations. *Laser Photonics Rev.* **2021**, *15*, 2000473. [CrossRef]
26. Shen, Y.; Harris, N.C.; Skirlo, S.; Prabhu, M.; Baehr-Jones, T.; Hochberg, M.; Sun, X.; Zhao, S.; Larochelle, H.; Englund, D.; et al. Deep learning with coherent nanophotonic circuits. *Nat. Photonics* **2017**, *11*, 441–446. [CrossRef]
27. Shokraneh, F.; Geoffroy-Gagnon, S.; Liboiron-Ladouceur, O. Towards Phase-Error-and Loss-Tolerant Programmable MZI-Based Optical Processors for Optical Neural Networks. In Proceedings of the 2020 IEEE Photonics Conference (IPC), Vancouver, BC, Canada, 28 September–1 October 2020.
28. Zhu, Y.; Zhang, G.L.; Li, B.; Yin, X.; Zhuo, C.; Gu, H.; Ho, T.; Schlichtmann, U. Countering variations and thermal effects for accurate optical neural networks. In Proceedings of the 2020 IEEE/ACM International Conference on Computer Aided Design (ICCAD), San Diego, CA, USA, 2–5 November 2020.
29. Aoyama, K.; Sawada, H. Acceleration method for learning fine-layered optical neural networks. In Proceedings of the 2021 IEEE/ACM International Conference on Computer Aided Design (ICCAD), Munich, Germany, 1–4 November 2021.
30. Sarantoglou, G.; Bogris, A.; Mesaritakis, C.; Theodoridis, S. Bayesian photonic accelerators for energy efficient and noise robust neural processing. *IEEE J. Sel. Top. Quantum Electron.* **2022**, *28*, 1–10. [CrossRef]
31. López Pastor, V.; Lundeen, J.; Marquardt, F. Arbitrary optical wave evolution with Fourier transforms and phase masks. *Opt. Express* **2021**, *29*, 38441. [CrossRef]
32. Pai, S.; Bartlett, B.; Solgaard, O.; Miller, D.A.B. Matrix Optimization on Universal Unitary Photonic Devices. *Phys. Rev. Appl.* **2019**, *11*, 064044. [CrossRef]
33. Burgwal, R.; Clements, W.R.; Smith, D.H.; Gates, J.C.; Kolthammer, W.S.; Renema, J.J.; Walmsley, I.A. Using an imperfect photonic network to implement random unitaries. *Opt. Express* **2017**, *25*, 28236. [CrossRef]
34. Suzuki, K.; Cong, G.; Tanizawa, K.; Kim, S.; Ikeda, K.; Namiki, S.; Kawashima, H. Ultra-high-extinction-ratio 2 × 2 silicon optical switch with variable splitter. *Opt. Express* **2015**, *23*, 9086. [CrossRef] [PubMed]
35. Miller, D.A.B. Perfect optics with imperfect components. *Optica* **2015**, *2*, 747–750. [CrossRef]
36. Wilkes, C.M.; Qiang, X.; Wang, J.; Santagati, R.; Paesani, S.; Zhou, X.; Miller, D.A.; Marshall, G.D.; Thompson, M.G.; O'Brien, J.L. 60 dB high-extinction auto-configured Mach-Zehnder interferometer. *Opt. Lett.* **2016**, *41*, 5318–5321. [CrossRef]
37. Leuthold, J.; Joyner, C.W. Multimode interference couplers with tunable power splitting ratios. *J. Light. Technol.* **2001**, *19*, 700–707. [CrossRef]
38. May-Arrioja, D.A.; LiKamWa, P. Reconfigurable 3-dB MMI splitter. In Proceedings of the 2008 Digest of the IEEE/LEOS Summer Topical Meetings, Acapulco, Mexico, 21–23 July 2008; pp. 39–40.
39. Rosa, Á.; Gutiérrez, A.; Brimont, A.; Griol, A.; Sanchis, P. High performace silicon 2 × 2 optical switch based on a thermo-optically tunable multimode interference coupler and efficient electrodes. *Opt. Express* **2016**, *24*, 191. [CrossRef]
40. Hassan, S.; Chack, D.; Pavesi, L. High extinction ratio thermo-optic based reconfigurable optical logic gates for programmable PICs. *Aip Adv.* **2022**, *12*, 055304. [CrossRef]
41. Fan Wang, J.Y. Optical Switch Based on Multimode Interference Coupler. *Ieee Photonic Tech. L* **2006**, *18*, 421–423. [CrossRef]
42. Perez, D.; Gutierrez, A.M.; Sanchez, E.; Dasmahapatra, P.; Capmany, J. Dual-Drive Directional Couplers for Programmable Integrated Photonics. In Proceedings of the 2019 IEEE Photonics Society Summer Topical Meeting Series (SUM), Ft. Lauderdale, FL, USA, 8–10 July 2019.

micromachines

Article

Efficiency Enhancing Technique for Rod Fiber Picosecond Amplifiers with Optimal Mode Field Matching

Danni Liu [1], Xiaojie Mao [2], Guojiang Bi [2], Tianqi Li [3,*], Dawei Zang [1,*] and Ninghui Sun [1]

1 Institute of Computing Technology, Chinese Academy of Sciences, Beijing 100190, China
2 Science and Technology on Solid-State Laser Laboratory, North China Research Institute of Electro-Optics, Beijing 100015, China
3 The Science and Technology on Inertial Laboratory, School of Instrumentation and Opto-Electronics Engineering, Beihang University, Beijing 100191, China
* Correspondence: tianqi_li@yeah.net (T.L.); zangdawei@ncic.ac.cn (D.Z.)

Abstract: A high power and high quality picosecond laser is crucial in MEMS fabrication regarding micromachines. Optimal seed beam coupling is an important precondition to enhance laser efficiency. However, empirical coupling limits its development. In this paper, the physical parameters related to coupling are determined. The relationships among them are established under optical mode matching constraints to satisfy optimal seed beam coupling. According to a theoretical analysis, the focal length cut-off and the optimal coupling position of the coupling lens are acquired. A maximum transmittance of 87.2% is acquired with a 6 W input seed power in the validation experiment. In further power amplification experiments, a diffraction-limited beam quality is achieved, with $M^2{}_X = 1.111$, $M^2{}_Y = 1.017$, an optical efficiency of 60.5% and a slope efficiency of 66%, benefiting from the previous theoretical guidance.

Keywords: rod fiber; picosecond amplifier; mode field matching; fiber coupling

Citation: Liu, D.; Mao, X.; Bi, G.; Li, T.; Zang, D.; Sun, N. Efficiency Enhancing Technique for Rod Fiber Picosecond Amplifiers with Optimal Mode Field Matching. *Micromachines* **2023**, *14*, 450. https://doi.org/10.3390/mi14020450

Academic Editors: He Yang, Xinyang Su and Yizhong Huang

Received: 5 December 2022
Revised: 8 February 2023
Accepted: 9 February 2023
Published: 15 February 2023

1. Introduction

The development from machine to micromachine depends on the MEMS fabrication. Etching [1], micro drilling [2], cutting [3] and thickness scanning [4] in MEMS fabrication need high power and a picosecond laser with excellent stability. A picosecond laser system is composed of an optical and a cooling structure, which is involved in stabilizing the laser system. According to the difference in the active regions in the optical structure, picosecond amplification systems can be divided into cryogenic Yb:YAG, thin-disk, Innoslab, single-crystal fiber (SCF), double clad active fiber (DCF) and photonic crystal fiber (PCF) systems. The former three solid-state lasers have the problem of being complicated systems with low stability, beam quality and conversion efficiency. Although the SCF is a simpler system with relatively high stability, beam quality and conversion efficiency compared to other solid-state lasers, it is still inferior to fiber lasers. The DCF system overcomes the above disadvantages of solid-state lasers. However, to satisfy single-mode operation, the mode field area of the DCF is limited to about 700 μm^2 [5]. The limited mode field area of DCF causes significant nonlinearity and damage under high power operation, which limits output power. Benefiting from an endless single-mode property, PCF can ensure single-mode operation while having a large mode area [6], which effectively controls fiber nonlinearity and potential material damage [7]. So far, the largest reported mode field diameter (MFD) of silica active PCF is 135 μm [8]. Nowadays, PCF, especially large mode area PCF (LMA-PCF), has been proven to be attractive. A comparison of the above different picosecond amplification systems is shown in Table 1.

Table 1. Comparison of different picosecond amplification systems.

Type	Stability		Optical Efficiency	M^2 Beam Quality
	Optical Structure	Cooling Structure		
Cryogenic Yb:YAG crystal [9,10]	Simple coupling system	Liquid nitrogen cooling	≤20%.	1.1–1.5
Thin-disk [11,12]	Complicated coupling system	Heat sink with water cooling	≤50%.	1.1–1.5
Innoslab [13–15]	Complicated coupling system	Heat sink with water cooling	≤40%.	1.1–1.5
SCF [16]	Simple coupling system	Heat sink with water or passive cooling	50–60%.	≤1.2
DCF [17]	Fiber coupler	Passive cooling	≥70%.	≤1.1
PCF [18]	Simple coupling system or fiber coupler	Heat sink with water cooling	≥60%.	≤1.2

Rod-type PCF is a kind of LMA-PCF, and is also called rod fiber. It was first proposed for use in ultrashort pulse amplification in 2005 [19]. The rod fiber is suitable for chirped pulse amplification (CPA), divided pulse amplification (DPA) and master oscillator power amplifier (MOPA). Based on the CPA technique, a 100 W average power femtosecond output was generated by a rod fiber amplification system [20]. However, due to the limited stretchability and compressibility of picosecond pulses induced by the narrow spectral bandwidth [21], the CPA technique is unsuitable for picosecond pulse amplification. The DPA technique uses a polarization-controlled free-space delay line or a birefringent crystal to divide and recombine pulses [22]. However, pulse distortion and polarization degradation induced by the pulse division and recombination cause power loss, which limits the efficiency of the DPA system. The MOPA technique has proven to be attractive because the fiber nonlinearity can be controlled effectively. According to the reported studies, rod fiber MOPA systems are developing towards higher power, beam quality and efficiency [23,24].

The optical efficiency enhancement method reported in the literature is mainly based on structure modification, assuming that the seed beam coupling is optimal. In actual experiments, optimal seed beam coupling is an important precondition [25]. However, traditional seed beam coupling depends on experiential adjustment. The empirically optimal seed beam coupling efficiency is easily affected by subjective factors. This study aims to provide the basic theory of optimal seed beam coupling, which guides the physical parameters and the installation position of seed beam coupling devices in actual applications.

In this paper, the theory basis for optimal seed coupling is presented, the related physical parameters are determined and the constraints of these parameters are provided. We begin with determining the optical mode matching condition that satisfies optimal seed beam coupling. A relationship among the related physical parameters can be established under an optimal optical mode matching situation. Based on the constraints of physical applications, the cut-off focal length and the optimal position of the coupling lens are determined. Section 3 describes a seed beam coupling experiment. The theoretical conclusion is verified by changing the single related variables. A passive transmission system is built to acquire the transmittance under different input seed powers, a maximum seed power transmittance of 87.2% is obtained with a 6 W input seed power and the optimal seed beam coupling theory is further enhanced from the perspective of mode field matching. In Section 4, the amplification experiment is implemented and an average power of 101.7 W is generated with a pump power of 162.4 W by using a rod fiber to amplify 6 ps pulses with a 30 MHz repetition rate and a coupled seed power of 5.23 W. The diffraction-limited beam quality with M^2_X = 1.111 and M^2_Y = 1.017 is also obtained. In addition, a maximum optical efficiency of 60.5% and a slope efficiency of 66% are acquired with a rod fiber length of 0.8 m.

2. Seed Beam Coupling Theory Based on Optical Mode Field Matching

2.1. The Influence Factors of Power Loss

The power loss of seed beam coupling with ultrafast pluses into rod fibers is mainly influenced by two factors: fiber loss and mode field mismatch.

For rod fibers, fiber loss is mainly caused by dissipation from the core to the clad, absorption from the core and reflection from the end face. According to the specification of the rod fibers (aeroGAIN-ROD-PM85), the two end facets of the fiber are coated with an antireflection coating and cut at an angle less than 0.5°. The total reflectivity of the two end facets is lower than 0.3%, and thus can be ignored. According to the instructions of aeroGAIN-ROD-PM85, the core/clad power ratio (CCR) can be expressed as:

$$R_{core/clad} = 10\lg\left(\frac{P_{core}}{P_{total} - P_{core}}\right) \tag{1}$$

From Equation (1), when the fiber is in passive operation, the fiber loss from dissipation and absorption can be measured by the CCR. The typical value of the CCR at 1030 nm is about −1 dB and it is relatively stable. Nevertheless, the mode field mismatching is mainly related to the parameters of laser path and coupling lens. It is very sensitive and changeable. If the parameters of the laser path and coupling lens are inappropriate, the power loss will increase significantly to even higher than 50% and the mode field of the passed beam will degrade, as shown in Figure 1. Mode field mismatch is the main influencing factor of power loss.

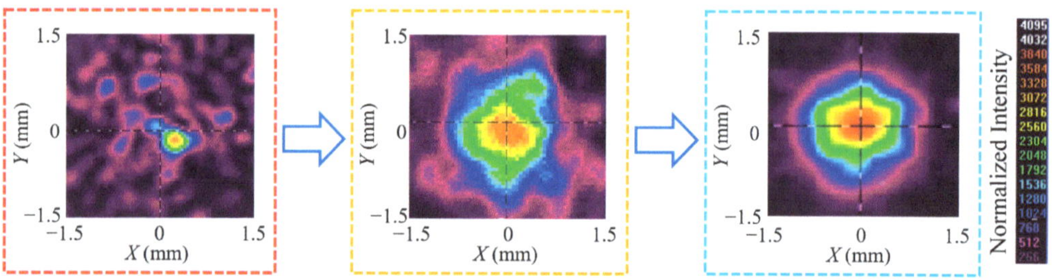

Unmatched Partially matched Matched

Figure 1. The mode field of coupling 5 W fiber laser into aeroGAINRODPM85 rod fiber.

Mode field mismatch can be classified into lateral and longitudinal mismatch. Lateral mismatch is decided by the pitch angle and the lateral position of the lenses. Longitudinal mismatch is decided by the longitudinal position and the focal length of the lenses. Lateral mismatch is easily detectable. Therefore, this study mainly focuses on longitudinal mismatch. In experiments, the longitudinal mismatch can be decreased by adjusting the longitudinal position of the lenses and by choosing lenses with different focal lengths; the theoretical analysis will prove that the former is more effective. The scheme of the theoretical analysis is shown in Figure 2.

2.2. Analysis of Optical Mode Field Matching

Mode field matching between the incident beam and the rod fiber is mainly measured by the coupling efficiency. The scheme of the coupling system is shown in Figure 3.

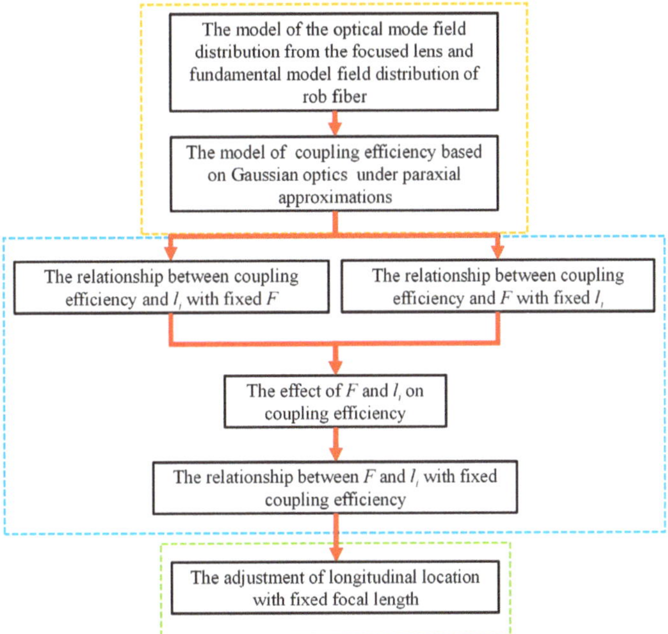

Figure 2. The scheme of the theoretical analysis.

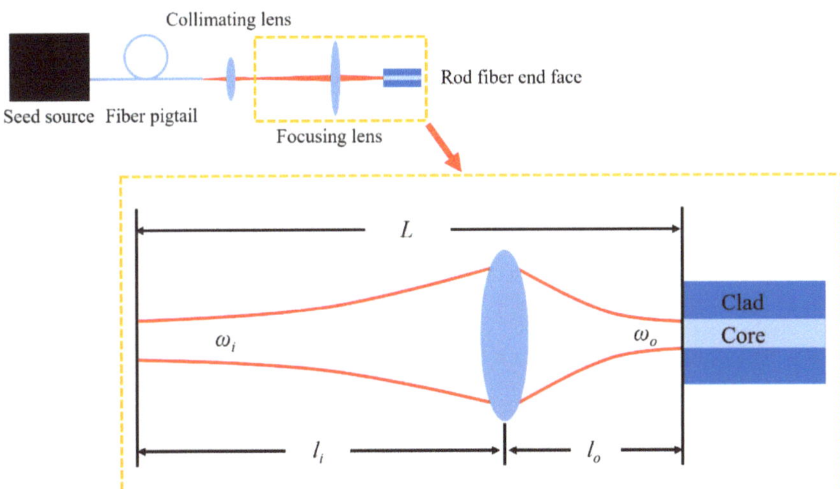

Figure 3. The scheme of lens coupling.

In the amplification system, the beam from the seed source is collimated by a collimating lens and focused by a coupling lens into the rod fiber core. In the actual experiment, to reduce the complexity of adjustment, the seed fiber end is placed at the focus point of the collimating lens. Therefore, the position and radius of the collimated beam waist are determined. Therefore, in the subsequent analysis, the collimated beam waist will be the starting point. According to the amplification system in this study, the following conditions can simplify the analysis while ensuring accuracy:

(1) The beam quality of the seed source is diffraction limited (M^2~1.1), so the seed beam can be approximated as a fundamental mode Gaussian beam.

(2) The length of the laser path (~1 m) is far more than the thickness of the lens (~3 mm), and the diameter of the lens (~20 mm) is also far more than the diameter of the beam on the lens (~1.7 mm), so the lens can be simplified as an ideal thin lens.

The beam waist radius before coupling and the beam radius after coupling are defined as ω_i and ω_o, respectively. The distance between the collimated beam waist and the rod fiber end face, the distance between the collimated beam waist and the coupling lens and the distance between the coupling lens and the focused beam waist are defined as L, l_i and l_o, respectively. The focal length of the coupling lens is defined as F. The wavelength of the incident beam is defined as λ.

L, l_i and l_o satisfy:

$$L = l_i + l_o. \tag{2}$$

The transformation matrix of the system can be expressed as:

$$T = \begin{pmatrix} A & B \\ C & D \end{pmatrix} = \begin{pmatrix} 1 & 0 \\ -\frac{1}{F} & 1 \end{pmatrix} \begin{pmatrix} 1 & l_i \\ 0 & 1 \end{pmatrix}. \tag{3}$$

The confocal parameter of the input beam is described as:

$$f_i = \frac{\pi \omega_i^2}{\lambda}. \tag{4}$$

The q parameter of the input beam waist satisfies:

$$q_i = jf_i \tag{5}$$

where j is an imaginary unit. The q parameter after system transformation is described as:

$$q_o = \frac{Aq_i + B}{Cq_i + D}. \tag{6}$$

The absolute value of the real part of q_o is the distance between the coupling lens and the output beam waist. The imaginary part of q_o is the confocal parameter of the output beam.

$$l = |\mathrm{Re}(q_o)| \tag{7}$$

$$f_o = \mathrm{Im}(q_o). \tag{8}$$

The output beam waist radius can be expressed as:

$$\omega = \sqrt{\frac{f_o \lambda}{\pi}}. \tag{9}$$

The q parameter of the output beam waist can be described as:

$$q = jf_o. \tag{10}$$

The output beam radius on the rod fiber end face can be described as:

$$\omega_o = \omega \sqrt{1 + \left(\frac{l_o - l}{f_o} \right)^2}. \tag{11}$$

The q parameter of the output beam on the rod fiber end face can be expressed as:

$$q_o = q + l_o - l. \tag{12}$$

The output beam mode field on the rod fiber end face can be described as:

$$U(r) \propto \exp\left\{-j\left[k(l_o - l) - \arctan\left(\frac{l_o - l}{f_o}\right)\right] - \frac{jkr^2}{2q_o}\right\}, \tag{13}$$

where $k = 2\pi/\lambda$ is the wave number of output beam.

Since the rod fiber is in single-mode operation, the rod fiber mode field can be approximated as a Gaussian distribution and described as:

$$U_P(r) \propto \exp\left[-\frac{r^2}{(D_{MF}/2)^2}\right], \tag{14}$$

where D_{MF} is the mode field diameter (MFD) of the rod fiber.

According to the coupling Gaussian beam into the rod fiber, the ideal condition that the beam axis should align to the rod fiber axis and the rod fiber end face should match the beam waist is satisfied. When the coupled beam waist is located at the rod fiber end face, l_o and l satisfy:

$$l_o = l. \tag{15}$$

Under this condition, ω, ω_o, q and q_o satisfy:

$$\omega_o = \omega \tag{16}$$

$$q_o = q. \tag{17}$$

Therefore, the output beam mode field on the rod fiber end face is reduced to the beam waist mode field:

$$U(r) \propto \exp\left(-\frac{r^2}{\omega^2}\right). \tag{18}$$

Under this circumstance, ω is the only factor affecting the mode field matching and ω is related to F and l_i.

The mode field coupling efficiency can be calculated by the mode field overlap integral [26]:

$$\eta = \frac{\left|\int_S U \bullet U_P^* dS\right|^2}{\int_S |U|^2 dS \bullet \int_S |U_P|^2 dS}, \tag{19}$$

where U_P^* is the complex conjugate of U_P.

According to the control variable principle, the analysis can be divided into two conditions:

(1) F is fixed and l_i is changeable.

The calculation parameters are $F = 57\text{mm}, 77\text{mm}, 97\text{mm}$, $\omega_i = 0.4$ mm, $D_{MF} = 65$ µm and $\lambda = 1030$nm. The coupling efficiency as a function of l_i, and the mode field of the coupled beam under $F = 77$mm and different l_i are shown in Figure 4.

According to Figure 4a, as l_i increases, coupling efficiency increases first and then decreases. From Figure 4c to Figure 4j, ω decreases with an increase in l_i. Compared Figure 4b with Figure 4g, $\omega = D_{MF}/2$ when l_i = 900 mm and F = 77 mm.The maximum coupling efficiency is obtained when l_i = 900 mm and F = 77 mm according to Figure 4a. According to these simulation results, the mode field matching is essentially the matching between ω and $D_{MF}/2$. Under different values of F, there always exists an l_i which can achieve the maximum coupling efficiency of 1 when $\omega = D_{MF}/2$. When F is larger, the corresponding l_i is larger to get maximum coupling efficiency. The coupling efficiency exceeds 0.975 in a certain range (±200 mm) near the value of l_i which results in the maximum coupling efficiency, which means that the effect of l_i on ω was not significant when F is fixed.

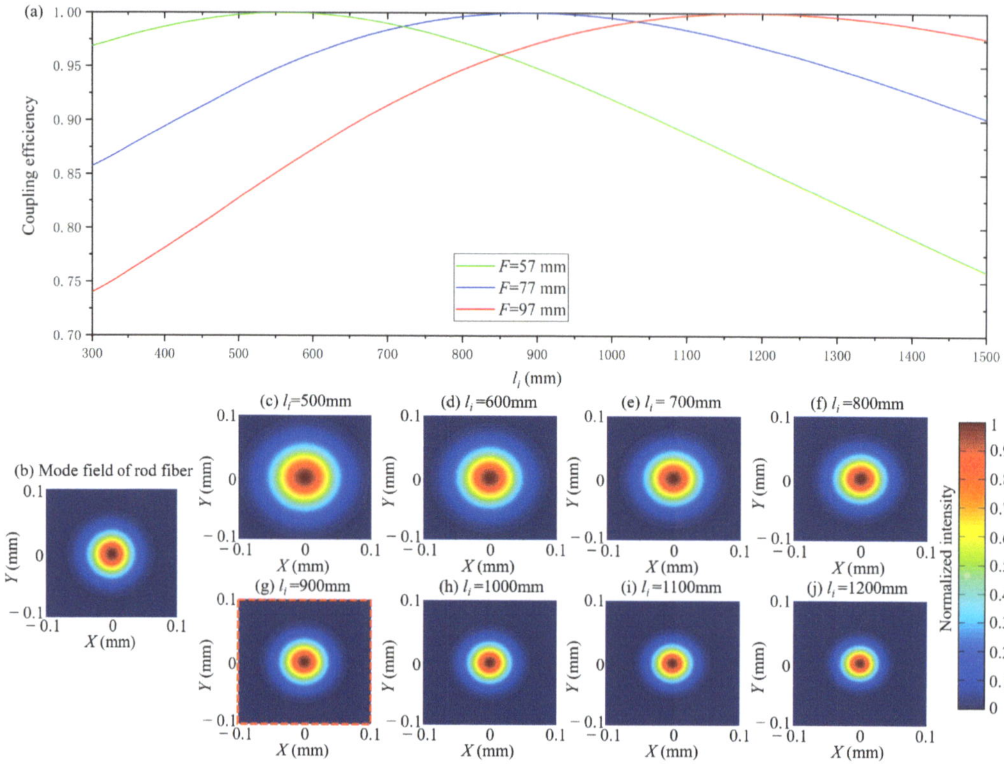

Figure 4. The coupling efficiency as a function of l_i with different F when $l_0 = l$ (**a**), the mode field of rod fiber (**b**), the mode field of coupled beam when l_i increased from 500 mm to 1200 mm with F = 77 mm (**c–j**).

(2) l_i is fixed and F is changeable.

The calculation parameters are l_i = 800 mm, 900 mm, 1000 mm, $\omega_i = 0.4$ mm, $D_{MF} = 65$ µm and $\lambda = 1030$nm. The coupling efficiency as a function of F and the mode field of the coupled beam when l_i = 900 mm at different values of F are shown in Figure 5.

According to Figure 5a, as F increases, the coupling efficiency increases first and then decreases. From Figure 4c to Figure 4j, the ω increases with an increasing F. Compared Figure 5b with Figure 5g, $\omega = D_{MF}/2$ when l_i = 900 mm and $F = 87$mm.The maximum coupling efficiency is obtained when l_i = 900mm and $F = 87$mm according to Figure 5a. These simulation results show that the mode field matching is essentially the matching between ω and $D_{MF}/2$ once again. Under different values of l_i, there is always an F that can achieve the maximum coupling efficiency of 1 when $\omega = D_{MF}/2$. When l_i is larger, the corresponding F is larger to get maximum coupling efficiency. The range of coupling efficiency changes significantly when F is increased from 0 mm to 200 mm, which means that the effect of F on ω is significant when l_i is fixed.

In summary, to acquire the optimum mode field matching, the coupled beam waist must be located at the rod fiber end face and the beam waist diameter must be equal to the MFD. The values of l_i and F for optimal mode field matching and the corresponding value of l_o can be calculated by these two conditions.

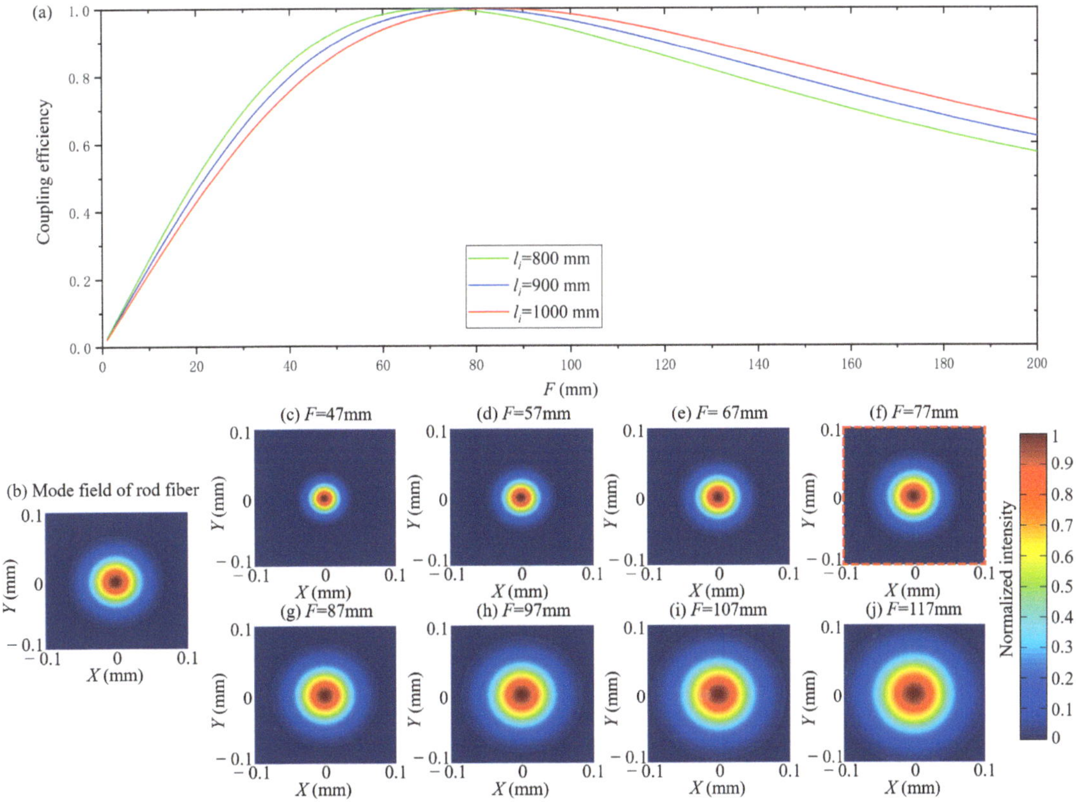

Figure 5. The coupling efficiency as a function of F with different l_i when $l_0 = l$ (**a**), the mode field of rod fiber (**b**), the mode field of coupled beam when F increased from 47 mm to 117 mm with l_i = 900 mm (**c–j**).

2.3. Seed Beam Coupling Theory

To further analyze the relation between l_i and F quantitatively, the following deduction is carried out.

Combining Equations (3), (5), (6) and (8), an equation which contains l_i, F, f_i and f_o can be acquired:

$$\frac{f_i}{f_o}F^2 - f_i^2 = (F - l_i)^2. \tag{20}$$

Substituting Equations (4) and (9) into Equation (20), the relation between l_i and F can be expressed as:

$$l_i = F \mp \frac{\omega_i}{\omega}\sqrt{\left(F - \frac{\pi\omega_i\omega}{\lambda}\right)\left(F + \frac{\pi\omega_i\omega}{\lambda}\right)}. \tag{21}$$

According to Equation (21), l_i can be a real number only if $F \geq \pi\omega_i\omega/\lambda$. Therefore, the lower bound of F satisfies:

$$F_c = \frac{\pi\omega_i\omega}{\lambda}. \tag{22}$$

When $F \geq F_c$, there are two conditions that should be considered:

(1) Choosing the plus sign, i.e., $l_i \geq F$. Under this condition, the input beam waist is located at or in front of the coupling lens focus.

(2) Choosing the minus sign, i.e., $l_i \leq F$. Under this condition, the input beam waist is located at or behind the coupling lens focus. In addition, when $\omega_i > \omega$, $l_i < 0$ may occur, which means that the input beam waist is located behind the coupling lens. When $l_i = 0$, the corresponding F can be expressed as:

$$F = \frac{\omega_i}{\sqrt{\omega_i^2 - \omega^2}} F_c. \tag{23}$$

Therefore, the relation between l_i and F can be further expressed as:

$$l_i = \begin{cases} F + \frac{\omega_i}{\omega}\sqrt{(F - F_c)(F + F_c)}, & (F \geq F_c \text{ and } F \leq l_i) \\ F - \frac{\omega_i}{\omega}\sqrt{(F - F_c)(F + F_c)}, & (F \geq F_c \text{ and } F \geq l_i) \end{cases}. \tag{24}$$

When $\omega_i \leq \omega$, the beam is expanded. When $\omega_i > \omega$, the beam is focused. It is easy to calculate the values of F and l_i which satisfy the maximum coupling efficiency and analyze the relation between F and l_i by Equation (24). The analysis in this study satisfies the conditions of $\omega_i > \omega$ and $F \leq l_i$. Therefore, in the following analysis, the first equation in Equation (24) will be used. The calculation parameters are $\omega_i = 0.4$ mm, $\omega = 32.5$ μm and $\lambda = 1030$nm. l_i as a function of F is shown in Figure 6.

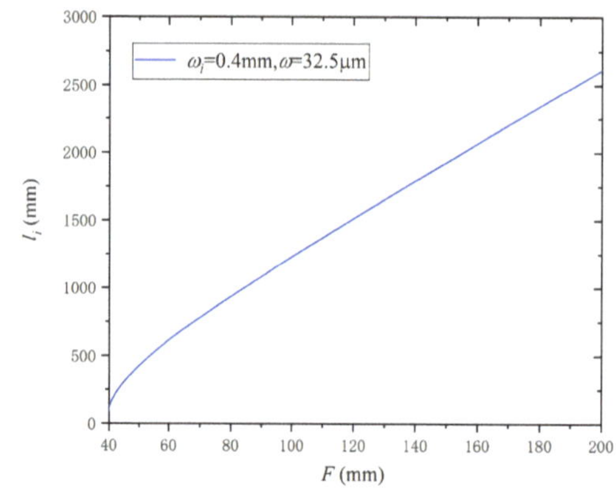

Figure 6. l_i as a function of F.

According to Figure 6, l_i increases as F increases, but l_i rises faster than F. This further illustrates the above conclusion that the effect of F on the coupling efficiency is more significant than that of l_i.

Through the above theoretical analysis, the optimal seed beam coupling should satisfy the following conclusions:

(1) The coupling lens focal length should satisfy $F \geq F_c$.

(2) When the focal length is determined, the l_i of coupling lens should satisfy Equation (24).

3. Experimental Verification of Seed Beam Coupling

The seed source is a fiber picosecond system, the wavelength is 1030 nm, the output average power is from 0 W to 6 W, the M^2 value is 1.1 and the MFD of the output fiber is 25 μm. The seed beam is collimated by a lens with a 15 mm focal length and the waist radius of the seed beam after collimating is 0.4 mm. The rod fiber used in the experiment is the aeroGAIN-ROD-PM85 produced by NKT Photonics. The MFD of the rod fiber is 65 μm, the length of the rod fiber is 803 ± 2 mm and the NA of the core of the rod fiber is about 0.006. According to the theoretical analysis, $F = 77$mm, l_i = 900 mm and $l_o = l$ = 82.33 mm were chosen as the laser path parameters for the experiment.

3.1. Verification of Optimal Seed Coupling Theory

In the experiment, the seed power transmittance as a function of l_i with fixed F was obtained. According to Figure 7, the maximum transmittance was acquired at l_i = 900 mm. The transmittance increases before l_i = 900 mm and drops after l_i = 900 mm. In addition, the growth is faster than the drop. Although the maximum transmittance at l_i = 900 mm is 87.2% due to fiber loss, the characteristics of the transmittance curve in the figure are consistent with the blue curve in Figure 4, which verifies the correctness and effectiveness of the theoretical analysis.

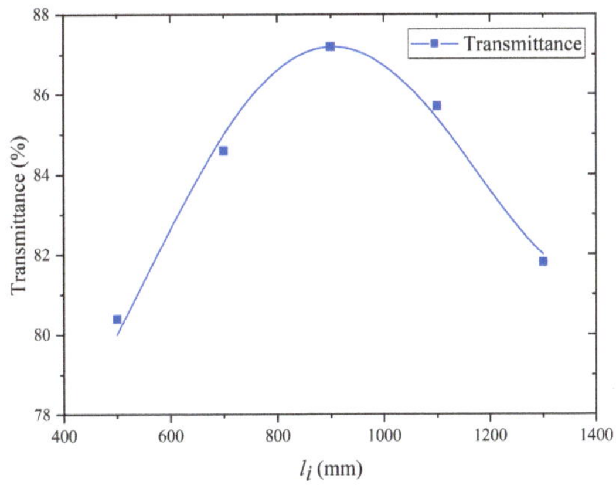

Figure 7. The seed power transmittance when l_i = 500 mm, l_i = 700 mm, l_i = 900 mm, l_i = 1100 mm and l_i = 1300 mm.

3.2. Seed Beam Coupling of an Amplification System

Under optimal seed beam coupling, to measure the seed power passing through the rod fiber, the transmittance and the coupled mode field under different output seed powers, the output seed power was gradually increased from 1 W to 6 W. The transient seed power and the transmittance with an increasing output seed power is shown in Figure 8a. Meanwhile, the power loss and the insertion loss with an increasing output seed power is shown in Figure 8b.

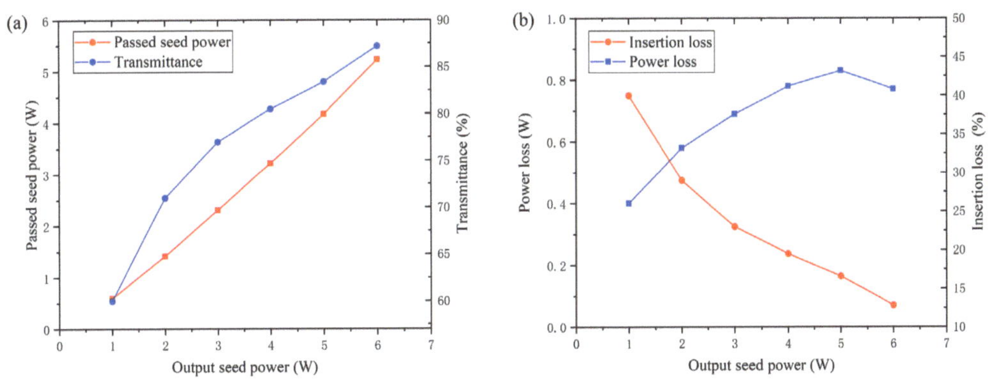

Figure 8. Transient seed power and the transmittance as function of the output seed power (**a**) and power loss and insertion loss as function of the output seed power (**b**).

According to Figure 7a, the transient seed power and the transmittance rise with the increase in the output seed power; the maximum transmittance of 87.2% was acquired at an output seed power of 6 W and the corresponding transient seed power was 5.23 W. From Figure 8b, the power loss increases first and reaches a maximum at 0.83 W at an output seed power of 5 W, then reduces with the further increase in output seed power. The insertion loss reduces with the increase in output seed power. There are two phenomena that need to be explained:

(1) The transmission rate exceeds 44%.

If the mode field of the seed beam waist and fiber core were fully matching, according to the instructions of the rod fiber, when the fiber is in passive operation, the power passing through the fiber core should be 44% of the total power. However, the results of this experiment show that the transmission rate exceeds 44%. The explanation is as follows. In this experiment, an aperture was not used to separate the power passing through the fiber core. Therefore, due to the limited optical field constraining ability of the rod fiber clad, the transmission power was composed of two parts of power. The first part is the power passing through fiber core and the second part is the power passing through the fiber clad. Although the transmission power consists of the power passing through the core and the clad, the proportion of the transmission power still cannot reach 100% of the output seed power, which means that the coupling still has mode field mismatch and the rod fiber has an absorption at 1030 nm.

(2) The insertion loss decreases with an increase in output seed power.

The reason for the variation in the insertion loss is that the seed beam mode field changes with the increase in output seed power and gradually matches the rod fiber mode field. The coupling mode fields at 2 W, 4 W and 6 W are shown in Figure 9; with the increase in output seed power, the mode field distribution of the transient beam has a tendency to be round, which confirms the above explanation.

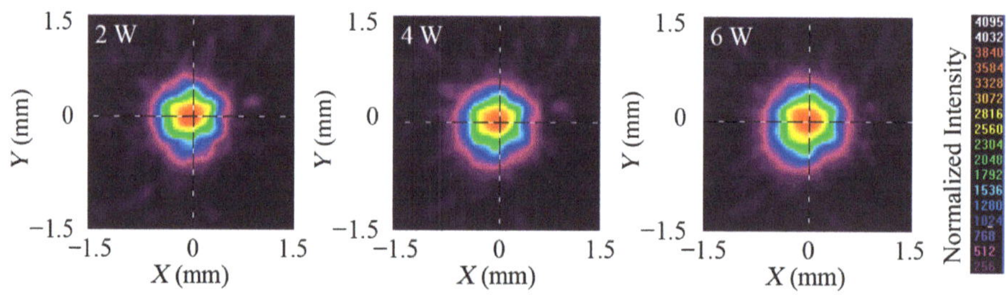

Figure 9. Variation in the coupled spots at output seed powers of 2 W, 4 W and 6 W.

Based on the experimental results and analysis, the theoretical analysis method has the ability to calculate the optimal mode field matching and enhance the coupling efficiency in experiments, and in the subsequent amplification experiments, 6 W will be selected as the output seed power.

4. Further Power Amplification Application

A scheme of the rod fiber amplification system is shown in Figure 10. For the convenience of adjustment, a counter-pumped structure was used in the experiment. The seed beam passes through a collimating lens (F = 15 mm), an isolator, dichroic mirror A (45° AR at 1030 nm and HR at 976 nm) and a coupling lens (F = 77 mm), respectively, then it is coupled to the rod fiber from the front-end. The pump beam passes through a telescope (both lenses are f = 15 mm) and dichroic mirror B (0° AR at 940 nm and HR at 1030 nm), respectively, then it is coupled to the rod fiber from the back-end. The infrared (IR) output is reflected into the power meter by dichroic mirror B and an IR mirror. The function of the beam splitter is to separate out a beam with low-power and reflect it into the CCD to observe the output spot. The seed source is a picosecond fiber laser system with a 6 W maximum average power, a central wavelength of 1030.24 nm, an autocorrelation pulse duration of about 5 ps, a repetition rate of 30 MHz, a spectral bandwidth of about 1.2 nm and an M^2 value of about 1.1. The MFD of the seed fiber is 25 μm. The central wavelength and the maximum output power of the pump diode are 976 nm and 200 W, respectively. The rod fiber is an aeroGAIN-ROD-PM85, produced by NKT Photonics. The MFD is 65 μm and length is 803 ± 2 mm. The two end caps are coated with antireflective film in order to decrease the Fresnel reflection loss. The rod fiber is clamped by a V-groove and sealed by a thermally conductive adhesive. The V-groove and pump diode are cooled by water at 20 °C.

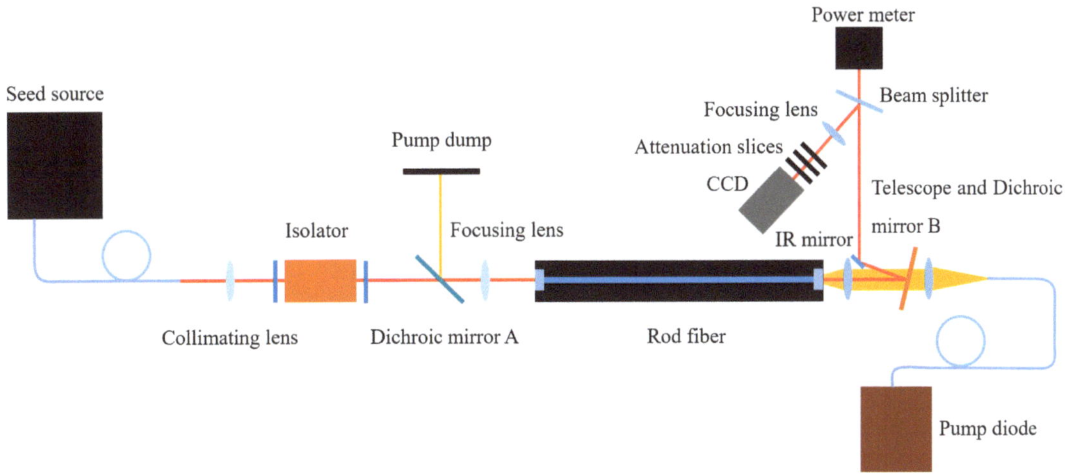

Figure 10. The scheme of the rod fiber amplification system.

According to the 87.2% transmittance with 6 W output seed power, a coupled seed power of 5.23 W is obtained. When the pump power was increased to 162.4 W, the output power and the optical efficiency as a function of pump power were measured and are shown in Figure 11a. A maximum optical efficiency of 60.5% was achieved with a 142.8 W pump power, and the corresponding output power was 91.6 W. When the pump power reached 162.4 W, a maximum output power of 101.7 W and a peak power of 500 kW were obtained; however, the corresponding optical efficiency was reduced to 59.4%. The slope efficiency of the amplifier was about 66.3%. In addition, the amplifier was unsaturated; however, considering that increasing the pump power may cause damage to the rod fiber and optical efficiency to decline, the pump power was not increased further in the experiment.

The beam diameters in the horizontal and vertical directions are subtly different, as shown in Figure 11b, and the M^2 values in the two directions are $M^2{}_X$ = 1.111 and $M^2{}_Y$ = 1.017 at 101.7 W, respectively, based on the D4σ method. Therefore, near diffraction-limited beam quality is acquired by this amplification system.

The spectral bandwidth of the seed is 1.2 nm (full-width at half-maximum, FWHM), but it was broadened to 1.8 nm at 101.7 W due to the self-phase modulation (SPM), as shown in Figure 11c. Attributed to the large mode area of the rod fiber, the maximum peak power of the output pulses is far below the threshold for stimulated Raman scattering (SRS); no SRS can be observed in the optical spectrum of the IR output. As shown in Figure 11d, the autocorrelation pulse width was measured as 6.4 ps (FWHM) at 101.7 W. The time-bandwidth product is 2.57, so the amplified pulses are chirped.

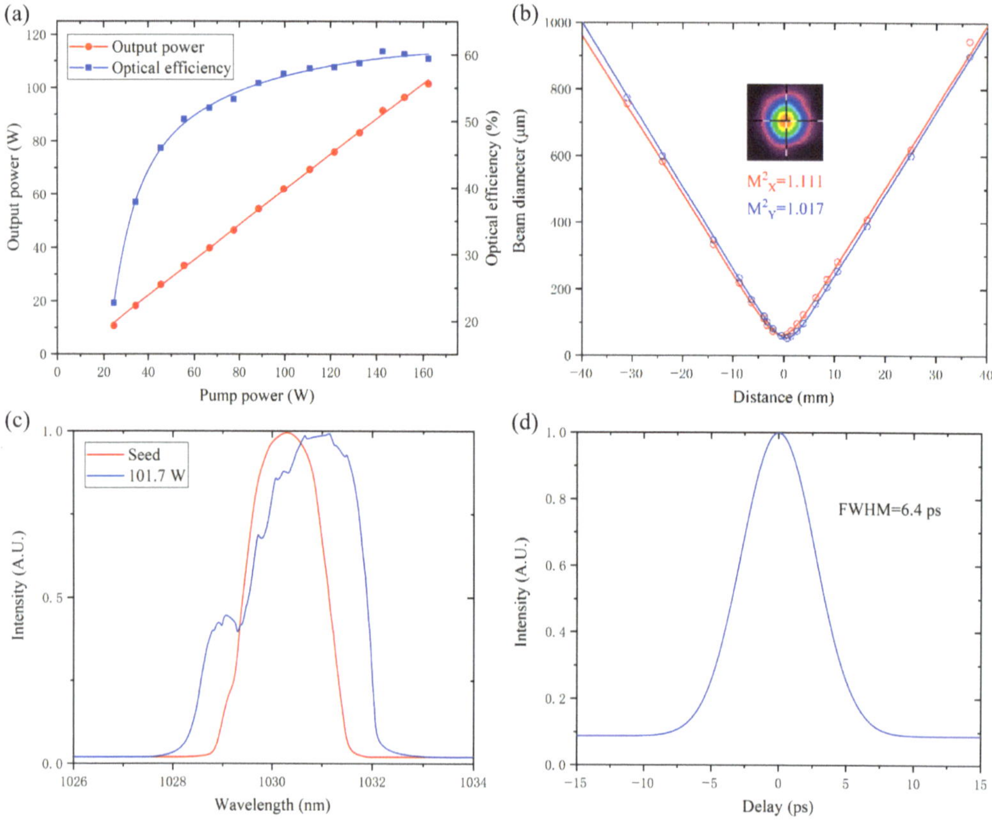

Figure 11. Output power and optical efficiency as a function of pump power (**a**); M^2 measurement at 101.7 W (**b**); optical spectrum from seed and 101.7 W output (**c**); autocorrelation signal of output pulses at 101.7 W (**d**).

5. Conclusions

In summary, to obtain a high coupling efficiency of the seed beam, thus enhancing the laser output power and beam quality of the rod fiber amplification system, a theoretical framework is established. The optical mode matching condition to satisfy optimal seed coupling is determined. The relationships among the main affecting factors of the coupling efficiency are determined. Among numerous variables, the focal length and position of the coupling lens are determined as the key variables for optimal seed coupling, and the cut-off focal length and the optimal position of the coupling lens are acquired.

According to the seed beam coupling experiment, the theoretical derivation is verified under single variable control. A passive transmission system is built and a maximum transmittance of 87.2% is obtained with an input seed power of 6 W. The optimal seed beam coupling theory is further enhanced considering the mode field mismatching.

The rod fiber amplification system achieves an average power of 101.7 W with a coupled seed power of 5.23 W and a pump power of 162.4 W. The beam quality of the amplification system is diffraction limited, with M^2 values in the horizontal and vertical directions of 1.111 and 1.017, respectively. A maximum optical efficiency of 60.5% and a slope efficiency of 66% are achieved with a rod fiber length of 0.8 m. Under the optimal mode field matching, the rod fiber amplification system obtains high beam quality and amplification efficiency, which both show good prospects and feasibility in MEMS fabrication.

Author Contributions: Conceptualization, D.L., T.L. and N.S.; methodology, D.L., T.L. and D.Z.; validation, T.L., X.M. and G.B.; formal analysis, D.L.; investigation, X.M. and G.B; resources, G.B.; data curation, T.L.; writing—original draft preparation, D.L.; writing—review and editing, D.L. and T.L.; visualization, T.L.; supervision, T.L.; project administration, X.M.; funding acquisition, G.B. and D.Z. All authors have read and agreed to the published version of the manuscript.

Funding: This work is partially supported in part by the Young Scientists Fund of the National Natural Science Foundation of China (Grant No. 61902373), by the Foundation of the Chinese Academy of Sciences, China (Grant No. JCPYJJ 22013), by the Strategic Priority Research Program of the Chinese Academy of Sciences, China (Grant No. XDB24050300, XDB44030300).

Data Availability Statement: Not applicable.

Conflicts of Interest: The authors declare no conflict of interest.

References

1. Li, Y.; Guo, X.; Wang, S.; Zhang, S.; Zhao, Y.; Guo, D.; Zhang, C.; Liu, S.; Cheng, G.J.; Liu, F. High-efficiency copper removal by nitrogen plasma-assisted picosecond laser processing. *Micromachines* **2022**, *13*, 1492. [CrossRef] [PubMed]
2. Wen, Q.; Wei, X.; Zhang, P.; Lu, J.; Jiang, F.; Lu, X. Enhanced microsphere-assisted picosecond laser processing for nanohole fabrication on silicon via thin gold coating. *Micromachines* **2021**, *12*, 611. [CrossRef] [PubMed]
3. Markauskas, E.; Zubauskas, L.; Voisiat, B.; Gečys, P. Efficient water-assisted glass cutting with 355 nm picosecond laser pulses. *Micromachines* **2022**, *13*, 785. [CrossRef] [PubMed]
4. Lye, C.S.; Wang, Z.; Lam, Y.C. Multi-foci laser separation of sapphire wafers with partial thickness scanning. *Micromachines* **2022**, *13*, 506. [CrossRef]
5. Zhao, Z.; Dunham, B.M.; Wise, F.W. Generation of 167 w infrared and 124 w green power from a 1.3-ghz, 1-ps rod fiber amplifier. *Opt. Express* **2014**, *22*, 25065–25070. [CrossRef]
6. Alkeskjold, T.T.; Laurila, M.; Scolari, L.; Broeng, J. Single-mode ytterbium-doped large-mode-area photonic bandgap rod fiber amplifier. *Opt. Express* **2011**, *19*, 7398–7409. [CrossRef]
7. Limpert, J.; Schreiber, T.; Nolte, S.; Zellmer, H.; Tünnermann, A.; Iliew, R.; Lederer, F.; Broeng, J.; Vienne, G.; Petersson, A.; et al. High-power air-clad large-mode-area photonic crystal fiber laser. *Opt. Express* **2003**, *11*, 818–823. [CrossRef]
8. Feng, X.; Ren, H.; Xu, F.; Shi, J.; Qi, S.; Hu, Y.; Tang, J.; Han, F.; Shen, D.; Yang, Z. Few-moded ultralarge mode area chalcogenide photonic crystal fiber for mid-infrared high power applications. *Opt. Express* **2020**, *28*, 16658–16672. [CrossRef]
9. Fu, X.; Hong, K.H.; Chen, L.J.; Kärtner, F.X. Performance scaling of high-power picosecond cryogenically cooled rod-type yb:Yag multipass amplification. *J. Opt. Soc. Am. B* **2013**, *30*, 2798–2809. [CrossRef]
10. Kowalewski, K.; Zembek, J.; Envid, V.; Brown, D.C. 201 w picosecond green laser using a mode-locked fiber laser driven cryogenic yb:Yag amplifier system. *Opt. Lett.* **2012**, *37*, 4633–4635. [CrossRef]
11. Teisset, C.; Schultze, M.; Bessing, R.; Haefner, M.; Metzger, T. *300 w Picosecond Thin-Disk Regenerative Amplifier at 10 khz Repetition Rate*; Advanced Solid-State Lasers Congress Postdeadline: Paris, France, 2013; p. JTh5A.1.
12. Ochi, Y.; Nagashima, K.; Maruyama, M.; Tsubouchi, M.; Yoshida, F.; Kohno, N.; Mori, M.; Sugiyama, A. Yb:Yag thin-disk chirped pulse amplification laser system for intense terahertz pulse generation. *Opt. Express* **2015**, *23*, 15057–15064. [CrossRef] [PubMed]
13. Lin, H.; Li, J.; Liang, X. 105 w, <10 ps, tem_00 laser output based on an in-band pumped nd:Yvo_4 innoslab amplifier. *Opt. Lett.* **2012**, *37*, 2634–2636. [PubMed]
14. Schmidt, B.E.; Hage, A.; Mans, T.; Légaré, F.; Wörner, H.J. Highly stable, 54mj yb-innoslab laser platform at 0.5kw average power. *Opt. Express* **2017**, *25*, 17549–17555. [CrossRef]
15. Xu, L.; Zhang, H.; Mao, Y.; Yan, Y.; Fan, Z.; Xin, J. High-average-power and high-beam-quality innoslab picosecond laser amplifier. *Appl. Opt.* **2012**, *51*, 6669–6672. [CrossRef] [PubMed]
16. Wang, N.N.; Li, F.; Wang, X.L.; Hu, X.H.; Wang, Y.S.; Zhao, W. Development of a 67.8 w, 2.5 ps ultrafast chirped-pulse amplification system based on single-crystal fiber amplifiers. *Appl. Opt.* **2020**, *59*, 8106–8110. [CrossRef]
17. Chan, H.Y.; Alam, S.U.; Xu, L.; Bateman, J.; Richardson, D.J.; Shepherd, D.P. Compact, high-pulse-energy, high-power, picosecond master oscillator power amplifier. *Opt. Express* **2014**, *22*, 21938–21943. [CrossRef]
18. Pierrot, S.; Liegeois, F.; Saby, J.; Cocquelin, B.; Hernandez, Y.; Salin, F.; Giannone, D. *All Fibre High Repetition Rate, High Power Picosecond Laser and uv Generation*; Advances in Optical Materials; Optical Society of America: Istanbul, Turkey, 2011; p. JWA5.
19. Limpert, J.; Deguil-Robin, N.; Manek-Hönninger, I.; Salin, F.; Röser, F.; Liem, A.; Schreiber, T.; Nolte, S.; Zellmer, H.; Tünnermann, A.; et al. High-power rod-type photonic crystal fiber laser. *Opt. Express* **2005**, *13*, 1055–1058. [CrossRef] [PubMed]
20. Manchee, C.P.K.; Möller, J.; Miller, R.J.D. Highly stable, 100 w average power from fiber-based ultrafast laser system at 1030 nm based on single-pass photonic-crystal rod amplifier. *Opt. Commun.* **2019**, *437*, 6–10. [CrossRef]

21. Zhao, Z.; Dunham, B.M.; Wise, F.W. Generation of 150 w average and 1 mw peak power picosecond pulses from a rod-type fiber master oscillator power amplifier. *J. Opt. Soc. Am. B* **2014**, *31*, 33–37. [CrossRef]
22. Lesparre, F.; Gomes, J.T.; Délen, X.; Martial, I.; Didierjean, J.; Pallmann, W.; Resan, B.; Druon, F.; Balembois, F.; Patrick Georges. Yb:Yag single-crystal fiber amplifiers for picosecond lasers using the divided pulse amplification technique. *Opt. Lett.* **2016**, *41*, 1628–1631. [CrossRef]
23. Otto, H.-J.; Stutzki, F.; Modsching, N.; Jauregui, C.; Limpert, J.; Tünnermann, A. 2 kw average power from a pulsed yb-doped rod-type fiber amplifier. *Opt. Lett.* **2014**, *39*, 6446–6449. [CrossRef] [PubMed]
24. Zhao, Z.; Sheehy, B.; Minty, M. Generation of 180 w average green power from a frequency-doubled picosecond rod fiber amplifier. *Opt. Express* **2017**, *25*, 8138–8143. [CrossRef] [PubMed]
25. Guang, Z.; Zhang, Y. Coupling ultrafast laser pulses into few-mode optical fibers: A numerical study of the spatiotemporal field coupling efficiency. *Appl. Opt.* **2018**, *57*, 9835–9844. [CrossRef] [PubMed]
26. Ma, J.; Zhao, F.; Tan, L.; Yu, S.; Han, Q. Plane wave coupling into single-mode fiber in the presence of random angular jitter. *Appl. Opt.* **2009**, *48*, 5184–5189. [CrossRef]

micromachines

MDPI

Article

Miniature Fourier Transform Spectrometer Based on Thin-Film Lithium Niobate

Lichao Zhang [1,2], Guangyang Gou [1,2], Jiamin Chen [1,2], Wangzhe Li [1,2], Weichao Ma [1,2], Ruoming Li [1,2], Junming An [3], Yue Wang [3], Yuanyuan Liu [3,4], Wei Yan [3], Tianjun Ma [1,2], Chunxiu Liu [1,2], Jianjun Cheng [5], Zhimei Qi [1,2,*] and Ning Xue [1,2,*]

1 State Key Laboratory of Transducer Technology, Aerospace Information Research Institute, Chinese Academy of Sciences, Beijing 100190, China
2 School of Electronic, Electrical and Communication Engineering, University of Chinese Academy of Sciences, Beijing 100049, China
3 Institute of Semiconductors, Chinese Academy of Sciences, Beijing 100083, China
4 Center of Materials Science and Optoelectronics Engineering, University of Chinese Academy of Sciences, Beijing 100049, China
5 School of Integrated Circuit, Quanzhou University of Information Engineering, Quanzhou 362000, China
* Correspondence: zhimei-qi@mail.ie.ac.cn (Z.Q.); xuening@mail.ie.ac.cn (N.X.)

Abstract: A miniature Fourier transform spectrometer is proposed using a thin-film lithium niobate electro-optical modulator instead of the conventional modulator made by titanium diffusion in lithium niobate. The modulator was fabricated by a contact lithography process, and its voltage-length and optical waveguide loss were 2.26 V·cm and 1.01 dB/cm, respectively. Based on the wavelength dispersion of the half-wave voltage of the fabricated modulator, the emission spectrum of the input signal was retrieved by Fourier transform processing of the interferogram, and the analysis of the experimental data of monochromatic light shows that the proposed miniaturized FTS can effectively identify the input signal wavelength.

Keywords: lithium niobate; Fourier transform spectrometer; electro-optical modulation

Citation: Zhang, L.; Gou, G.; Chen, J.; Li, W.; Ma, W.; Li, R.; An, J.; Wang, Y.; Liu, Y.; Yan, W.; et al. Miniature Fourier Transform Spectrometer Based on Thin-Film Lithium Niobate. *Micromachines* **2023**, *14*, 458. https://doi.org/10.3390/mi14020458

Academic Editors: He Yang, Xinyang Su and Yizhong Huang

Received: 19 January 2023
Revised: 4 February 2023
Accepted: 6 February 2023
Published: 15 February 2023

1. Introduction

Due to the ability to detect the characteristics of optical signals, spectrometers are widely used in many fields, such as space exploration [1], nanotechnology [2], bioengineering [3], modern medicine [4], meteorological monitoring [5], resource exploration [6], environmental monitoring [7] and life sciences [8]. With the development of science and technology, there is an increasingly urgent need for spectrometers to move from the traditional benchtop to miniaturization.

In recent years, the development of integrated optics has promoted the development of a miniature Fourier transform spectrometer (FTS) based on the electro-optical effect in lithium niobate (LN) to achieve a tunable optical pathlength difference (*OPD*) has received great attention.

In 1893, German physicist Pockels discovered the linear electro-optical effect, which revealed that the refractive index of certain crystals varies linearly with an applied voltage. As a material with a linear electro-optic effect, LN is a good material platform for miniature FTS because of its relatively large electro-optic coefficient, wide optical transparency window (0.4~5 μm), and low power consumption. In 2002, Howard constructed the first FTS based on bulk LN crystal [9]. Then a miniature FTS was introduced by Bentini et al. in 2007, which was achieved by utilizing high-energy ion implantation in LN [10]. In 2014, Li et al. proposed a miniature FTS, which is based on a conventional electro-optical modulator made by diffusing titanium into LN [11]. However, the electro-optical modulation efficiency of titanium diffused LN waveguide is low. Around 2016, thin film lithium niobate (TFLN) wafers were commercially produced, giving rise to a series of studies based on TFLN [12–16]. TFLN waveguides are mostly formed by dry etching, allowing for high electro-optical modulation efficiency. There are many explorations of an electro-optical modulator based on TFLN, but there are few studies combining it with FTS. In 2019, Grange et al. combined the principles of LN electro-optical modulation and standing wave interference to propose a hybrid integrated FTS based on a TFLN-silicon nitride platform [17].

In this work, we present a miniature FTS based on a TFLN electro-optical modulator which was fabricated by a contact lithography process. Due to the half-wave voltage V_π dependence on wavelength, the emission spectrum of the input signal is retrieved by the Fourier transform of the measured temporal interferogram. This work is of great value for the miniaturization of FTS.

This paper is organized as follows: Section 2 describes the detailed design of FTS and simulation of the modulator; then the fabrication, measurement, and analysis of results are presented in Section 3.

2. Design and Simulation

In this section, the general description of the device is given and then simulations are performed based on the relevant parameters to determine the dimensions of the modulation area.

The schematic diagram of the miniature FTS based on lithium niobate on insulator (LNOI) described in this paper is shown in Figure 1a, which consists of an electro-optical modulator, an arbitrary waveform generator (AWG), a photodetector (PD), a polarization controller (PC), a data acquisition card (DAQ), a signal processing unit, and input and output optical fibers. The core component is an electro-optical modulator based on LNOI, which consists of a Mach-Zehnder interferometer (MZI) and traveling wave electrodes (TWE), as shown in Figure 1b. Firstly, the optical signal is coupled into the MZI through the input fiber and propagates in the two arms of the MZI. At the same time, a modulating voltage is generated by the voltage generator and is applied to the TWEs on both sides of the waveguide, which leads to an opposite change in the refractive index of the two LN arms, thus modulating the *OPD* between both arms. Subsequently, the two optical signals interfere at the Y-branch combiner, and the interferometric optical signal is captured by the PD through the output fiber, and then is converted into an electrical signal at the output to the DAQ. Simultaneously, the modulating voltage is also sampled by the DAQ. Based on the Fourier transform principle and the relevant algorithms [18], the Fourier transform is performed on the acquired signals through the signal processing unit, and the emission spectrum of the input signal can be obtained. Among them, the V_π of the electro-optical modulator has a great influence on the performances of the miniature FTS.

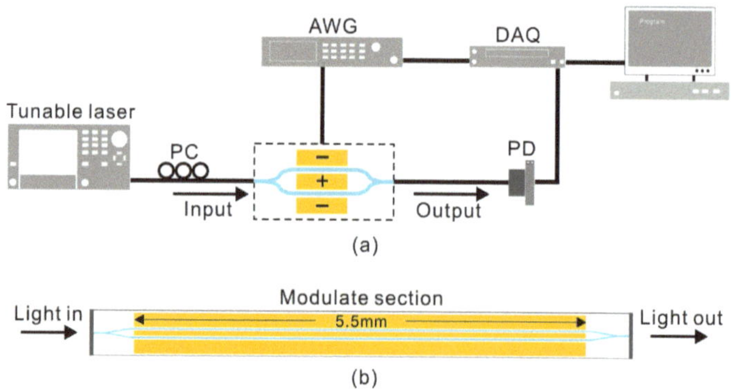

Figure 1. (**a**) The schematic of the FTS; (**b**) The schematic of the TFLN electro-optical modulator.

The configuration and dimensions of the electro-optical modulator are shown in Figure 1b. The chip size is 7.5 mm × 0.25 mm, the modulating arm length of the interferometer is 5.5 mm, and the width of both waveguides is 1.1 μm. Considering that the input and output fiber waist diameters are 4 μm, the input port of the waveguide is designed with a tapered shape with the width gradually decreasing from 6 μm to 1.1 μm in order to make the input optical signal better coupled into the waveguide through the fiber, and the length of the taper is 500 μm. The output port of the waveguide is similarly designed. The two modulating arms are flanked by TWEs.

Figure 2a shows the cross-sectional structure of the modulated region. From top to bottom, are the SiO_2 cladding layer, the two gold electrodes, the x-cut LNOI ridge waveguide, the SiO_2 buried layer, and the silicon substrate. The optical signals are confined in the ridge waveguide for transmission, and the sidewall roughness of the ridge waveguide is greatly affected by fabrication conditions. Considering that contact lithography is used in this research, in the simulations, the upper width of the ridge waveguide is set to 1.1 μm. When the modulating voltage is applied to the gold electrodes on both sides of the ridge waveguide, it will result in a change in the refractive index of the ridge waveguide. The electric field applied to the two arms of the interferometer in opposite directions causes opposite changes in the refractive index of the two arms, resulting in the *OPD* at the end of the interferometer. For the electro-optical modulator, the V_π and optical waveguide loss are greatly influenced by the gap D of the two gold electrodes, the etching depth h of the ridge waveguide, and the height of gold electrodes H, as shown in Figure 2a. Related studies have shown [19] that the larger the gap D and the larger the etching depth h, the larger the V_π and the smaller the optical loss caused by metal absorption. In order to obtain a relatively low V_π and relatively small optical waveguide loss, with reference to the results of the literature [19] and a series of related simulations using COMSOL Multiphysics, in our research, the gap D is set as 5 μm, the etching depth h equals 300 nm and the height of gold electrodes H equals 900 nm.

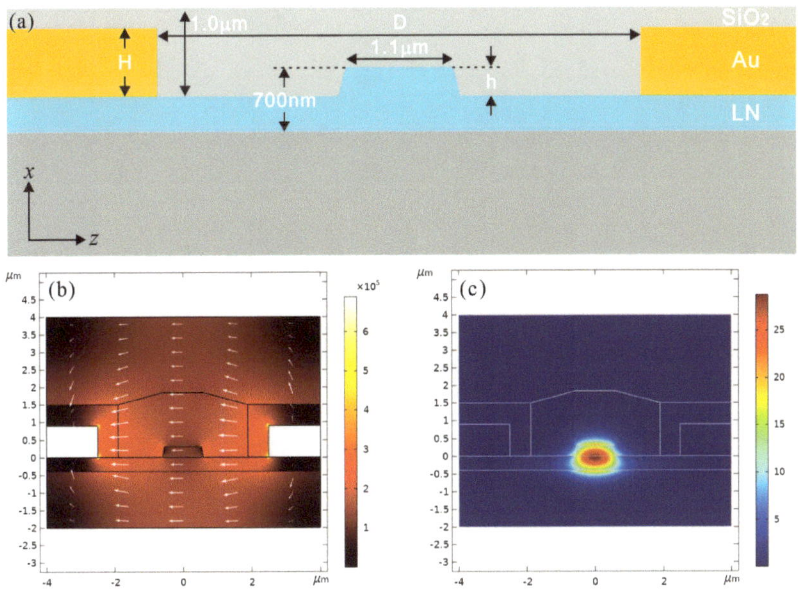

Figure 2. (**a**) The cross−sectional structure and dimensions of the modulation area; (**b**) The electrostatic field of two electrodes; (**c**) The optical field of the transverse electric (TE) mode.

Based on the determined dimensions, the simulations were carried out by COMSOL Multiphysics, and the results are shown in Figure 2b,c, including the optical field of the simulated transverse electric (TE) mode and the electrostatic field of the gold electrodes, with a calculated electro-optic overlap integral factor of 0.5401. Then the V_π and optical waveguide loss of the electro-optical modulator can be calculated based on COMSOL simulations. At λ = 1550 nm, the corresponding V_π is 4.08 V, and the optical waveguide loss is 0.38 dB/cm.

3. Fabrication of the Device and Analysis of Measurement Results

3.1. Fabrication and Measurement of the Device

We use contact lithography to prepare the electro-optic modulator chip, and the fabrication process is shown in Figure 3. First, an 80 nm thick Chromium layer (Cr) is sputtered on the top surface of the exposed LNOI by magnetron sputtering, which is used as a hard mask for the etching process. A 1.2-μm-thick negative photoresist (PR) is then used to define the waveguide pattern. Next, the patterns are transferred into a Cr layer using an inductively coupled plasma (ICP) etcher, and subsequently into the TFLN by an optimized dry ICP etching process based on Ar and SF_6. The physical and chemical reactions are carried out simultaneously during the TFLN etching process, and the etching speed is 0.32 nm/s. Then the residual Cr mask is removed using cerium ammonium nitrate. The metal electrode (15 nm-thick Cr/900 nm-thick Au) is then fabricated and formed by a standard lift-off process. A 1 μm thick silica cladding is then deposited by plasma enhanced chemical vapor deposition (PECVD). Then windows are made in the silica cladding to facilitate the connection of the bottom electrode to the top electrode afterward, and finally, a second lift-off process is performed to produce the top electrode.

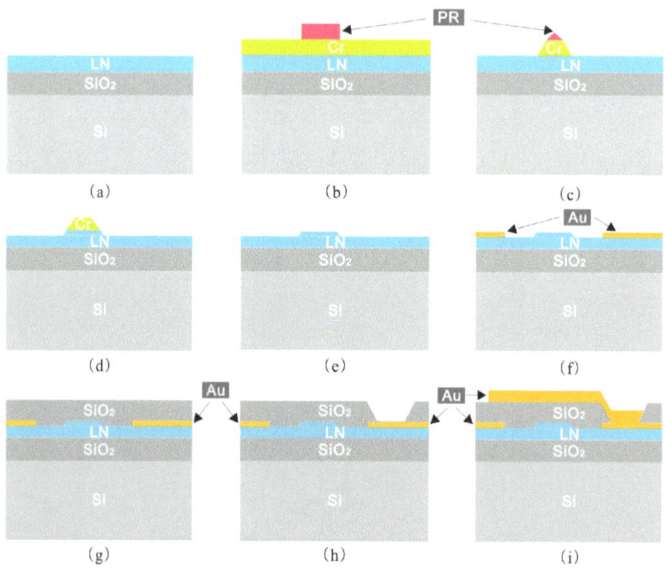

Figure 3. Schematic diagram of the fabrication process of the TFLN electro-optical modulator. (**a**) An x-cut LNOI substrate is (**b**) coated with 80 nm thick Cr by magnetron sputter. Then a 1.2-μm-thick negative PR is used to define the waveguide pattern. (**c**) The patterns are then transferred into the Cr layer using ICP etcher, and (**d**) subsequently into the LN thin film using Ar and SF6. (**e**) The residue mask materials are removed in cerium ammonium nitrate. (**f**) Bottom metal electrodes (15 nm-thick Cr/900 nm-thick Au) are formed using a standard lift-off process. (**g**) A 1 μm thick silica cladding layer is then deposited by PECVD. (**h**) Via windows are opened by wet etching. (**i**) A second lift-off process is performed to produce the top electrodes.

Figure 4 shows the microscope image of a part of the fabricated modulator and the scanning electron microscope (SEM) image of a cross section of the waveguide. The upper width of the ridge waveguide is measured to be about 1.1 μm, the sloping sidewall of the waveguide is about 75° from horizontal, and the bottom width of the ridge waveguide is about 1.2 μm.

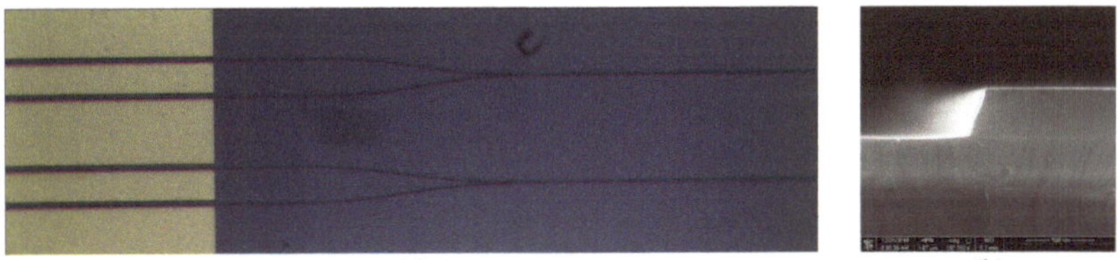

(a) (b)

Figure 4. (**a**) Microscopic image of the part of the modulator; (**b**) The SEM image of the etched waveguide cross section.

The measurement setup is shown in Figure 5. The input optical signal is generated by a tunable laser and coupled into the electro-optical modulator through the input fiber and PC. After modulation by the modulating voltage, the optical signals in the two arms interfere at the Y-branch combiner, and the interfering optical signals are captured by the PD through the output fiber and then are converted into electrical signals, and the DAQ is set at the end to capture the electrical signals. Table 1 compares the fabricated TFLN modulator in this work with the one obtained in the [19]. They have similar structures and working principles.

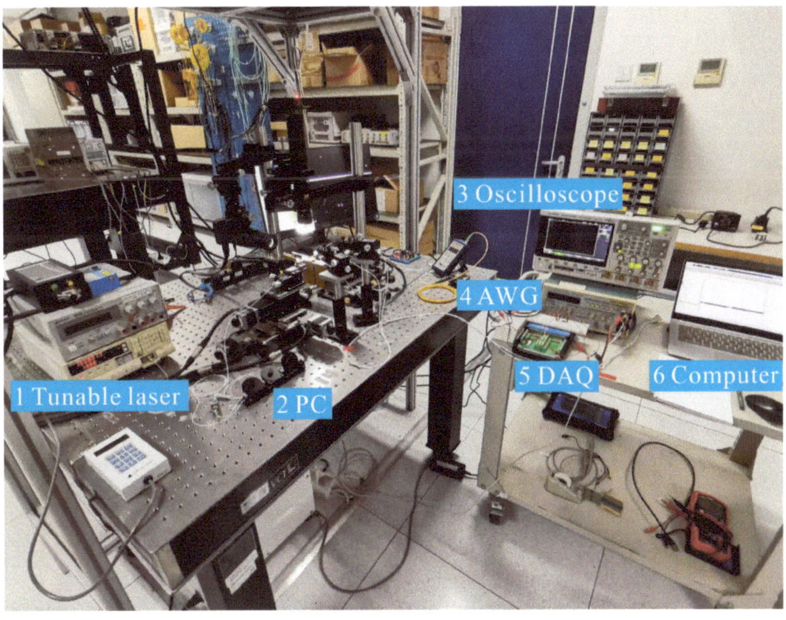

Figure 5. The measurement setup.

Table 1. Performance comparison of TFLN modulators.

Voltage-Length Product	Pattern Definition	Half-Wave Voltage	Gap	Etching Depth	Propagation Loss
1.75 V·cm	Photolithography	3.5 V	5 µm	0.3 µm	0.7 dB/cm
2.26 V·cm	Photolithography	4.51 V	5 µm	0.3 µm	1.01 dB/cm

3.2. Analysis of Measurement Results

In this section, the principle of optical signal identification based on FTS is described, and then the identification of the input optical signal based on the experimental data is conducted.

The change of *OPD* in the proposed miniature FTS is achieved by the electro-optical effect in LN. The opposite refractive index change caused by the modulating voltage V(t) will produce *OPD* although the physical paths of the two arms are the same. The *OPD* between the two arms can be expressed as

$$OPD = \frac{n_e^3(\lambda)\gamma_{33}(\lambda)\Gamma(\lambda)L}{D}V(t) \tag{1}$$

where D is the gap between the two gold electrodes, L is the arm length, $n_e(\lambda)$, $\gamma_{33}(\lambda)$, $\Gamma(\lambda)$ are the parameters related to modulator configuration and optical signal wavelength λ. The half-wave voltage $V_\pi(\lambda)$ of the electro-optical modulator can be expressed as

$$V_\pi(\lambda) = \frac{\lambda D}{2n_e^3(\lambda)\gamma_{33}(\lambda)\Gamma(\lambda)L} \tag{2}$$

Substituting Equation (2) into Equation (1), we can get

$$OPD = \frac{\lambda V(t)}{2V_\pi(\lambda)} \tag{3}$$

Defining $g(\lambda) = 1/(2V_\pi(\lambda))$, Equation (3) can be rewritten as

$$OPD = \lambda V(t) g(\lambda) \tag{4}$$

For a fabricated electro-optical modulator, L and D are all constant, and the parameters $n_e(\lambda)$, $\gamma_{33}(\lambda)$ and $\Gamma(\lambda)$ have definite values when the input optical signal wavelength λ is determined. Then according to Equation (2), $V_\pi(\lambda)$, and $g(\lambda)$ can be determined, and *OPD* is only related to the applied modulating voltage $V(t)$ according to Equation (1).

The modulating voltage will cause the change of *OPD*, thus causing the change of the interferometric light intensity I. I varies with the modulating voltage V and can be expressed as follows, according to the FTS theory [11],

$$I(V) = \int_0^\infty A(\lambda)\cos\left[2\pi\frac{OPD}{\lambda}\right]d\lambda \tag{5}$$

where $A(\lambda)$ is the light intensity in the emission spectrum of the input signal. Equations (4) and (5) can be rewritten as

$$I(V) = \int_0^\infty A(\lambda)\cos[2\pi V(t)g(\lambda)]d\lambda \tag{6}$$

To facilitate the Fourier transform, Equation (6) is rewritten as [11]

$$I(V) = \int_0^\infty A'(g)\cos[2\pi V(t)g]dg \tag{7}$$

Fourier transform of Equation (7) gives

$$A(g) = \int_{-\infty}^{\infty} I(V) \exp(-j2\pi g V(t)) dV \tag{8}$$

It is worth noting that due to the presence of a breakdown electric field (~10 V/μm) in LN, the range of modulating voltage is limited, then Equation (8) can be rewritten as

$$A_0(g) = \int_{-V_{max}}^{V_{max}} I(V) \exp(-j2\pi g V) dV \tag{9}$$

where V_{max} is the maximum modulating voltage, and $A_0(g)$ is the spectral intensity distribution with g. Since the collected data of the modulating voltage $V(t)$ and the interferometric light intensity I are discrete sequences, the discrete Fourier transform is used, which is expressed as [20]

$$A_0(g) = \sum_{i=0}^{N_s-1} I_i \exp(-j2\pi g V_i) dV_i, \qquad dV_i = (V_{i+1} - V_{i-1})/2 \tag{10}$$

$A_0(g)$ can be obtained through the discrete Fourier transform. Obtaining $A(\lambda)$ from $A_0(g)$ also requires determining the dispersion relation between the half-wave voltage $V_\pi(\lambda)$ and the wavelength λ.

For the fabricated FTS, the dependence of V_π on wavelength λ can be determined by experimentally measuring a series of input optical signals of different λ, and the study shows that the function has monotonicity [20]. By using the measured dispersion of $V_\pi(\lambda)$ and the formula $g = [2V_\pi(\lambda)] - 1$, the plot of $A_0(g)$ against g can be converted into the wanted laser power spectrum $A(\lambda)$. A more detailed discussion of the above theory can be found in the literature [11]. Based on the above principles, specific signal processing algorithms are written to analyze the experimental data of the fabricated miniature FTS, then the identification of the input optical signal can be conducted.

In order to obtain the half-wave voltage $V_\pi(\lambda)$ as a function of wavelength λ, a series of optical signals with different wavelengths (1528.8 nm, 1535.8 nm, 1545.3 nm, 1553.3 nm, 1566.7 nm) are fed in the TFLN electro-optic modulator, and the V_π corresponding to a certain wavelength can be obtained by the intensity-versus-voltage interferogram, then a series of the values of V_π and λ can be obtained, as shown by the black dots in Figure 6. The half-wave voltage dispersion can be obtained by fitting the measured multiple wavelengths to the half-wave voltage data, as shown in Figure 6. The fitting function is expressed as

$$V_\pi(\lambda) = 0.00853 \times \lambda - 8.80257 \tag{11}$$

In order to verify the spectral detection capability of the miniature FTS proposed in this paper, the light signal with a wavelength of 1561.828 nm was fed into the miniature FTS as the light source to be measured, and the intensity-versus-voltage interferogram was obtained, as shown in Figure 7a. It can be read from Figure 7a that the half-wave voltage is 4.51 V, corresponding to a voltage-length product of 2.26 V·cm. After Fourier transforming of the interferogram, the plot of normalized $A_0(g)$ against g was obtained (as shown in Figure 7b), on which the plot of $A(\lambda)$ against λ can be obtained utilizing the formula $g = [2V_\pi(\lambda)] - 1$ and the Equation (11). It can be deduced that the main peak is located at 1561.46 nm, which is very close to the wavelength of the signal to be measured (1561.828 nm), indicating that the proposed miniature FTS can accurately identify the wavelength of the input signal.

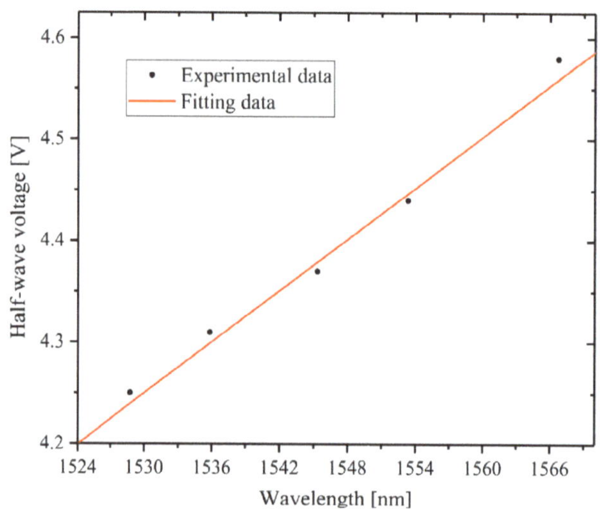

Figure 6. The half-wave voltage dispersion.

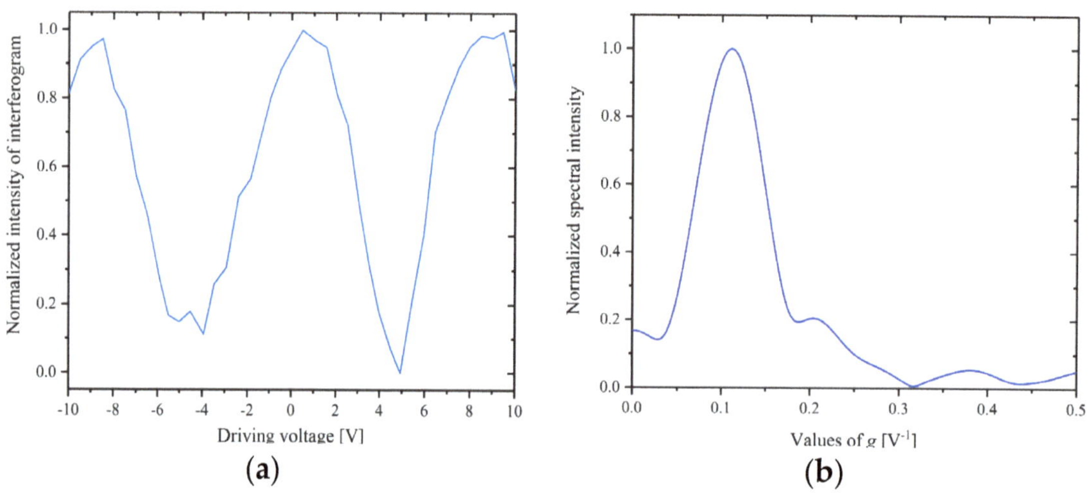

Figure 7. (**a**) The normalized intensity−versus−voltage interferogram; (**b**) The normalized spectral intensity distributed with g.

4. Conclusions

A miniature FTS is proposed using a shallowly etched TFLN electro-optical modulator instead of the conventional modulator made by titanium diffusion in LN. The modulator was fabricated by a contact lithography process and ICP dry etching process, which can reduce the cost of the device. Its voltage-length product and optical waveguide loss were 2.26 V·cm and 1.01 dB/cm, respectively, and the measured results agree well with simulated results, which indicates that the simulation is a good guide to improve the modulator performance. The half-wave voltage as a function of wavelength was experimentally determined. Based on the wavelength dispersion of half-wave voltage, the emission spectrum of the input signal was retrieved by Fourier transform processing of the interferogram. A signal with a wavelength of 1561.828 nm was input to the proposed miniaturized FTS, and the signal spectrum with a peak located at 1561.46 nm was obtained with the help of the half-wave voltage dispersion equation and Fourier transform. This shows that the proposed miniaturized FTS can effectively identify the input signal wavelength. We believe that the miniaturized FTS presented in this paper can be a promising candidate for compact FTS.

Author Contributions: Conceptualization, N.X. and Z.Q.; Data curation, L.Z., G.G. and J.C. (Jiamin Chen); Software, L.Z., T.M. and J.C. (Jianjun Cheng); Validation, W.L., W.M. and R.L.; Formal analysis, L.Z. and G.G.; Investigation, L.Z.; Resources, G.G. and C.L.; Writing—original draft preparation, L.Z.; Writing—review and editing, L.Z., N.X. and Z.Q.; Supervision, J.A., Y.W., Y.L. and W.Y. All authors have read and agreed to the published version of the manuscript.

Funding: This research was funded by National Key R&D Program of China (2021YFB3200600, 2020YFC2004500); National Natural Science Foundation of China (NSFC) (62273321, 62073307, 62204243, 61774157, 81771388, 61675203, 61931018, 61871365); CAMS Innovation Fund for Medical Sciences (2019-I2M-5-019); Chinese Academy of Sciences Joint Fund for Equipment Pre Research (8091A140106).

Data Availability Statement: Data are available upon reasonable request from the corresponding author.

Conflicts of Interest: The authors declare no conflict of interest.

Abbreviations

FTS	Fourier transform spectrometer
LN	lithium niobate
OPD	optical pathlength difference
TFLN	thin film lithium niobate
LNOI	lithium niobate on insulator
AWG	arbitrary waveform generator
PD	photodetector
PC	polarization controller
DAQ	data acquisition card
MZI	Mach-Zehnder interferometer
TWE	traveling wave electrodes
TE	transverse electric
PR	photoresist
ICP	inductively coupled plasma
PECVD	plasma enhanced chemical vapor deposition
SEM	scanning electron microscope

References

1. Rohner, U.; Whitby, J.A.; Wurz, P. Compact fourier-transform imaging spectrometer for small satellite massions. *Meas. Sci. Technol.* **2003**, *14*, 2159–2164. [CrossRef]
2. DeCarlo, P.F.; Kimmel, J.R.; Trimborn, A.; Northway, M.J.; Jayne, J.T.; Aiken, A.C.; Gonin, M.; Fuhrer, K.; Horvath, T.; Docherty, K.S. Field-deployable, high-resolution, time-of-flight aerosol mass spectrometer. *Anal. Chem.* **2006**, *78*, 8281–8289. [CrossRef] [PubMed]
3. Huang, Y.; Lan, Y.; Thomson, S.J.; Fang, A.; Hoffmann, W.C.; Lacey, R. Development of soft computing and applications in agricultural and biological engineering. *Comput. Electron. Agric.* **2010**, *71*, 107–127. [CrossRef]
4. Ivanov, S.V.; Novoderezhkin, V.I.; Panchenko, V.Y.; Solomatin, V.S.; Kholodnykh, A.I. Laser infrared spectrometer for atmosphere gas analysis and medicine. *Opt. Eng.* **1994**, *33*, 3202. [CrossRef]
5. Almeida, T.S.; Brancher, M.; Lisboa, H.D.M.; Franco, D.; Maranhão, T.A.; Borges, D.L. Direct analysis of particulate matter (PM10) for the determination of Be, Cd and Pb using high resolution-continuum source electrothermal atomic absorption spectrometry: Assessment of the potential correlation between analyte content and meteorological parameters. *Spectrochim. Acta B* **2020**, *172*, 105951.
6. Piercey, S.J.; Devine, M.C. Analysis of powdered reference materials and known samples with a benchtop, field portable X-ray fluorescence (pXRF) spectrometer: Evaluation of performance and potential applications for exploration lithogeochemistry. *Geochem. Explor. Environ. Anal.* **2014**, *14*, 139–148. [CrossRef]
7. Cai, F.; Lu, W.; Shi, W.; He, S. A mobile device-based imaging spectrometer for environmental monitoring by attaching a lightweight small module to a commercial digital camera. *Sci. Rep.* **2017**, *7*, 15602. [CrossRef] [PubMed]
8. Wilson, R.; Tapp, H. Mid-infrared spectroscopy for food analysis: Recent new applications and relevant developments in sample presentation methods. *TrAC Trends Anal. Chem.* **1999**, *18*, 85–93. [CrossRef]

9. Howard, J. Electro-optically modulated polarizing Fourier-transform spectrometer for plasma spectroscopy applications. *Appl. Opt.* **2002**, *41*, 197–208. [CrossRef] [PubMed]
10. Bentini, G.G.; Bianconi, M.; Cerutti, A.; Nubile, A.; Cerabolini, P.; Chiarini, M.; Pennestrì, G.; Dinicolantonio, W.; Guzzi, R. A New Miniaturised Optical System for Chemical Species Spectroscopic Detection Based on a Scanning Integrated Mach–Zehnder Microinterferometer on LiNbO3. *Orig. Life Evol. Biosph.* **2007**, *36*, 597–603. [CrossRef] [PubMed]
11. Li, J.; Lu, D.; Qi, Z. Miniature Fourier transform spectrometer based on wavelength dependence of half-wave voltage of a LiNbO3 waveguide interferometer. *Opt. Lett.* **2014**, *39*, 3923–3926. [CrossRef] [PubMed]
12. Rollinson, J.; Hella, M.; Toroghi, S.; Rabiei, P.; Wilke, I. Thin-film lithium niobate modulators for non-invasive sensing of high-frequency electric fields. *J. Opt. Soc. Am. B* **2021**, *38*, 336–341. [CrossRef]
13. Weigel, P.O.; Mookherjea, S. Design of folded hybrid silicon carbide-lithium niobate waveguides for efficient second-harmonic generation. *J. Opt. Soc. Am. B* **2018**, *35*, 593–600. [CrossRef]
14. Yu, Z.J.; Sun, X.K. Gigahertz acousto-optic modulation and frequency shifting on etchless lithium niobate integrated platform. *ACS Photonics* **2021**, *8*, 798–803. [CrossRef]
15. Yu, Z.J.; Tong, Y.Y.; Tsang, H.K.; Sun, X.K. High-dimensional communication on etchless lithium niobate platform with photonic bound states in the continuum. *Nat. Commun.* **2020**, *11*, 2602–2611. [CrossRef] [PubMed]
16. Yu, M.J.; Barton, D.; Cheng, R.; Reimer, C.; Kharel, P.; He, L.Y.; Shao, L.B.; Zhu, D.; Hu, Y.W.; Grant, H.R.; et al. Integrated femtosecond pulse generator on thin-film lithium niobate. *Nature* **2022**, *612*, 252–258. [CrossRef] [PubMed]
17. Pohl, D.; Escalé, M.R.; Madi, M.; Kaufmann, F.; Brotzer, P.; Sergeyev, A.; Guldimann, B.; Giaccari, P.; Alberti, E.; Meier, U.; et al. An integrated broadband spectrometer on thin-film lithium niobate. *Nat. Photonics* **2019**, *14*, 24–29. [CrossRef]
18. Li, J.; Lu, D.; Qi, Z. A modified equation for the spectral resolution of Fourier transform spectrometers. *J. Light. Technol.* **2015**, *33*, 19–24. [CrossRef]
19. Liu, Y.; Li, H.; Liu, J.; Tan, S.; Lu, Q.Y.; Guo, W.H. Low V_π thin-film lithium niobate modulator fabricated with photolithography. *Opt. Express* **2021**, *29*, 6320–6329. [CrossRef] [PubMed]
20. Li, J.; Lu, D.; Qi, Z. A prototype stationary Fourier transform spectrometer for near-infrared absorption spectroscopy. *Appl. Spectrosc.* **2015**, *69*, 1112–1117. [CrossRef] [PubMed]

micromachines

MDPI

Article

High−Accuracy Film−Integrated Optical Sensor for Real−Time Intraocular Pressure Monitoring

Xiaobin Xu [1], Zixuan Liu [1,*], Liqiang Wang [2], Yifei Huang [2] and He Yang [1,*]

1 School of Instrumentation and Optoelectronic Engineering, Beihang University, Beijing 100191, China
2 Department of Ophthalmology, Chinese PLA General Hospital, Medical School of Chinese PLA, Beijing 100853, China
* Correspondence: zy2017110@buaa.edu.cn (Z.L.); yanghe@buaa.edu.cn (H.Y.)

Abstract: Intraocular pressure (IOP) is a key indicator to evaluate the risk and status of glaucoma, which is one of the main causes of irreversible blindness. However, the IOP value is susceptible to circadian changes and is difficult to be measured real−time. In this paper, we designed a thin−film integrated optical IOP sensor based on the interferometry principle, which could read out the IOP value by interference patterns and monitor the value changes real−time at the same time. The theoretical and experimental results indicated that our sensor exhibited a sensitivity of 0.19 μm/mmHg and an average accuracy of 0.84 mmHg over the pressure range of 0–45 mmHg, which is comparable with the other reported optical systems but with the advantage of easier fabrication process and low−cost. Our sensor device implies great potential in the application of human physiological index measurement and other chip−integrated medical sensing instruments.

Keywords: glaucoma; intraocular pressure (IOP); implantable IOP sensor; interferometry

Citation: Xu, X.; Liu, Z.; Wang, L.; Huang, Y.; Yang, H. High−Accuracy Film−Integrated Optical Sensor for Real−Time Intraocular Pressure Monitoring. *Micromachines* **2023**, *14*, 353. https://doi.org/10.3390/mi14020353

Academic Editors: Mehmet Remzi Dokmeci and Romeo Bernini

Received: 18 December 2022
Revised: 20 January 2023
Accepted: 30 January 2023
Published: 31 January 2023

1. Introduction

Glaucoma is one of the main causes of irreversible blindness [1]. It is estimated that about 111.8 million people worldwide will suffer from glaucoma by 2040 [2]. In addition to the examination of the optic nerve head and retinal nerve fiber layer changes, intraocular pressure (IOP) is another important indicator in the diagnosis of glaucoma with higher values compared with the normal data [3,4]. Therefore, it is significant to study precise clinical methods of IOP measurement for the treatment of glaucoma. Until now, Goldman applanation tonometry is one of the widely used methods for IOP measurement by the method of pressing the central corneal to a given area [5]. However, due to the different central corneal thicknesses, this method is susceptible to causing errors [6]. In addition, studies showed that the majority (67.2%) of the peak 24−h IOP values in glaucoma patients occurred at night; only 32.8% of patients experienced peak IOP during daytime [7]. Therefore, the variation of IOP peak values is easy to mislead the doctors' diagnosis. In order to improve glaucoma care, it is important and meaningful to develop reliable IOP measurement and monitoring devices.

For the past few years, various methods based on different principles have been studied to realize real−time and reliable IOP monitoring. According to the method of pressure−sensing, IOP sensors could be classified into wearable and implantable IOP sensors [8]. The wearable IOP sensors were integrated into contact lenses to detect the pressure based on the changes of corneal curvature [8]. However, the accuracy of this measurement method was severely limited for the corneal biomechanical properties and thickness variations in the central corneal [9]. We could compare the sensitivity of implantable IOP sensors and wearable IOP sensors in the same principle, as shown in Table 1. Obviously, the average sensitivity of implantable IOP sensors is higher than that of wearable IOP sensors.

Table 1. Comparison of the sensitivity of inductive couple telemetry−based wearable IOP sensors and implantable IOP sensors.

Wearable IOP Sensors		Implantable IOP Sensors	
Ref.	**Sensitivity**	**Ref.**	**Sensitivity**
[10]	~2.2 MHz $mmHg^{-1}$	[11]	243 kHz $mmHg^{-1}$
[12]	23 kHz $mmHg^{-1}$	[13]	1.14 MHz $mmHg^{-1}$
[14]	8 kHz $mmHg^{-1}$	[15]	119.88 kHz $mmHg^{-1}$
[16]	57 kHz $mmHg^{-1}$	[17]	120 kHz $mmHg^{-1}$
[18]	35.1 kHz $mmHg^{-1}$	[19]	156 kHz $mmHg^{-1}$

In comparison, implantable IOP monitoring devices could obtain more accurate and reliable measurement results for the direct measurement methods as the IOP sensors were implanted into eyeball structures by surgery. When the IOP varied, the flexible material was deformed, directly changing the signal standing for the intensity fluctuation of IOP. Furthermore, various measurement schemes based on electrical, microfluidic, and optical technologies have been studied. For example, Po−Jui Chen et al. reported an IOP sensor based on LCR circuit [20] with the results of 7000 ppm/mmHg sensitivity and 1 mmHg resolution [8]. Amit Todani et al. proposed an active approach based on capacitance variations [21], which gave a measurement accuracy of 0.81 mmHg. In this structure, the IOP variations changed following the value of the capacitor, which would be detected by the external reading device. Recently, some new approaches using microfluidic technology emerged with the advantage of simple structure and low cost. For example, Araci et al. reported an implantable IOP sensor integrated into an intraocular lens for IOP measurement by testing the changes in a gas−liquid interface, giving the results of an average error of ±0.5 mm Hg [22]. Though the microfluidic technology was simple and low−cost, it was still difficult to read out the pressure optically through a hazy cornea.

Recent advances in continuous IOP monitoring were optical methods, such as spectral reflectance [23], optical grating technology [24], and membrane interferometry [25] for the advantages of optical sensors in IOP measurement. Jeong Oen Lee et al. proposed an IOP sensor with an optical cavity [23] giving the higher accuracy of 0.29 mm Hg, but it required gold−nanodot arrays to strengthen the optical resonance, leading to more complicated and higher−cost fabrication. Jayer Fernandes et al. reported a new approach using optical grating technology with a pressure range of 0–50 mmHg, but the measurement accuracy is not high enough. To reduce the fabrication complexity and improve the measurement accuracy, Alex Phan et al. proposed another IOP device based on the interferometry principle showing the advantages of easier fabrication, lower cost but higher accuracy, which opened new avenues for reliable IOP monitoring [25]. However, only few works were reported demonstrating the integrated optical sensor device using this method, which still needs to be further explored.

In this work, we experimentally demonstrated an IOP monitoring system based on the optical interferometry principle by integrating the flexible polydimethylsiloxane (PDMS) membranes. Compared with the work in [25], we designed a novel wedge−cavity structure to produce vertical interference stripes, showing the relationship between the pressure and the moving distance of center fringe visually. What is more, we innovatively selected PDMS as the material of pressure−sensitive membrane. Our results showed that the IOP sensor gave an accuracy as high as 0.84 mmHg over the range of 0–45 mmHg. Simultaneously, our IOP sensor could examine the IOP values by monitoring the interference fringes distance and shape variations on the screen easily and in real−time, which provided high convenience and accuracy in the diagnosis of glaucoma. Thus, our IOP sensor shows great opportunity in the application of human physiological index measurement, optical medical instruments, water pressure−based sensing measurement, and other micro− and nano−medical integrated devices.

2. Principle and Design

2.1. System Operating Principle

As shown in Figure 1, the IOP measurement system consists of an implantable sensor and an optical reading part. The optical reading part includes a camera, a filter (CWL = 632.8 ± 0.2 nm), a beam splitter, and a LED (λ = 635 nm, P_{power} = 4 mW). In order to generate high−quality interference patterns in our sensor, the monochromatic light should have a long temporal coherence length. After comparison, the monochromatic light at 632.8 nm wavelength with 330 μm temporal coherence length was selected. The implantable IOP sensor is a hermetically hollow wedge cavity that includes top glass, spacer, and bottom glass substrate with a drilled hole in the center covered by polymer membrane, as demonstrated in the inset of Figure 1.

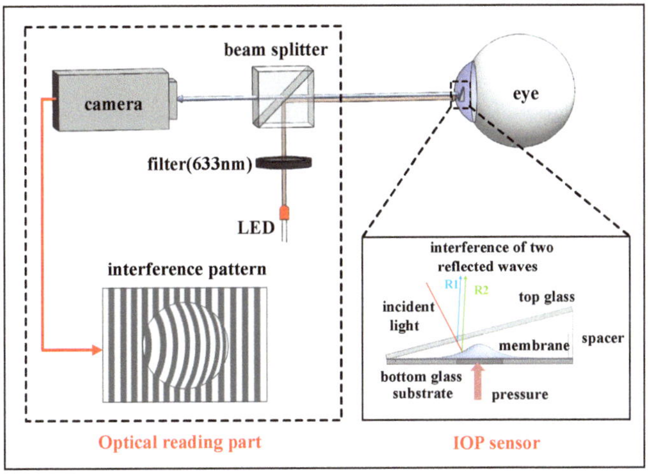

Figure 1. Optical intraocular pressure measurement system with implantable sensor and optical reading part.

The working principle of our IOP sensor is described as the following: the monochromatic light at the wavelength of 633 nm was divided into two beams. One beam was reflected at the undersurface of the top glass and the other was reflected at the surface of the membrane. Then, the two beams interfered at the undersurface of the top glass, producing fringes depending on the gap distance and the phase difference. Figure 2 shows the detailed interference working scheme. The function between phase difference $\Delta\varphi(x, y)$ and gap distance $d(x, y)$ between the top glass and membrane could be expressed as [25]:

$$\Delta\varphi(x,y) = \frac{4\pi}{\lambda}d(x,y) + \pi \tag{1}$$

where λ is the wavelength of the incident light, π is the phase shift occurring at the top surface of the diaphragm, $d(x, y) = h - \Delta d$ means the gap distance, h is the spacer thickness and Δd is the deflection of the diaphragm. The polymer membrane is flexible, serving as a pressure−sensitive element by detecting the value of Δd. With the increase of the pressure, the deflection of the membrane increases, corresponding to the increase of Δd shown in Figure 2.

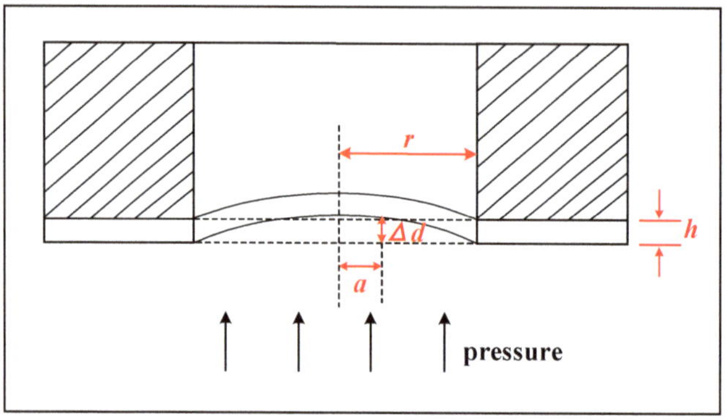

Figure 2. Schematic of membrane deformation under pressure.

According to the principle of elasticity, if the maximum deflection under pressure is not more than 30% of the membrane thickness and, at the same time, the membrane thickness is not larger than 20% of the diameter, the deflection Δd of a diaphragm with fixed edges under pressure (P) at any radial distance (a) could be expressed as [26]:

$$\Delta d = \frac{3}{16}\frac{(1-\mu^2)(r^2-a^2)^2}{Eh^3}P \tag{2}$$

where Δd, μ, r, a, E, h, and P are the deflection, Poisson's ratio of the diaphragm, radius of the diaphragm, radial distance, modulus of elasticity, thickness, and pressure, respectively. Based on Equations (1) and (2), the precise pressure applied to the membrane could be calculated.

2.2. Design of the IOP Sensor

The detailed structure of IOP sensor is shown in Figure 3a. In order to guarantee the high sensitivity, accuracy, and safety of IOP sensor, the membrane material should be flexible and biocompatible. PDMS is a kind of biocompatible material that is widely used in biomedical devices and its elasticity modulus of PDMS is 750 K, which is much lower than other flexible materials. What is more, PDMS membrane is also low−cost and easy−fabricated. Considering the flexibility and biocompatibility of the membrane, PDMS was selected as the membrane material to detect the pressure. Following Equation (2), as the diameter of the hole was 1 mm, the membrane thickness should not be more than 200 μm. To select the most appropriate membrane, we compared the deflection results from two membranes with different thicknesses of 100 μm and 200 μm under the same pressure, respectively. Figure 3b shows the simulated curve of the membrane deflection that the defection is more sensitive with the PDMS thickness at 100 μm compared with that of 200 μm−thick PDMS membrane. According to Equation (2), the thinner membrane had a larger deflection under the same pressure, which was consistent with our simulation results. In our experiments, by setting the pressure at 22.5 mmHg, the interference patterns resulting from the membrane thickness at 100 μm and 200 μm also showed huge differences, as shown in Figure 3c,d separately. Compared with the two figures, 100 μm−thick PDMS membrane had sharper deflection and the stripes became circles. While, the 200 μm−thick PDMS membrane had a measurable moving distance of stripes. For the convenience of our experiment, the higher deflection sensitivity from the membrane was not easy for the algorithm to distinguish the transverse displacement and results error. Thus, the PDMS membrane with the thickness of 200 μm was selected in our experiment to ensure the accuracy of the results. Considering the above all, our sensor was finally constructed with the PDMS membrane at 200 μm thickness, the spacer height at ~85 μm, the laser−drilled hole diameter at 1 mm, and the size of the holding substrate of 5 mm × 5 mm.

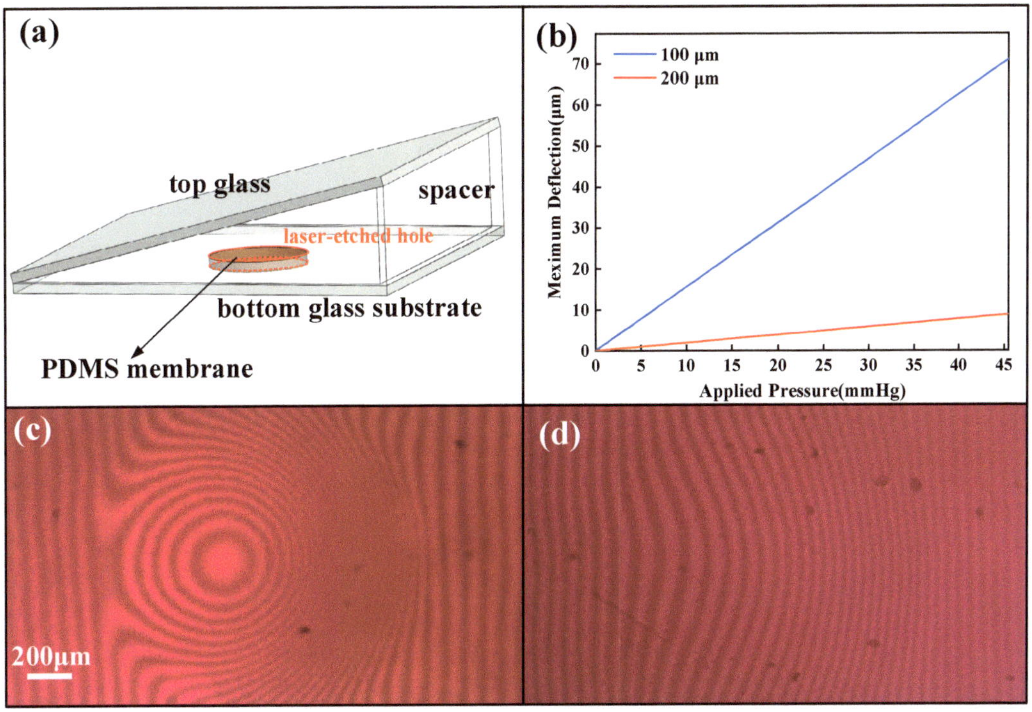

Figure 3. (**a**) The structure of IOP sensor. (**b**) Simulation of 100 μm− and 200 μm−thick membranes deflection under pressure. (**c**) Interference pattern of 100 μm−thick membrane under the pressure of 22.5 mmHg. (**d**) Interference pattern of 200 μm−thick membrane under the same pressure.

2.3. Algorithm

To analyze the interference patterns, we developed a unique image−processing algorithm to reconstruct the 3D model of membrane deflection and calculate the pressure according to Equations (1) and (2). To test the performance of the algorithm, the interference pattern generated under the pressure of 3 Kpa (22.5 mmHg) was selected for the algorithm analysis, as shown in Figure 4a. Firstly, the interference pattern was transformed into the frequency domain by Fourier Transformation. Then, we could obtain the phase diagram from the filtered figures (Figure 4b). Using Equations (1) and (2), the 3D model of the membrane defection and the pressure distribution diagram on the membrane was constructed, as illustrated in Figure 4c,d separately. Lastly, the pressure applied could be determined with a high degree polynomial fitting. Figure 4d showed that the pressure was not uniformly distributed across the membrane and the highest pressure with the value of 2.9 Kpa was located at the center of the membrane. Since the edge of the membrane was fixed, the pressure on the edge of the membrane was nearly 1 Kpa (~7.5 mmHg) less than that at the center of the membrane.

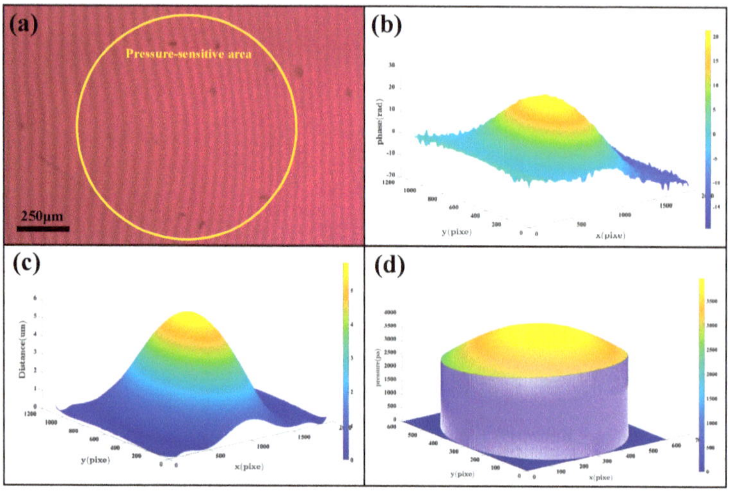

Figure 4. (**a**) Interference pattern at 22.5 mmHg. (**b**) Phase diagram of Interference pattern. (**c**) 3D model of the membrane. (**d**) The distribution diagram of pressure on the membrane.

3. Results and Discussion

To test the performance of our designed sensor, we built the experimental sensing setup shown in Figure 5. The IOP sensor was fixed in a pressure chamber which was customized to mimic the environment inside the human eye. An infusion bag was connected to the pressure chamber to alter the inner pressure by adjusting the height. A standard pressure sensor was used to record the pressure changes. The camera was mounted to a microscope to capture the interference patterns of the IOP sensor. When varied pressure was applied to the sensor, the interference patterns were captured and processed by MATLAB image analysis algorithms.

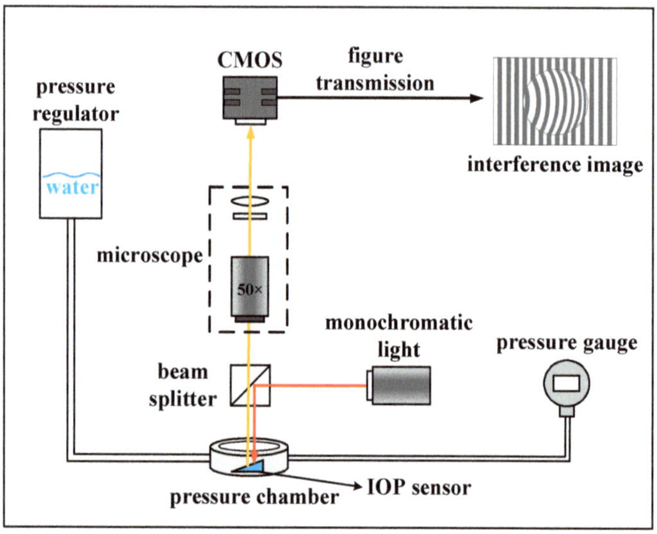

Figure 5. The setup of the experiments of sensor performance testing.

The accuracy, range, and linearity of the IOP sensor were measured inside the controlled pressure chamber filled with water. As shown in Figure 6, the fringes moved far away from the center area when the pressure inside the chamber increased from 0 to 45 mmHg at the steps of 3.75 mmHg. Figure 6a shows the interference patterns at the pressure load of 0 mmHg. When no pressure was loaded inside the pressure chamber, the interference fringe stayed vertically. As the hydrostatic pressure increased, fringes deformed sharply towards the right, which was demonstrated in Figure 6b,c.

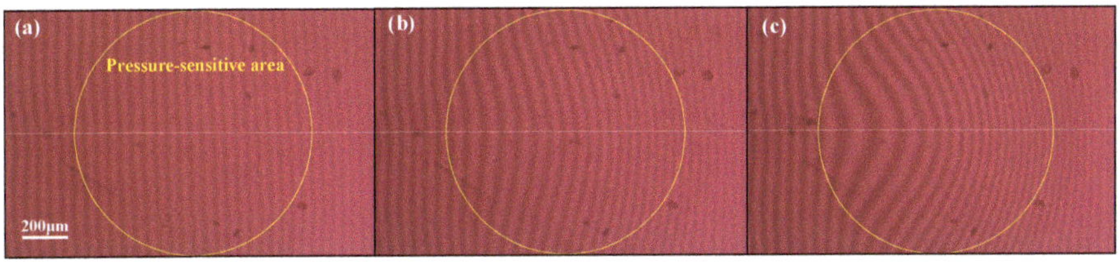

Figure 6. Images of interference patterns at pressure load of (**a**) 0 mmHg, (**b**) 22.5 mmHg, (**c**) 45 mmHg.

We also investigated the relationship between the moving distance of the center fringe of the membrane and the pressure applied. In the experiment, we found that when the pressure was applied at the step of 3.75 mmHg, the central fringe moved at the corresponding step of ~19.27 μm toward the right. Fitting the relationship between the moving distance and the applied pressure, we could obtain a linear fitting curve with a specific value of 0.99604, as shown in Figure 7a.

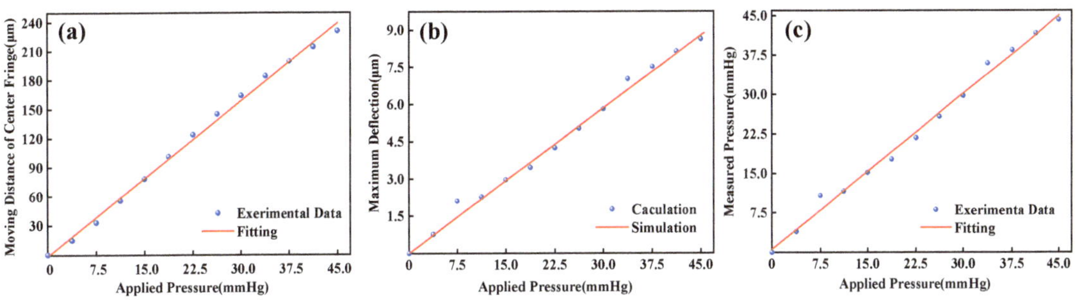

Figure 7. (**a**) Moving distance of center fringe as a function of applied pressure. (**b**) Maximum deflection of the sensor diaphragm and the ideal curve. (**c**) Measured pressure versus applied pressure using the IOP sensor.

At the same time, we adopted our image processing algorithm to analyze the interference patterns and calculated the maximum deflection of the PDMS membrane, which was plotted by dots in Figure 7b. The simulated curve of the 200 μm−thick membrane deflection was also plotted by the red line in Figure 7b. The simulated line was fully following the experimental results with the average error of ±0.16 μm considering the dimensional error of laser−etched hole radius and membrane thickness. Thus, the sensitivity of our sensor of 0.19 μm/mmHg could be determined. Further work will be performed to improve the sensor sensitivity by modifying the size of the sensor, such as reducing the thickness of the membrane and shrinking the radius of the laser−etched hole.

Figure 7c shows the relationship between the pressure value analyzed by the image processing algorithm and the applied pressure from the infusion bag, where the red line is fitted by a third−order polynomial curve on the experimental data. The experimental data was plotted by dots which had a linear relationship with calculated deflection data shown in Figure 7b. On the other hand, it showed that the IOP sensor had highly linear responses, and the average error from the standard pressure−gauge reading was 0.84 mmHg, which further verify the high quality or reliability of our constructed sensor. A performance comparison with previous reports is shown in Table 2.

Table 2. Comparison of our implantable IOP sensor with previous reports.

Ref.	Working Principle	Sensitivity	Accuracy	Merit and Demerit
[20]	Inductive	7000 ppm/mmHg	Not mentioned	Merit: high sensitivity Demerit: non−portability of the external reading device
[21]	Capacitive	Not mentioned	0.81 mmHg	Merit: long−term implantation of IOP detector in animal experimentations will not produce inflammation. Demerit: the size is too large to be implanted in humans' eye
[22]	Microfluidic	106 μm/mm Hg	0.5 mmHg	Merit: simple and low−cost Demerit: it is difficult to read out the pressure optically through a hazy cornea
[23]	Spectral reflectance	Not mentioned	0.29 mmHg	Merit: high accuracy Demerit: more complicated and higher−cost fabrication
[25]	interferometry	31 nm/mmHg	0.3 mmHg	Merit: high accuracy and sensitivity Demerit: more complicated fabrication
This work	interferometry	0.19 μm/mmHg	0.84 mmHg	Merit: easy fabrication process and low cost

In further studies, we would make more efforts to improve the portability of the IOP monitoring system and increase its accuracy. In this study, the factors affecting the accuracy of IOP sensors include manufacturing error and image definition. To increase the accuracy of IOP sensor, we will use MEMS technology to reduce manufacturing errors during the fabrication process of IOP sensor. In experiments, the accuracy of the image processing algorithm is susceptible to image definition. We could improve image definition by using a monochromatic light source with longer temporal coherence length and higher pixel cameras. To allow patients to realize IOP self−monitoring anywhere and at any time, the optical reading part would be integrated into an adapter that could be installed on smartphones. In the future, the IOP sensor could be integrated on smaller substrate and bonded to intraocular lenses to guarantee safety and biocompatibility.

4. Conclusions

In this work, we constructed an optical implantable sensor system based on the interferometry principle for continuous and real−time IOP monitoring. The sealed wedge cavity and pressure−sensitive membrane in the structure were designed to improve the detection sensitivity of IOP. As the variation of the IOP could be detected in real−time by monitoring the phase change from the interference patterns, no external power source was required. Our designed IOP monitoring system shows a sensitivity as high as 0.19 μm/mmHg and an average accuracy of 0.84 mmHg. Thus, it shows great reliability in the diagnosis and treatment of glaucoma patients as an implantable medical device, providing a warning of danger in the range of high IOP value. Our work also implied that the optical IOP monitoring systems would make more contributions in relieving the glaucoma patient's indisposition and providing more sensitive and accurate IOP information to advance the development of the optical sensors.

Author Contributions: Conceptualization, X.X.; methodology, Z.L. and H.Y.; software, Z.L.; formal analysis, Z.L. and H.Y.; investigation, Z.L. and H.Y.; funding acquisition, H.Y.; project administration, X.X. and H.Y.; supervision, L.W. and Y.H.; writing—original draft, Z.L. and H.Y.; writing—review and editing, H.Y. All authors have read and agreed to the published version of the manuscript.

Funding: This work was supported by the Fundamental Research Funds for the Central Universities (KG16165601).

Data Availability Statement: Data presented in this study are available on request from the corresponding author.

Conflicts of Interest: The authors declare no conflict of interest.

References

1. Flaxman, S.R.; Bourne, R.R.A.; Resnikoff, S.; Ackland, P.; Braithwaite, T.; Cicinelli, M.V.; Das, A.; Jonas, J.B.; Keeffe, J.; Kempen, J.H.; et al. Global causes of blindness and distance vision impairment 1990–2020: A systematic review and meta−analysis. *Lancet Glob. Health* **2017**, *5*, e1221–e1234. [CrossRef] [PubMed]
2. Tham, Y.C.; Li, X.; Wong, T.Y.; Quigley, H.A.; Aung, T.; Cheng, C.Y. Global prevalence of glaucoma and projections of glaucoma burden through 2040: A systematic review and meta−analysis. *Ophthalmology* **2014**, *121*, 2081–2090. [CrossRef] [PubMed]
3. De Moraes, C.G.V.; Juthani, V.J.; Liebmann, J.M.; Teng, C.C.; Tello, C.; Susanna, R., Jr.; Ritch, R. Risk factors for visual field progression in treated glaucoma. *Arch. Ophthalmol.* **2011**, *129*, 562–568. [CrossRef]
4. Bengtsson, B.; Leske, M.C.; Hyman, L.; Heijl, A.; Early Manifest Glaucoma Trial, G. Fluctuation of intraocular pressure and glaucoma progression in the early manifest glaucoma trial. *Ophthalmology* **2007**, *114*, 205–209. [CrossRef] [PubMed]
5. Wang, Y.W.L.L.H. Application progress of tonometer in clinic. *Chin. J. Mod. Nurs.* **2017**, *23*, 292–296.
6. Bhan, A.; Browning, A.C.; Shah, S.; Hamilton, R.; Dave, D.; Dua, H.S. Effect of corneal thickness on intraocular pressure measurements with the pneumotonometer, Goldmann applanation tonometer, and Tono−Pen. *Investig. Ophthalmol. Vis. Sci.* **2002**, *43*, 1389–1392.
7. Mosaed, S.; Liu, J.H.K.; Weinreb, R.N. Correlation between office and peak nocturnal intraocular pressures in healthy subjects and glaucoma patients. *Am. J. Ophthalmol.* **2005**, *139*, 320–324. [CrossRef]
8. Yang, C.; Huang, X.; Li, X.; Yang, C.; Zhang, T.; Wu, Q.; Liu, D.; Lin, H.; Chen, W.; Hu, N.; et al. Wearable and Implantable Intraocular Pressure Biosensors: Recent Progress and Future Prospects. *Adv. Sci.* **2021**, *8*, 2002971. [CrossRef]
9. Kotecha, A.; Elsheikh, A.; Roberts, C.R.; Zhu, H.; Garway−Heath, D.F. Corneal thickness− and age−related biomechanical properties of the cornea measured with the ocular response analyzer. *Investig. Ophthalmol. Vis. Sci.* **2006**, *47*, 5337–5347. [CrossRef]
10. Kim, J.; Kim, M.; Lee, M.S.; Kim, K.; Ji, S.; Kim, Y.T.; Park, J.; Na, K.; Bae, K.H.; Kyun Kim, H.J.N.C. Wearable smart sensor systems integrated on soft contact lenses for wireless ocular diagnostics. *Nat. Commun.* **2017**, *8*, 14997. [CrossRef]
11. Chen, P.J.; Saati, S.; Varma, R.; Humayun, M.S.; Tai, Y.C.J.J.o.M.S. Wireless Intraocular Pressure Sensing Using Microfabricated Minimally Invasive Flexible−Coiled LC Sensor Implant. *J. Microelectromech. Syst.* **2010**, *19*, 721–734. [CrossRef]
12. Chen, G.Z.; Chan, I.S.; Lam, D.J.S.; Physical, A.A. Capacitive contact lens sensor for continuous non−invasive intraocular pressure monitoring. *Sens. Actuators A Phys.* **2013**, *203*, 112–118. [CrossRef]
13. Chen, P.J.; Rodger, D.C.; Saati, S.; Humayun, M.S.; Tai, Y.C. Microfabricated Implantable Parylene−Based Wireless Passive Intraocular Pressure Sensors. *J. Microelectromech. Syst.* **2008**, *17*, 1342–1351. [CrossRef]
14. Chen, G.Z.; Chan, I.S.; Leung, L.K.; Lam, D.C. Soft wearable contact lens sensor for continuous intraocular pressure monitoring. *Med. Eng. Phys.* **2014**, *36*, 1134–1139. [CrossRef] [PubMed]
15. Katuri, K. *Design and Optimization of Passive Wireless Intraocular Pressure Sensor*; North Carolina State University: Raleigh, NC, USA, 2012.
16. Kouhani, M.H.M.; Weber, A.; Li, W. Wireless intraocular pressure sensor using stretchable variable inductor. In Proceedings of the 2017 IEEE 30th International Conference on Micro Electro Mechanical Systems (MEMS), Las Vegas, NV, USA, 22–26 January 2017; pp. 557–560.
17. Akar, O.; Akin, T.; Najafi, K. A wireless batch sealed absolute capacitive pressure sensor. *Sens. Actuators A Phys.* **2001**, *95*, 29–38. [CrossRef]
18. Kouhani, M.H.M.; Wu, J.; Tavakoli, A.; Weber, A.J.; Li, W. Wireless, passive strain sensor in a doughnut−shaped contact lens for continuous non−invasive self−monitoring of intraocular pressure. *Lab Chip* **2020**, *20*, 332–342. [CrossRef]
19. Li, B.C.W. Parylene−based fold−and−bond wireless pressure sensor. In Proceedings of the 8th Annual IEEE International Conference on Nano/Micro Engineered and Molecular Systems, Suzhou, China, 7–10 April 2013.
20. Chen, P.J.; Rodger, D.C.; Saati, S.; Humayun, M.S.; Tai, Y.C. Implantable parylene−based wireless intraocular pressure sensor. In Proceedings of the IEEE 2008 IEEE 21st International Conference on Micro Electro Mechanical Systems, Tucson, AZ, USA, 13–17 January 2008.
21. Todani, A.; Behlau, I.; Fava, M.A.; Cade, F.; Cherfan, D.G.; Zakka, F.R.; Jakobiec, F.A.; Gao, Y.; Dohlman, C.H.; Melki, S.A. Intraocular pressure measurement by radio wave telemetry. *Investig. Ophthalmol. Vis. Sci.* **2011**, *52*, 9573–9580. [CrossRef]

22. Araci, I.E.; Su, B.; Quake, S.R.; Mandel, Y. An implantable microfluidic device for self−monitoring of intraocular pressure. *Nat. Med.* **2014**, *20*, 1074–1078. [CrossRef]
23. Lee, J.O.; Park, H.; Du, J.; Balakrishna, A.; Chen, O.; Sretavan, D.; Choo, H. A microscale optical implant for continuous in vivo monitoring of intraocular pressure. *Microsyst. Nanoeng.* **2017**, *3*, 17057. [CrossRef]
24. Fernandes, J.; Kwon, Y.H.; Kim, J.J.; Liu, H.; Jiang, H. High Contrast Grating Based Strain Sensor for Intraocular Applications. *J. Microelectromech. Syst.* **2018**, *27*, 599–601. [CrossRef]
25. Phan, A.; Truong, P.; Trumpp, J.; Talke, F.E. Design of an Optical Pressure Measurement System for Intraocular Pressure Monitoring. *IEEE Sens. J.* **2018**, *18*, 61–68. [CrossRef]
26. Giovanni, M. *Flat and Corrugated Diaphragm Design Handbook*; Marcel Dekker: New York, NJ, USA, 1982; pp. 130–154.

micromachines

MDPI

Article

Tailoring the Topological Charge of a Superposition of Identical Parallel Laguerre–Gaussian Beams

Victor V. Kotlyar [1,2], Alexey A. Kovalev [1,2], Elena S. Kozlova [1,2,*] and Alexandra A. Savelyeva [1,2]

1 Image Processing Systems Institute of the RAS, Branch of FSRC "Crystallography & Photonics" of the RAS, 151 Molodogvardeyskaya St., Samara 443001, Russia

2 Technical Cybernetics Department, Samara National Research University, Moskovskoye Shosse 34, Samara 443086, Russia

* Correspondence: kozlova.elena.s@gmail.com

Abstract: In optical computing machines, data can be transmitted by optical vortices, and the information can be encoded by their topological charges. Thus, some optical mechanisms are needed for performing simple arithmetic operations with the topological charges. Here, a superposition of several parallel identical Laguerre–Gaussian beams with single rings is studied. It is analytically and numerically shown that if the weighting coefficients of the superposition are real, then the total topological charge of the superposition is equal to the topological charge of each component in the initial plane and in the far field. We prove that the total topological charge of the superposition can be changed by the phase delay between the beams. In the numerical simulation, we demonstrate the incrementing and decrementing the topological charge. Potential application areas are in optical computing machines and optical data transmission.

Keywords: optical computing machine; topological charge; Laguerre–Gaussian beam; superposition of parallel beams

Citation: Kotlyar, V.V.; Kovalev, A.A.; Kozlova, E.S.; Savelyeva, A.A. Tailoring the Topological Charge of a Superposition of Identical Parallel Laguerre–Gaussian Beams. *Micromachines* **2022**, *13*, 2227. https://doi.org/10.3390/mi13122227

Academic Editor: He Yang, Xinyang Su and Yizhong Huang

Received: 22 November 2022
Accepted: 12 December 2022
Published: 15 December 2022

1. Introduction

Optical vortices constitute a great family of light fields, which is actively studied over 30 years [1]. The studies include various aspects, including direct generation in lasers [2], interaction with matter [3], and propagation and focusing [4]. Light fields with optical vortices are usually characterized by the orbital angular momentum (OAM) [5] and the topological charge (*TC*) [6]. In a number of studies, the topological charge (*TC*) of a superposition of parallel optical vortices (OVs), and in particular, parallel Laguerre–Gaussian (LG) beams, was studied. This problem has been of interest since 2000, when the number and location of OVs in a superposition of two parallel Gaussian beams with embedded OVs were studied in [7]. In [7], a transcendental equation was obtained analytically for determining the position of OVs. However, it is applicable only for cases when the vortices in both beams are of the first order. It was also shown that when two beams are separated by a certain critical distance, negative-order vortices appear along with positive-order vortices. Later, in [8], using the analysis of forks in the interference pattern of two parallel LG beams, it was shown that varying the distance between the beams changes the arrangement of screw dislocations in the superposition. In addition, the same authors [9] showed that the number of vortices in the superposition of two parallel LG beams can change during propagation in space, although the total *TC* remains unchanged. In [10], the superposition of two off-axis optical vortices, but with orthogonal polarization, was studied. Instead of the dynamics of phase singularities, this paper studied the dynamics of polarization singularities and the position of C-points as a function of the distance between vortices, their *TC*, and the phase delay between them. In [11], the interference of two off-axis Gaussian beams with different curvatures of the wave front was also considered. The conditions for vortex dipoles (two OVs of opposite orders) formation were obtained. In [12], the coherent and the incoherent superposition of two parallel partially coherent OVs were studied. It was shown that the type of superposition, the distance between the beams, the propagation distance, and the coherence parameter affect the number and location of coherence vortices. The number of vortices, however, is determined only numerically. Study [13] considered OVs that are formed in a superposition of off-axis vortices during a nonlinear process of three-wave mixing. The number of vortices and their *TC*s were established in some particular cases. In a recent article [14], the interaction of parallel Bessel–Gaussian beams was considered. The dependence of formation, annihilation, and splitting of OVs on the displacement of the beams from the optical axis, on their *TC*, and on the phase difference between them was studied. It was shown that the total *TC* of such a composite field is not necessarily equal to the sum of the *TC*s of the composite beams. In [15], it was shown how to calculate the *TC* of a superposition of only two parallel LG beams. In particular, in [15], it was analytically shown that if two beams have the same *TC*, for example, m, then the superposition of such beams with arbitrary distance between them will also have a *TC* equal to m.

In this paper, we generalize the results on a superposition of a finite number of parallel one-ring identical LG beams. Moreover, we will show that if the weight coefficients of such a superposition are real (that is, all LG beams have the same phase, but can have different amplitudes), then the superposition *TC* is equal to the *TC* of each beam, that is, m. It has already been proven earlier that the power-normalized OAM of such a superposition is also equal to the OAM of one LG beam in the superposition, that is, also m [16].

2. *TC* of the Superposition of Identical Parallel LG Beams in the Initial Plane

We consider a superposition of N parallel identical one-ring LG beams in the initial plane:

$$E_m(x,y) = \sum_{n=1}^{N} c_n \left(re^{i\varphi} - r_n e^{i\varphi_n}\right)^m \exp\left(-r^2 - r_n^2 + 2rr_n\cos(\varphi - \varphi_n)\right), \quad (1)$$

where (x, y) and (r, φ) are the Cartesian and the polar coordinates ($x = r\cos\varphi$, $y = r\sin\varphi$), respectively.

The *TC* of each beam in (1) is equal to m, the waist radius is included in the radial variable r: r/w. We assume that the weight coefficients c_n in (1) are real numbers. Polar coordinates of beam centers are (r_n, φ_n). The *TC* of superposition (1) will be calculated using the Berry formula [6]:

$$TC = \frac{1}{2\pi}\lim_{r\to\infty}\mathrm{Im}\int_0^{2\pi} d\varphi \frac{\partial E(r,\varphi)/\partial\varphi}{E(r,\varphi)}. \quad (2)$$

where lim means the limit at $r \to \infty$, and Im is the imaginary part of a complex number.

Substituting (1) into (2), we obtain:

$$\begin{aligned} TC &= \frac{1}{2\pi}\lim_{r\to\infty}\mathrm{Im}\int_0^{2\pi}\left[\sum_{n=1}^{N} c_n\left(re^{i\varphi} - r_n e^{i\varphi_n}\right)^m e^{-r_n^2+2rr_n\cos(\varphi-\varphi_n)}\right. \\ &\left.\times\left(\frac{imre^{i\varphi}}{re^{i\varphi}-r_ne^{i\varphi_n}} - 2rr_n\sin(\varphi-\varphi_n)\right)\right]\left[\sum_{n=1}^{N} c_n\left(re^{i\varphi} - r_n e^{i\varphi_n}\right)^m e^{-r_n^2+2rr_n\cos(\varphi-\varphi_n)}\right]^{-1} d\varphi \\ &= m - \frac{2r}{2\pi}\mathrm{Im}\int_0^{2\pi}\frac{\sum_{n=1}^{N} c_n r_n\sin(\varphi-\varphi_n)\left(re^{i\varphi}\right)^m e^{-r_n^2+2rr_n\cos(\varphi-\varphi_n)}}{\sum_{n=1}^{N} c_n\left(re^{i\varphi}\right)^m e^{-r_n^2+2rr_n\cos(\varphi-\varphi_n)}}d\varphi \\ &= m - \frac{2r}{2\pi}\mathrm{Im}\int_0^{2\pi}\frac{\sum_{n=1}^{N} c_n r_n\sin(\varphi-\varphi_n)e^{-r_n^2+2rr_n\cos(\varphi-\varphi_n)}}{\sum_{n=1}^{N} c_n e^{-r_n^2+2rr_n\cos(\varphi-\varphi_n)}}d\varphi = m. \end{aligned} \quad (3)$$

In (3), the imaginary part of the last integral is equal to zero, since it is real. It follows from (3) that the *TC* of the superposition of parallel identical single-ring LG beams with numbers $(0, m)$ is equal to m in the initial plane.

3. *TC* of the Superposition of Parallel Identical LG Beams with Different Weight Coefficients in the Far Field

We propose that there is a superposition of N identical one-ring LG beams displaced from the optical axis in the initial plane. Then, the complex amplitude in the initial plane is equal to:

$$E_m(x,y) = \sum_{n=1}^{N} c_n \left\{ \frac{\sqrt{2}}{w_0} [(x-a_n) + i(y-b_n)] \right\}^m \exp\left[-\frac{(x-a_n)^2 + (y-b_n)^2}{w_0^2} \right], \quad (4)$$

where (x, y) are Cartesian coordinates in the initial plane, w_0 is Gaussian beam waist, (a_n, b_n) are coordinates of beam centers, and c_n are superposition coefficients. In contrast to (1), in (4), the superposition beams are taken with complex weight coefficients c_m and the Gaussian beam waist radius is explicitly distinguished.

In the far zone, the displacement of each LG beam turns out to be a slope of the wavefront, that is, in the far zone, the LG beams become axial, but with slopes. Therefore, the complex amplitude of the entire superposition in the far zone has the form:

$$E_m(r,\varphi,z >> z_0) = \exp\left(-\frac{r^2}{w_0^2}\right) \left(\frac{\sqrt{2}}{w_0} r e^{i\varphi}\right)^m \sum_{n=1}^{N} c_n \exp(ika_n r\cos\varphi + ikb_n r\sin\varphi). \quad (5)$$

where (r, φ) are far-field polar coordinates. According to the formula of M.V. Berry [6], *TC* is equal to

$$\begin{aligned} TC = & \frac{1}{2\pi} \lim_{r\to\infty} \mathrm{Im} \int_0^{2\pi} \frac{\sum_{n=1}^{N} c_n \frac{\partial}{\partial\varphi} [\exp(ika_n r\cos\varphi + ikb_n r\sin\varphi)]}{\sum_{n=1}^{N} c_n \exp(ika_n r\cos\varphi + ikb_n r\sin\varphi)} d\varphi \\ & + \frac{1}{2\pi} \lim_{r\to\infty} \mathrm{Im} \int_0^{2\pi} \frac{\frac{\partial}{\partial\varphi}\left[\left(re^{i\varphi}\right)^m\right]}{\left(re^{i\varphi}\right)^m} d\varphi. \end{aligned} \quad (6)$$

Reducing the common factors in the numerator and denominator, we obtain further:

$$\begin{aligned} TC = & m + \frac{1}{2\pi} \lim_{r\to\infty} \mathrm{Im} \int_0^{2\pi} \sum_{n=1}^{N} c_n \frac{\partial}{\partial\varphi} [\exp(ika_n r\cos\varphi + ikb_n r\sin\varphi)] \\ & \times \left[\sum_{n=1}^{N} c_n \exp(ika_n r\cos\varphi + ikb_n r\sin\varphi) \right]^{-1} d\varphi. \end{aligned} \quad (7)$$

The second term in (7) is the *TC* of some additional field of the form (without a Gaussian envelope).

$$E_{\text{add}}(x,y) = \sum_{n=1}^{N} c_n \exp(ika_n x + ikb_n y). \quad (8)$$

Since the numbers a_n, b_n, c_n are arbitrary, Formula (8) can describe a wide class of light fields. In particular, the additional field can be a vortex and therefore give an additional *TC*. For example, if we take $N = 4$, $c_1 = -i$, $c_2 = 1$, $c_3 = i$, $c_4 = -1$, $a_1 = -a_3 = a$, $a_2 = a_4 = 0$, $b_1 = b_3 = 0$, $b_2 = -b_4 = a$, we then obtain an additional field in the form

$$E_{\text{add}}(x,y) = 2\sin(kax) + 2i\sin(kay), \quad (9)$$

which is near the center, approximately

$$E_{\text{add}}(x \approx 0, y \approx 0) = 2ka(x+iy), \quad (10)$$

that is, it contains a vortex of the first order. If all coefficients c_n are real in superposition (4), then we can show that

$$E^*_{\text{add}}(x,y) = E_{\text{add}}(-x,-y). \tag{11}$$

From (11), it follows that if there is a zero amplitude at some point (x_{null}, y_{null}), then the amplitude is also zero at the point ($-x_{\text{null}}$, $-y_{\text{null}}$), and near zero, the amplitude is complex conjugate. That is, there is a "conjugate" vortex for each vortex in the field (8). The *TC*s of these "conjugate" vortices compensate each other, and therefore the *TC* of the field (8) with real coefficients c_m is equal to zero. Expression (11) is proved simply:

$$E^*(u,v) = \sum_{n=1}^{N} c_n \exp(-ixu - iyv) = \sum_{n=1}^{N} c_n \exp(ix(-u) + iy(-v)) = E(-u,-v). \tag{12}$$

There is also a physical reason why the field (8) cannot have other *TC* than zero. Indeed, the amplitude of the form (8) is formed in the Fourier plane (in the focus of a spherical lens) by a light field, which in the initial plane consists of *N* point sources with different amplitudes, but the same phase. A light field whose amplitude is a real function can only create OVs in pairs with $+p$ and $-p$ *TC*s. This also follows from the fact that an OV that has passed through the amplitude mask does not change its *TC* [17,18].

If the *TC* of the superposition (1) in the initial plane and in the far zone is the same and equals *m*, then in any other plane it is equal to *m*, if the coefficients c_n are real.

4. Modeling

For example, Figure 1 shows the intensities and phases of three superpositions of off-axis single-ring LG beams with the following parameters: the wavelength is $\lambda = 532$ nm; the waist radius of all beams is $w_0 = 0.5$ mm; the number of LG beams is $N = 4$; the *TC* of each of them is $m = 3$; the centers of these beams (in Cartesian coordinates) are $(a_1, b_1) = (r_0, 0)$, $(a_2, b_2) = (0, r_0)$, $(a_3, b_3) = (-r_0, 0)$, $(a_4, b_4) = (0, -r_0)$, where $r_0 = 3w_0$; the superposition coefficients for LG beams are $c_1 = c_2 = c_3 = c_4 = 1$ (Figure 1a,d), $c_1 = -i$, $c_2 = 1$, $c_3 = i$, $c_4 = -1$ (Figure 1b,e), and $c_1 = -i$, $c_2 = -1$, $c_3 = i$, $c_4 = 1$ (Figure 1c,f,g); the computational domain is restricted by $|x|, |y| \leq R$, where $R = 5$ mm (Figure 1a–f) and $R = 10$ mm (Figure 1g); the radius of the circle for calculating the *TC* is $R_1 = 4.5$ mm (Figure 1d–f) and $R_1 = 9.5$ mm (Figure 1g); the grid size in each direction is $N = 1024$. When all superposition coefficients are the same (Figure 1a,d), the phase distribution becomes asymmetric, but the total *TC* of four LG beams turns out to be the same as for each of them: $TC = 3.0042 \approx 3$. If we choose the coefficients, as in Figure 1b,e, then despite their dissimilarity, the phase distribution is symmetrical around the center and the total *TC* changes and turns out to be equal to $TC = 4.0003 \approx 4$. For the other coefficients (Figure 1c,f,g), the total *TC* also changes and is equal to $TC = 5.9234 \approx 6$ (Figure 1f). However, according to Equation (2), the *TC* is computed over an infinite-radius circle, and thus accounts all the vortices in the light field (or, equivalently, the 2π phase jumps in the beam periphery). Computation indicates that the figure size $2R = 10$ mm is sufficient to account for all phase jumps in Figure 1d,e, but insufficient for Figure 1f. Figure 1g illustrates the phase distribution in a wider area ($2R = 20$ mm). In the periphery, there are four phase jumps by 2π and two phase jumps by -2π (denoted by '+' and by '–', respectively). Thus, the *TC* should be equal to $(4 \times 2\pi - 2 \times 2\pi)/2\pi = 2$. Numerical computation confirms it and yields the value $TC = 1.9975 \approx 2$, which, in contrast with Figure 1b,e, is less than the *TC* of the constituent LG beams. Thus, the superposition coefficients $c_1 = -i$, $c_2 = 1$, $c_3 = i$, $c_4 = -1$ (Figure 1b,e) increment the *TC* of the whole superposition, whereas the coefficients $c_1 = -i$, $c_2 = -1$, $c_3 = i$, $c_4 = 1$ (Figure 1c,f,g) decrement this *TC*.

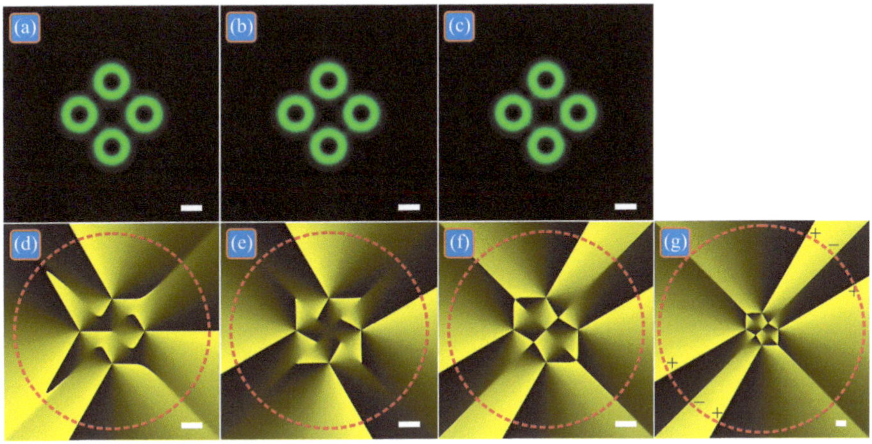

Figure 1. Intensities and phases of three superpositions of off-axis single-ring LG beams, in which the total *TC* is the same as that of each beam (**a**,**d**), or differs from the *TC* of each beam by +1 (**b**,**e**) and by −1 (**c**,**f**,**g**) due to the presence of complex weight coefficients in the superposition. Black and green intensity means zero and maximal values, respectively. Black and yellow phase means 0 and 2π, respectively. Here and in all other figures, red dashed circle on the phase distributions denotes the circle of the *TC* computation, whereas white scale marks in the right bottom denote 1 mm.

When propagating in free space, four LG beams expand and begin to interfere with each other. Figure 2 shows the intensities and phases of the beams from Figure 1 with the same parameters, but at the Rayleigh distance $z = z_0 = kw_0^2/2 \approx 1.476$ m. When all superposition coefficients are equal to each other (Figure 2a,d), the total *TC* of four LG beams remains equal to three: $TC = 2.9968 \approx 3$. For the beam in Figure 2b,e, the total *TC* remains equal to four: $TC = 3.9903 \approx 4$. For the beam in Figure 2c,f, the total *TC* is equal to six ($TC = 5.9148 \approx 6$), but, again, choosing a wider domain (Figure 2g) yields the value $TC = 1.9713 \approx 2$.

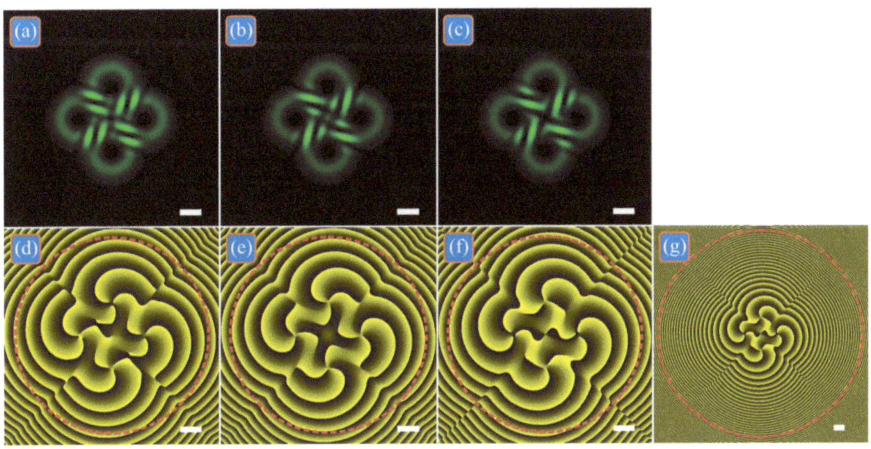

Figure 2. Intensities (**a**–**c**) and phases (**d**–**g**) of three superpositions of off-axis single-ring LG beams with the initial distribution shown in Figure 1 at the Rayleigh distance. The total *TC* of the superpositions is the same as that of each constituent beam (**a**,**d**), or differs from the *TC* of each beam by +1 (**b**,**e**) and by −1 (**c**,**f**,**g**) due to the presence of complex weight coefficients in the superposition.

In the far zone, all four LG beams mix with each other, and the distributions of their intensity and phase are shown in Figure 3. All calculation parameters are the same as in Figure 1, but the propagation distance $z = 3z_0 \approx 4.429$ m, computational area is limited by $|x|, |y| \leq R$, where $R = 7.5$ mm, circle radius for calculating TC is $R_1 = 7$ mm. When all superposition coefficients are the same (Figure 3a,d), the total TC of four LG beams remains equal to three: $TC = 2.9875 \approx 3$. For the beam in Figure 3b,e, the total TC is incremented and is equal to four: $TC = 3.9760 \approx 4$. For the beam in Figure 3c,f, the total TC is decremented and is equal to two: $TC = 2.0036 \approx 2$.

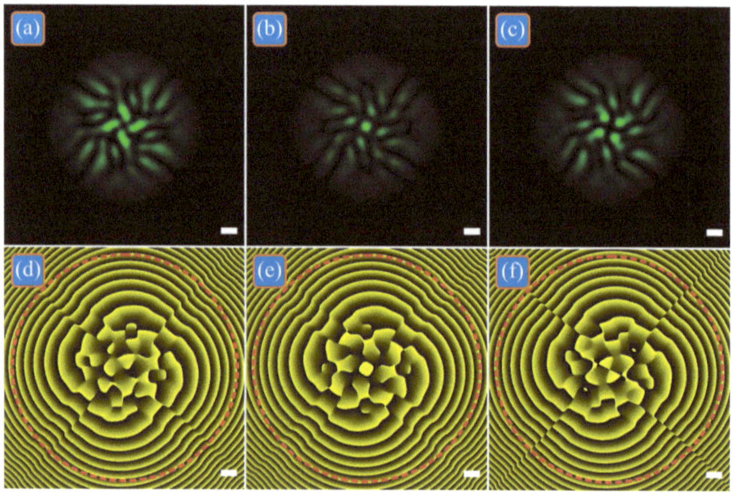

Figure 3. Intensities (**a**–**c**) and phases (**d**–**f**) of three superpositions of off-axis single-ring LG beams with the initial distribution shown in Figure 1 at triple Rayleigh distance (far zone). The total TC of the superpositions is the same as that of each constituent beam (**a**,**d**), or differs from the TC of each beam by +1 (**b**,**e**) and by -1 (**c**,**f**) due to the presence of complex weight coefficients in the superposition.

It should be noted that all LG beams in the superpositions shown in Figures 1–3 have the same power. However, it follows from the theory above that if all superposition coefficients are real then the TC does not change even in the case of a superposition of LG beams with different power (LG beams are added in phase or in antiphase). Therefore, Figure 4 shows two such superpositions. In one of them, the LG beams on the horizontal axis are twice as powerful as the LG beams on the vertical axis: $c_1 = c_3 = 1$, $c_2 = c_4 = 1/2^{1/2}$ (Figure 4a–d). In another superposition, the beam power decreases in a circle: $c_1 = 1$, $c_2 = 3^{1/2}/2$, $c_3 = 2^{1/2}/2$, $c_4 = 1/2$ (Figure 4e–h). Other calculation parameters are the same as in Figure 1, but the propagation distance $z = 0$ (Figure 4a,b,e,f) and $z = z_0/2$ (Figure 4c,d,g,h). The numerically calculated TC for both beams at both distances along the optical axis turned out to be equal to three: $TC = 3.0037$ (Figure 4a,b), $TC = 2.9995$ (Figure 4c,d), $TC = 3.0035$ (Figure 4e,f), $TC = 2.9993$ (Figure 4g,h).

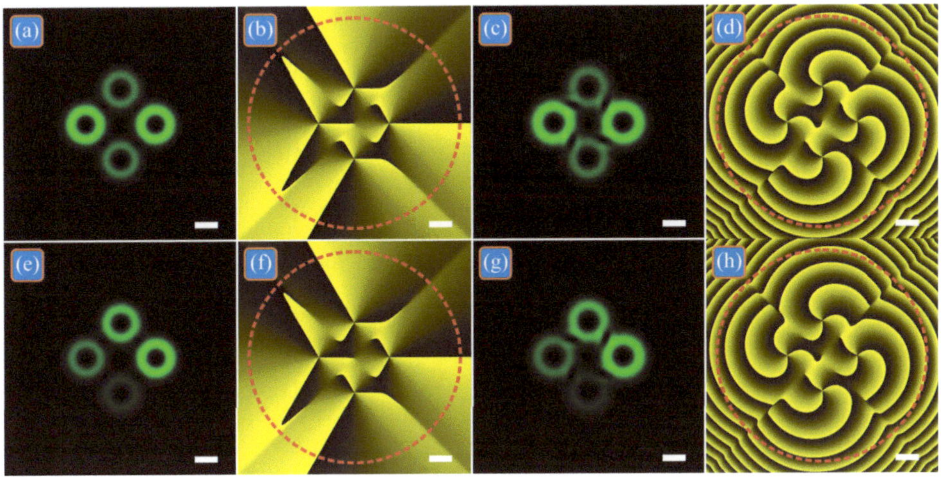

Figure 4. Intensities (**a**,**c**,**e**,**g**) and phases (**b**,**d**,**f**,**h**) of two superpositions of off-axis single-ring LG beams with different powers in the initial plane (**a**,**b**,**e**,**f**) and at the half of the Rayleigh distance (**c**,**d**,**g**,**h**). The superposition coefficients are $c_1 = c_3 = 1$, $c_2 = c_4 = 1/2^{1/2}$ (**a**–**d**) and $c_1 = 1$, $c_2 = 3^{1/2}/2$, $c_3 = 2^{1/2}/2$, $c_4 = 1/2$ (**e**–**h**).

5. Discussion

In this section, we briefly consider the applications of the obtained results as well as the implementation issues. The potential application areas are optical data transmission and performing simple arithmetic operations in optical computing machines where data are carried by vortex light beams and are encoded by the topological charges [19,20].

In wireless optical data transmission, the results can be used by generating considered superpositions of LG beams using a spatial light modulator (SLM) and by identifying the incoming signals by the superposition's topological charge using a Shack–Hartmann wavefront sensor. According to Figure 3, simply adding phase delays between the LG beams allows the *TC* of the whole superposition to change, whereas using the superposition of the LG beams instead of separate LG beams increases the resistance to the turbulence-induced distortions, adds degrees-of-freedoms in data encoding, and improves the data security, since the *TC* of a single LG mode can be determined simply by the radius of the light ring, while the *TC* of the superposition depends not only on the radii of the constituent beams, but also on the phase delays between them.

In optical computing, for designing compact devices, the LG beams should be localized in guiding microstructures rather than propagating in free space. However, the implementation of LG beam propagation in on-chip devices is now challenging for typical semiconductor-manufacturing processes, since most of integrated optical waveguides do not have circular symmetry in their transverse geometries in contrast to optical fibers. However, such techniques are developed now. For instance, a $\pm$1st-order vortex can be implemented in a rectangular-shaped waveguide as a superposition of the TE01 and TE10 modes with a phase delay of $\pi/2$ [21]. In [22], an integrated cross-shaped waveguide structure to support high-order OAM modes up to the fourth order was proposed. This waveguide was designed for a wavelength of 1550 nm and has transverse sizes of just 1.626 μm × 1.504 μm. The obtained results show that simple free-space mixing of the LG beams, coming out from such structures, can generate light fields with incremented or decremented *TC*, although we did not investigate the influence of the mode purity of the LG beams.

6. Conclusions

Thus, we have proven that the superposition of identical single-ring Laguerre–Gaussian beams with numbers (0, m) that are parallel to the optical axis and have the same phase but different power in the initial plane has a topological charge equal to m, regardless of the distance between the beams and of the power of each beam. Furthermore, only if the beams have a different phase in the initial plane (in the waist plane), then the superposition TC changes. The theoretical consideration is confirmed by the simulation results, which demonstrate how tuning the phase delays between the superposition coefficients allows the TC of the superposition to be conserved, or incremented, or decremented. The potential application areas are optical data transmission and optical computing.

Author Contributions: Conceptualization, V.V.K. and A.A.K.; methodology, V.V.K. and A.A.K.; software, A.A.K. and A.A.S.; validation, V.V.K. and A.A.K.; formal analysis, V.V.K. and A.A.K.; investigation, V.V.K., A.A.K. and E.S.K.; resources, A.A.K. and E.S.K.; data curation, V.V.K. and A.A.K.; writing—original draft preparation, V.V.K., A.A.K. and E.S.K.; writing—review and editing, V.V.K., A.A.K. and E.S.K.; visualization, A.A.K., E.S.K. and A.A.S.; supervision, V.V.K.; project administration, A.A.K.; funding acquisition, A.A.K. All authors have read and agreed to the published version of the manuscript.

Funding: The work was funded by the Russian Science Foundation under grant #22-12-00137.

Institutional Review Board Statement: Not applicable.

Informed Consent Statement: Not applicable.

Data Availability Statement: Not applicable.

Conflicts of Interest: The authors declare no conflict of interest. The funders had no role in the design of the study; in the collection, analyses, or interpretation of data; in the writing of the manuscript; or in the decision to publish the results.

References

1. Shen, Y.; Wang, X.; Xie, Z.; Min, C.; Fu, X.; Liu, Q.; Gong, M.; Yuan, X. Optical vortices 30 years on: OAM manipulation from topological charge to multiple singularities. *Light Sci. Appl.* **2019**, *8*, 90. [CrossRef] [PubMed]
2. Lin, X.J.; Feng, Q.C.; Zhu, Y.; Ji, S.H.; Xiao, B.; Xu, H.Y.; Li, W.S.; Cai, Z.P. Diode-pumped wavelength-switchable visible Pr3+:YLF laser and vortex laser around 670 nm. *Opto-Electron. Adv.* **2021**, *4*, 210006. [CrossRef]
3. Zhang, D.S.; Li, X.Z.; Fu, Y.; Yao, Q.H.; Li, Z.G.; Sugioka, K. Liquid vortexes and flows induced by femtosecond laser ablation in liquid governing formation of circular and crisscross LIPSS. *Opto-Electron. Adv.* **2022**, *5*, 210066. [CrossRef]
4. Zhang, Y.; Liu, X.; Lin, H.; Wang, D.; Cao, E.; Liu, S.; Nie, Z.; Jia, B. Ultrafast Multi-Target Control of Tightly Focused Light Fields. *Opto-Electron. Adv.* **2022**, *5*, 210026. [CrossRef]
5. Allen, L.; Beijersbergen, M.; Spreeuw, R.; Woerdman, J. Orbital angular momentum of light and the transformation of Laguerre-Gaussian laser modes. *Phys. Rev. A* **1992**, *45*, 8185–8189. [CrossRef]
6. Berry, M.V. Optical vortices evolving from helicoidal integer and fractional phase steps. *J. Opt. A Pure Appl. Opt.* **2004**, *6*, 259–268. [CrossRef]
7. Molina-Terriza, G.; Recolons, J.; Torner, L. The curious arithmetic of optical vortices. *Opt. Lett.* **2000**, *25*, 1135–1137. [CrossRef]
8. Pyragaite, V.; Smilgevicius, V.; Stabinis, A.; Maslinska, V. The interference of higher order Laguerre-Gaussian beams. *Lith. J. Phys.* **2007**, *47*, 21–26. [CrossRef]
9. Pyragaite, V.; Stabinis, A. Free-space propagation of overlapping light vortex beams. *Opt. Commun.* **2002**, *213*, 187–191. [CrossRef]
10. Lopez-Mago, D.; Perez-Garcia, B.; Yepiz, A.; Hernandez-Aranda, R.I.; Gutiérrez-Vega, J.C. Dynamics of polarization singularities in composite optical vortices. *J. Opt.* **2013**, *15*, 044028. [CrossRef]
11. Naik, D.N.; Viswanathan, N.K. Generation of singular optical beams from fundamental Gaussian beam using Sagnac interferometer. *J. Opt.* **2016**, *18*, 095601. [CrossRef]
12. Cheng, K.; Lü, B. Composite coherence vortices in coherent and incoherent superpositions of two off-axis partially coherent vortex beams. *J. Mod. Opt.* **2008**, *55*, 2751–2764. [CrossRef]
13. Sukhorukov, A.P.; Kalinovich, A.A.; Molina-Terriza, G.; Torner, L. Superposition of noncoaxial vortices in parametric wave mixing. *Phys. Rev. E* **2002**, *66*, 036608. [CrossRef] [PubMed]
14. Zhang, Y.; Wu, Z.; Yang, K.; Li, P.; Wen, F.; Gu, Y. Splitting, generation, and annihilation of phase singularities in non-coaxial interference of Bessel-Gaussian beams. *Phys. Scr.* **2021**, *96*, 125105. [CrossRef]
15. Kotlyar, V.V.; Kovalev, A.A.; Amiri, P.; Soltani, P.; Rasouly, S. Topological charge of two parallel Laguerre-Gaussian beams. *Opt. Express* **2021**, *29*, 42962–42977. [CrossRef]

16. Kovalev, A.A.; Kotlyar, V.V. Orbital angular momentum of superposition of identical shifted vortex beams. *J. Opt. Soc. Am. A* **2015**, *32*, 1805–1810. [CrossRef]
17. Kotlyar, V.V.; Kovalev, A.A. *Topological Charge of Optical Vortices*; Novaya Technika: Samara, Russia, 2021; ISBN 978-5-88940-157-5. (In Russian)
18. Kotlyar, V.V.; Kovalev, A.A.; Volyar, A.V. Topological charge of a linear combination of optical vortices: Topological competition. *Opt. Express* **2020**, *28*, 8266–8281. [CrossRef]
19. Zhang, Q.; Ni, J.; Qiu, C.W. Vortex 4.0 on chip. *Light Sci. Appl.* **2020**, *9*, 103. [CrossRef]
20. Akulshin, A.M.; Novikova, I.; Mikhailov, E.E.; Suslov, S.A.; McLean, R.J. Arithmetic with optical topological charges in stepwise-excited Rb vapor. *Opt. Lett.* **2016**, *41*, 1146–1149. [CrossRef]
21. Lee, I.J.; Kim, S. Directional coupler design for orbital angular momentum mode-based photonic integrated circuits. *Opt. Express* **2020**, *28*, 30085. [CrossRef]
22. Lee, I.J.; Kim, S. On-Chip Guiding of Higher-Order Orbital Angular Momentum Modes. *Photonics* **2019**, *6*, 72. [CrossRef]

micromachines

MDPI

Article

Metasurfaces Assisted Twisted α-MoO_3 for Spinning Thermal Radiation

Yasong Sun [1,2], Derui Zhang [1,2,3], Biyuan Wu [1,2,3], Haotuo Liu [3,4], Bing Yang [5] and Xiaohu Wu [3,*]

1 Basic Research Center, School of Power and Energy, Northwestern Polytechnical University, Xi'an 710072, China
2 Center of Computational Physics and Energy Science, Yangtze River Delta Research Institute of NPU, Northwestern Polytechnical University, Taicang 215400, China
3 Shandong Institute of Advanced Technology, Jinan 250100, China
4 School of Energy Science and Engineering, Harbin Institute of Technology, Harbin 150001, China
5 Centre for Advanced Laser Manufacturing (CALM), School of Mechanical Engineering, Shandong University of Technology, Zibo 255000, China
* Correspondence: xiaohu.wu@iat.cn

Abstract: Spinning thermal radiation has demonstrated applications in engineering, such as radiation detection and biosensing. In this paper, we propose a new spin thermal radiation emitter composed of the twisted bilayer α-MoO_3 metasurface; in our study, it provided more degrees of freedom to control circular dichroism by artificially modifying the filling factor of the metasurface. In addition, circular dichroism was significantly enhanced by introducing a new degree of freedom (filling factor), with a value that could reach 0.9. Strong-spin thermal radiation resulted from the polarization conversion of circularly polarized waves using the α-MoO_3 metasurface and selective transmission of linearly polarized waves by the substrate. This allowed for extra flexible control of spinning thermal radiation and significantly enhanced circular dichroism, which promises applications in biosensing and radiation detection. As a result of their unique properties, hyperbolic materials have applications not only in spin thermal radiation, but also in areas such as near-field thermal radiation. In this study, hyperbolic materials were combined with metasurfaces to offer a new idea regarding modulating near-field radiative heat transfer.

Keywords: spin thermal radiation; metasurface; twisted α-MoO_3

Citation: Sun, Y.; Zhang, D.; Wu, B.; Liu, H.; Yang, B.; Wu, X. Metasurfaces Assisted Twisted α-MoO_3 for Spinning Thermal Radiation. *Micromachines* **2022**, *13*, 1757. https://doi.org/10.3390/mi13101757

Academic Editors: He Yang, Xinyang Su and Yizhong Huang

Received: 5 September 2022
Accepted: 15 October 2022
Published: 17 October 2022

1. Introduction

In recent years, thermal radiation has attracted considerable attention from researchers due to its high potential for applications in areas such as energy harvesting [1–4] and coherent heat sources [5–7]. According to wave-particle duality, the nature of thermal radiation is electromagnetic waves. Therefore, thermal radiation possesses various properties of electromagnetic waves, such as superposition and coherence properties, spectral properties and polarization properties [8–10]. Greffet et al. demonstrated that periodic microstructures could emit a coherent and linearly polarized wave [5], which offers significant promise for controlling the spectral, coherent and polarization properties of thermal radiation [11–13]. Spin polarized (circularly polarized) wave has gained extensive attention in chiral optics [14–16] and spin-controlled nanophotonics [17–19]; spin angular momentum is used to engineer spin-dependent nanoscale light-matter interactions. Recently, studies regarding chiral microstructures have demonstrated the feasibility of spin thermal radiation for engineering, including thermal detection [20–22].

In general, spin thermal radiation can be generated by breaking rotational symmetry and mirror symmetry simultaneously. Circular dichroism (CD) is defined as the difference in the absorption between left-hand circular polarization (LCP) and right-hand circular polarization (RCP); CD is an important parameter when measuring spin thermal radiation [21,22]. At present, many approaches have been proposed to improve CD [23,24]. It is possible to break mirror symmetry using an applied magnetic field (due to the spin-orbit interaction of electrons) resulting in spin thermal radiation [25]. Nevertheless, this approach requires additional incentives and is not conducive to practical application.

Hyperbolic materials (HMs) have attracted much attention due to their unique properties [26,27]. HMs have a wide range of promising applications in broadband enhanced local density of states (LDOS) [28], spontaneous emission [29–31], hyperbolic lensing [32–34], negative refraction [35,36], super absorption [37] and Förster energy transfer [38–40]. As a natural biaxial hyperbolic crystal with in-plane anisotropy, α-MoO_3 has a unique advantage in exciting spin thermal radiation. Hexagonal boron nitride (hBN) is another hyperbolic material with out-of-plane anisotropy, which is also capable of exciting spin thermal radiation. Generally, spin thermal radiation requires more anisotropy. Compared to the uniaxial hyperbolic material hBN, α-MoO_3 is a natural biaxially hyperbolic material with both in-plane and out-of-plane anisotropy, enabling it to facilitate spin thermal radiation. In addition, α-MoO_3 has a wider hyperbolic band, carrying larger electromagnetic wave energy, which offers the possibility of enhancing circular dichroism. [41]. Wu et al. studied the spin thermal radiation properties of single-layer α-MoO_3 [42] and double-layer twisted α-MoO_3 structures [43]. Although the structures mentioned above can excite spin thermal radiation properties, the CD obtained by optimizing the rotation angle and thickness parameters was always very limited.

Another way to achieve spin thermal radiation is to create a structure with chiral surface morphology or with the help of chiral metamaterials. Dyakov et al. proposed a photonic crystal slab waveguide with chiral morphology that can excite spin thermal radiation without an external magnetic field [44]. Kong et al. proposed a novel chiral metamaterial structure with Γ-shaped aligned nanocrystals to achieve significant CD [24]. To date, many two-dimensional (2D) or three-dimensional (3D) chiral microstructures have been designed that enhance spin thermal radiation significantly [45,46]. Although chiral metamaterials can effectively improve CD, subwavelength nanostructures tend to increase the complexity of structural fabrication. Metasurfaces, as two-dimensional derivatives of metamaterials composed of a single or a few patterned layer planar structures, reduce the fabrication requirement. In recent years, metasurfaces have attracted much attention from researchers and have a high potential for important applications [47,48]. More importantly, thermal radiation devices based on metasurfaces possess more freedom of regulation. Recently, metasurfaces based on α-MoO_3 rectangular strips, which only need to be etched on a single layer of slab, have attracted interest. Huang et al. [49] studied hyperbolic phonon polarization excitons (HPhPs) of van der Waals semiconductors coupled to terahertz and LWIR radiation based on gratings etched directly on α-MoO_3 semiconductor flat plates, ultimately obtaining quality factors as high as 300. However, the spin thermal radiation of α-MoO_3 microstructures is still seldom studied.

This paper describes our study of the spin thermal radiation properties of the metasurface-assisted twisted bilayer α-MoO_3. First, the effects of the thicknesses of the two layers and the rotation angle on the CD value were investigated. In addition, a new degree of freedom (filling factor) was introduced. It was found that the structure can greatly enhance spin thermal radiation, and also provide more degrees of freedom to control the spin thermal radiation instead of limiting it to a specific angle. Furthermore, this paper explains the physical mechanism of CD dependence on the filling factor from the perspective of polarization conversion. This study achieved strong spin thermal radiation, which allows greater freedom in tuning the spin thermal radiation.

2. Theory and Method

Figure 1 shows the proposed metasurface structure, which consists of a periodic α-MoO_3 rectangular strip and an α-MoO_3 substrate. As shown in Figure 1, d_1 and d_2 represent the thicknesses of rectangular strips and substrate, respectively. δ represents the relative rotation angle between the rectangular strips and the substrate. When the rectangular strips had a rotation angle with respect to the substrate, the overall symmetry of the structure broke. w represents the spacing of the rectangular strips, Λ is the period, and the incident light was directed along the z-axis. For the α-MoO_3 substrate, the crystal axes [100], [001] and [010] were along the x, y and z directions, respectively. Thus, the permittivity tensor of the α-MoO_3 substrate can be denoted by $\varepsilon = diag(\varepsilon_x, \varepsilon_y, \varepsilon_z)$, where ε_x, ε_y and ε_z can be represented by the Lorentz model as [50]:

$$\varepsilon_{\rm m} = \varepsilon_{\infty,\rm m}\left(1 + \frac{\omega_{\rm LO,m}^2 - \omega_{\rm TO,m}^2}{\omega_{\rm TO,m}^2 - \omega^2 - j\omega\Gamma_{\rm m}}\right) \quad (1)$$

where w is the angular frequency. The values of the other parameters are shown in Table 1 [51].

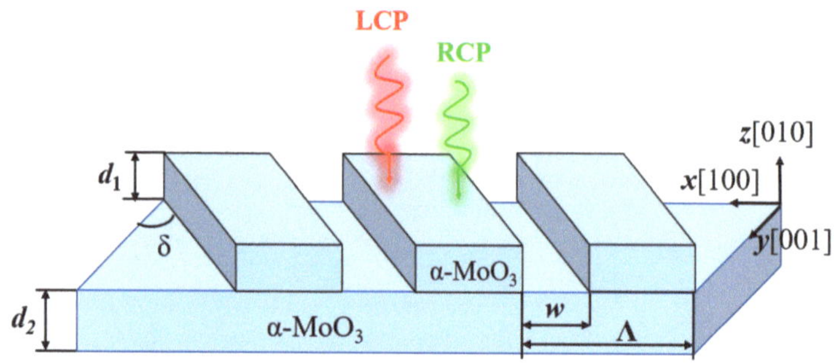

Figure 1. The metasurface structure with spin thermal radiation; both substrate and rectangular strips are α-MoO_3.

Table 1. Values and parameters of the permittivity.

Physical Parameter	Value	Physical Parameter	Value
$\varepsilon_{\infty,x}$	4	$\omega_{TO,x}$	1.5457×10^{14} rad/s
$\varepsilon_{\infty,y}$	5.2	$\omega_{TO,y}$	1.8322×10^{14} rad/s
$\varepsilon_{\infty,z}$	2.4	$\omega_{TO,z}$	1.8058×10^{14} rad/s
$\omega_{LO,x}$	1.8322×10^{14} rad/s	Γ_x	7.5398×10^{11} rad/s
$\omega_{LO,y}$	1.6041×10^{14} rad/s	Γ_y	7.5398×10^{11} rad/s
$\omega_{LO,z}$	1.8925×10^{14} rad/s	Γ_z	3.7699×10^{11} rad/s

We first analyzed the top α-MoO_3 rectangular strips using the effective medium theory [52]. The effective permittivity can be expressed as:

$$\begin{aligned} \varepsilon_{\rm eff,xx} &= \left(\frac{f}{\varepsilon_{\alpha-{\rm MoO_3},x}} + 1 - f\right)^{-1} \\ \varepsilon_{\rm eff,yy} &= \varepsilon_{\alpha-{\rm MoO_3},y} f + 1 - f \\ \varepsilon_{\rm eff,zz} &= \varepsilon_{\alpha-{\rm MoO_3},z} f + 1 - f \end{aligned} \quad (2)$$

where f is the filling factor and its value is $f = w/\Lambda$.

When the top rectangular strips had a rotation angle δ with respect to the substrate, rotation broke the diagonal tensor form of the original dielectric function; the permittivity tensor of α-MoO_3 follows the following transformation form [53]:

$$\varepsilon = \begin{pmatrix} \cos\delta & -\sin\delta & 0 \\ \sin\delta & \cos\delta & 0 \\ 0 & 0 & 1 \end{pmatrix} \begin{pmatrix} \varepsilon_{\rm eff,xx} & 0 & 0 \\ 0 & \varepsilon_{\rm eff,y} & 0 \\ 0 & 0 & \varepsilon_{\rm eff,z} \end{pmatrix} \begin{pmatrix} \cos\delta & \sin\delta & 0 \\ -\sin\delta & \cos\delta & 0 \\ 0 & 0 & 1 \end{pmatrix} \quad (3)$$

The new permittivity tensor was obtained after the calculation as follows:

$$\varepsilon = \begin{bmatrix} \varepsilon_{\text{eff,xx}}\cos^2\delta + \varepsilon_{\text{eff,yy}}\sin^2\delta & \left(\varepsilon_{\text{eff,xx}} - \varepsilon_{\text{eff,yy}}\right)\sin\delta\cos\delta & 0 \\ \left(\varepsilon_{\text{eff,xx}} - \varepsilon_{\text{eff,yy}}\right)\sin\delta\cos\delta & \varepsilon_{\text{eff,xx}}\sin^2\delta + \varepsilon_{\text{eff,yy}}\cos^2\delta & 0 \\ 0 & 0 & \varepsilon_{\text{eff,zz}} \end{bmatrix} \quad (4)$$

In this study, the transfer matrix method (TMM) was used to calculate the transmission of the above structures [43].

A large area of α-MoO_3 flakes was first grown using the physical vapor deposition method. This was then transferred to a silicon substrate and a combination of electron beam lithography and reactive ion etching was used to etch one-dimensional nanoribbons with different periods and angles on the flakes. Electron beam lithography was performed using a Poly (methyl methacrylate) (PMMA) photoresist and ion etching was performed using a mixture of oxygen, argon and CHF_3 at 50 W for 10 min, after which we obtained the α-MoO_3 1D grating structure [49].

3. Results and Discussion

CD is a key parameter for measuring spin thermal radiation's radiative properties. In this study, we primarily considered the transmission of the structure. Therefore, CD could be calculated using:

$$CD = |T_{LCP} - T_{RCP}|, \quad (5)$$

where T_{LCP} and T_{RCP} are the transmission of the LCP and RCP waves, respectively.

Based on [44], it is known that the thickness and the relative rotation angle significantly influence the spin radiation properties of the structure. The variation in CD with thickness and the relative rotation angle was first calculated for any wavelength (here, the wavelength was fixed at 12 μm) and f = 0 (bilayer slabs), as shown in Figure 2. The CD value tended to increase and then decrease as the angle of rotation increased. CD reached a maximum value of 0.0178 at $d_1 = d_2 = 0.175$ μm. Although CD can be controlled by changing the rotation angle, the CD was still very weak. Results indicate that there was almost no excitation of spin thermal radiation at f = 0; therefore, the bilayer slabs had some limitations regarding exciting spin thermal radiation.

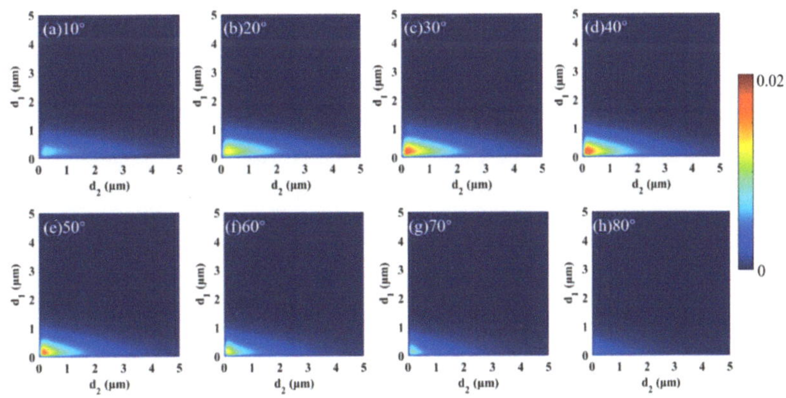

Figure 2. When the wavelength was fixed at 12 μm, and f = 0, CD varied with d_1 and d_2 for different rotation angles: (**a**) 10°, (**b**) 20°, (**c**) 30°, (**d**) 40°, (**e**) 50°, (**f**) 60°, (**g**) 70° and (**h**) 80°.

Based on the above study, we introduced the filling factor f. Next, the effect of f on CD is discussed in detail. Here, the wavelength was the same as that in Figure 2. Variation in CD with d_1 and d_2 as well as the rotation angle are provided in Figure 3. Here, the grating period of the grating was 3 μm. Notably, the maximum value of the color bar is 1, whereas that of Figure 2 is 0.02. Compared to when $f = 0$, CD has been significantly enhanced. CD could reach 0.6848 at $d_1 = 0.65$ μm, $d_2 = 0.525$ μm and a 20° rotation angle, which is tens of times higher than that at $f = 0$. The results illustrate that the metasurface structure greatly enhanced spin thermal radiation. In addition, we used the same method to optimize the structure; it was found that CD could reach 0.9 when $f = 0.7$, $d_1 = 6.25$ μm, $d_2 = 0.5$ μm and $\delta = 40°$, which exceeded the results in previous studies [45,46].

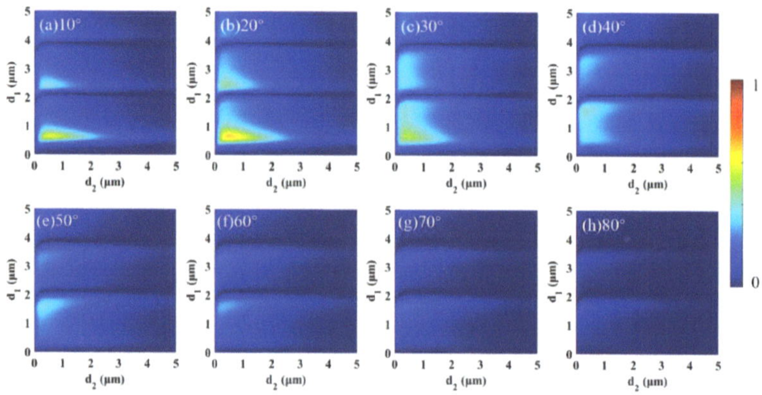

Figure 3. When the wavelength was fixed at 12 μm and $f = 0.1$, CD varied with d_1 and d_2 for different rotation angles: (**a**) 10°, (**b**) 20°, (**c**) 30°, (**d**) 40°, (**e**) 50°, (**f**) 60°, (**g**) 70° and (**h**) 80°.

Next, to further illustrate the effect of f on CD, we calculated the variation in the maximum value of CD with the rotation angle when f increased from 0 to 0.6 at every 0.1 interval. Figure 4a,b show results for wavelengths of 12 μm and 11 μm, respectively. In Figure 4a, it can be seen that the overall trend of CD increased with an increase in f, implying that the value of f can enhance the spin thermal radiation in a wide range, which is more beneficial to practical applications. When the wavelength was 11 μm, it can be seen in Figure 4b that, although the CD decreased somewhat at $f = 0.1$ and $f = 0.2$, it still showed an overall increasing trend at larger f. We conducted similar studies at other wavelengths, with results similar to those of 12 μm and 11 μm, namely that CD was enhanced as f increased. This suggests that the metasurface structure not only enhances spin thermal radiation, but also has a greater degree of freedom in the excitation of thermal radiation.

To better understand the physical mechanism, we discuss the polarization conversion of circularly polarized waves at a fixed wavelength of 12 μm. Figure 5a shows TE (transverse electric wave) and TM (transverse magnetic wave) components in the transmitted wave varying with the rotation angle for different spin direction circularly polarized waves incidence when $f = 0$ and $d_1 = d_2 = 0.175$ μm. LCP-TM represents the TM wave component in the transmitted wave for LCP wave incidence; RCP-TM, RCP-TE and LCP-TE have similar definitions. When $f = 0$, the proposed structure can be considered a bilayer slab structure. It can be seen in Figure 5a that regardless of whether LCP or RCP waves were incidents, the TM wave component in the transmitted wave decreased with increasing rotation angle, whereas the TE wave component gradually increased. However, the overall TE wave component was low; therefore, the TM wave component played a major role in spin thermal radiation at this time. Thus, CD mainly originated from the difference in TM wave components in the transmitted waves at the incidence of LCP and RCP waves. Clearly, the difference between TM wave components in the transmitted wave for LCP and RCP incidence was small at any rotation angle. Combined with Figure 4a, it was found that CD was always at a low level at $f = 0$, which coincides with the result in Figure 5a. The phenomenon in Figure 5b is more obvious in Figure 5a; $f = 0.6$, $d_1 = 4.8$ μm and $d_2 = 0.4$ μm. TE wave components of LCP and RCP waves were almost zero, whereas the difference in the TM wave components reached a maximum at a rotation angle of 40°, which corresponds almost exactly to when f = 0.6 in Figure 4a. These results further indicate that the difference in the TM wave was the key to influencing spin thermal radiation.

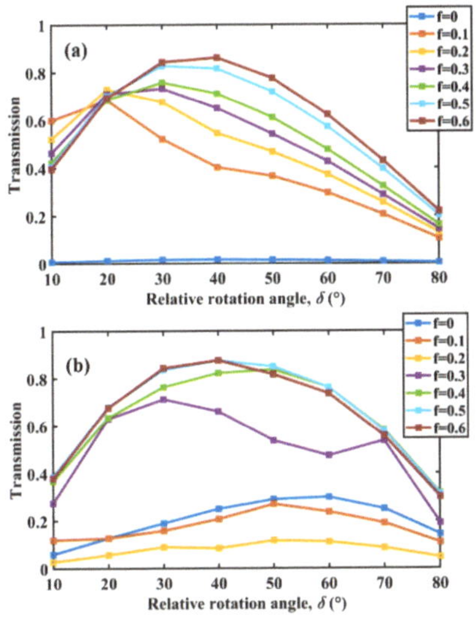

Figure 4. Maximum value of CD as a function of the rotation angle under different f when the wavelength was (**a**) 12 μm and (**b**) 11 μm, respectively.

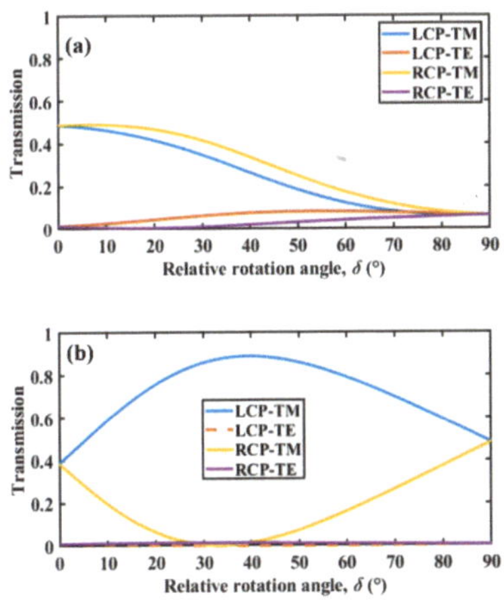

Figure 5. TE wave and TM wave components in the transmitted wave as a function of the rotation angle for LCP and RCP waves: (**a**) f = 0 and (**b**) f = 0.6.

To further illustrate the above mechanism, we now discuss the polarization conversion for the monolayer α-MoO_3 slab, shown in Figure 6. In Figure 6a, it can be seen that differences in TM and TE wave components in the transmitted wave for LCP and RCP waves were basically the same, and both were relatively low overall. When the wavelength was 12 μm, the permittivity of α-MoO_3 in the x and y directions were $\varepsilon_x = -45.51 - 7.99i$ and $\varepsilon_y = -0.4 - 0.04i$, respectively. As the real part of ε_x is negative and has a large absolute value, the α-MoO_3 exhibited metal-like properties in the x direction. After quantitative calculation, the transmission was only 0.135 when the TE wave related to ε_x was incident at a 0.175 μm thick monolayer α-MoO_3 slab, whereas the transmission for the TM wave related to ε_y could reach 0.99. Thus, the effect of the difference in the TM wave component on CD was further confirmed.

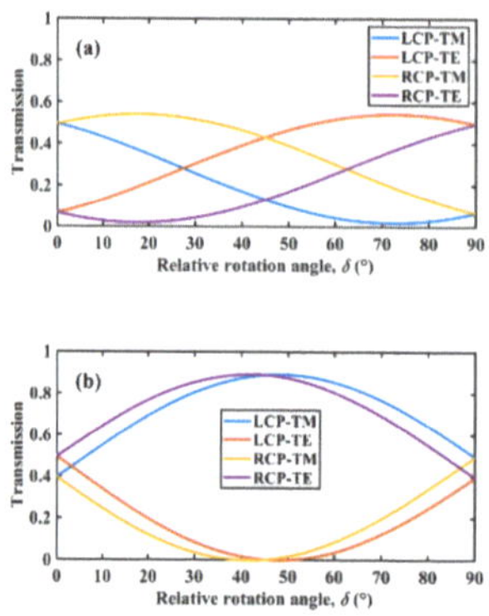

Figure 6. TE wave and TM wave components in the transmitted wave as a function of the rotation angle for LCP and RCP waves: (**a**) single layer slab (f = 0) and (**b**) single-layer rectangular strips structure (f = 0.6).

Next, the polarization conversion for single-layer rectangular strips was studied. According to the effective medium theory, the permittivity in x and y directions can be written as $\varepsilon_x = 1.69 - 0.004i$ and $\varepsilon_y = 0.44 - 0.01i$. Therefore, both TE and TM waves can theoretically be transmitted in a single-layer rectangular strips structure. Figure 6b illustrates that the TM wave component in the transmitted wave for LCP wave incidence tended to increase and then decrease with an increase in the rotation angle, whereas the TM wave component in the transmitted wave for the RCP wave incidence first decreased and then increased. Thus, the TM component was significantly different in the transmitted wave for LCP and RCP waves. However, there was also a large difference in the TE wave component of the transmitted wave at LCP and RCP incidence, which means that the main role of the rectangular strips structure in the top layer was to achieve polarization conversion. According to the polarization conversion results, we then placed the rectangular strips on a 0.4 μm thick substrate and found that the transmission of TE waves was only 0.018, which further indicated that the TE wave component could not pass through the substrate and had little effect on CD. In contrast, the transmission of TM waves could reach 0.969. These results suggest that the role of the substrate was to achieve selective transmission to TE and TM waves. In addition, the trend of the TM wave component difference with rotation angle illustrated in Figure 6b was essentially the same as that in Figure 5b, indicating that the difference in the TM wave component played a decisive role in CD.

4. Conclusions

In summary, we systematically investigated the spin thermal radiation in a twisted bilayer α-MoO_3 metasurface. With the introduction of the filling factor f, the spin thermal radiation was greatly enhanced and more flexibly excited. The numerical results show that CD could reach 0.9 via optimizing the filling factor, thickness and rotation angle. Based on analysis of bilayer and single layer structures, it was found that the spin thermal radiation of the structure originated from the polarization conversion of the top periodic rectangular strips structure and the selective transmission of the substrate. Specifically, the difference in the TM wave component of the transmitted wave for LCP and RCP waves incidence effected the structure's CD. The TM wave component in the transmitted wave was affected by the filling factor; therefore, the spin thermal radiation of the structure proposed in this paper could be flexibly tuned by the filling factor. We believe that this study has potential applications in biosensing and radiation detection.

Author Contributions: Conceptualization, X.W.; data curation, H.L.; funding acquisition, Y.S. and X.W.; software, D.Z. and H.L.; supervision, Y.S. and B.Y.; visualization, X.W.; writing—original draft, D.Z.; writing—review and editing, B.W. and H.L. All authors have read and agreed to the published version of the manuscript.

Funding: This study was supported by the National Natural Science Foundation of China (Nos. 52106099, 51976173), the Natural Science Foundation of Shandong Province (No. ZR2020LLZ004), the Natural Science Foundation of Jiangsu Province (No. BK20201204), the Basic Research Program of Taicang (No. TC2019JC01), and Fundamental Research Funds for the Central Universities (No. D5000210779).

Data Availability Statement: Data presented in this study are available on request from the corresponding author.

Conflicts of Interest: The authors declare no conflict of interest. The funders had no role in the design of the study; in the collection, analyses, or interpretation of data; in the writing of the manuscript; or in the decision to publish the results.

References

1. Rephaeli, E.; Fan, S. Absorber and emitter for solar thermo-photovoltaic systems to achieve efficiency exceeding the Shockley-Queisser limit. *Opt. Express* **2009**, *17*, 15145–15159. [CrossRef] [PubMed]
2. Fan, S. An alternative'Sun'for solar cells. *Nat. Nanotechnol.* **2014**, *9*, 92–93. [CrossRef] [PubMed]
3. Lenert, A.; Bierman, D.M.; Nam, Y.; Chan, W.R.; Celanović, I.; Soljačić, M.; Wang, E.N. A nanophotonic solar thermophotovoltaic device. *Nat. Nanotechnol.* **2014**, *9*, 126–130. [CrossRef] [PubMed]
4. Bierman, D.M.; Lenert, A.; Chan, W.R.; Bhatia, B.; Celanović, I.; Soljačić, M.; Wang, E.N. Enhanced photovoltaic energy conversion using thermally based spectral shaping. *Nat. Energy* **2016**, *1*, 16068. [CrossRef]
5. Greffet, J.-J.; Carminati, R.; Joulain, K.; Mulet, J.-P.; Mainguy, S.; Cheng, Y. Coherent emission of light by thermal sources. *Nature* **2002**, *416*, 61–64. [CrossRef]
6. Guo, Y.; Cortes, C.L.; Molesky, S.; Jacob, Z. Broadband super-Planckian thermal emission from hyperbolic metamaterials. *Appl. Phys. Lett.* **2012**, *101*, 131106. [CrossRef]
7. Wang, L.; Zhang, Z.M. Measurement of Coherent Thermal Emission Due to Magnetic Polaritons in Subwavelength Microstructures. *J. Heat Transf.* **2013**, *135*, 091505. [CrossRef]
8. Bermel, P.; Boriskina, S.V.; Yu, Z.; Joulain, K. Control of radiative processes for energy conversion and harvesting. *Opt. Express* **2015**, *23*, A1533–A1540. [CrossRef]
9. Li, W.; Fan, S. Nanophotonic control of thermal radiation for energy applications [Invited]. *Opt. Express* **2018**, *26*, 15995–16021. [CrossRef]
10. Miller, D.A.B.; Zhu, L.; Fan, S. Universal modal radiation laws for all thermal emitters. *Proc. Natl. Acad. Sci. USA* **2017**, *114*, 4336–4341. [CrossRef]
11. Sai, H.; Kanamori, Y.; Yugami, H. High-temperature resistive surface grating for spectral control of thermal radiation. *Appl. Phys. Lett.* **2003**, *82*, 1685–1687. [CrossRef]
12. Biswas, R.; Ding, C.G.; Puscasu, I.; Pralle, M.; McNeal, M.; Daly, J.; Greenwald, A.; Johnson, E. Theory of subwavelength hole arrays coupled with photonic crystals for extraordinary thermal emission. *Phys. Rev. B* **2006**, *74*, 045107. [CrossRef]
13. Costantini, D.; Lefebvre, A.; Coutrot, A.-L.; Moldovan-Doyen, I.; Hugonin, J.-P.; Boutami, S.; Marquier, F.; Benisty, H.; Greffet, J.-J. Plasmonic Metasurface for Directional and Frequency-Selective Thermal Emission. *Phys. Rev. Appl.* **2015**, *4*. [CrossRef]
14. Lodahl, P.; Mahmoodian, S.; Stobbe, S.; Rauschenbeutel, A.; Schneeweiss, P.; Volz, J.; Pichler, H.; Zoller, P. Chiral quantum optics. *Nature* **2017**, *541*, 473–480. [CrossRef] [PubMed]
15. Kwon, D.-H.; Werner, P.L.; Werner, D.H. Optical planar chiral metamaterial designs for strong circular dichroism and polarization rotation. *Opt. Express* **2008**, *16*, 11802–11807. [CrossRef]
16. Tang, Y.; Cohen, A.E. Optical Chirality and Its Interaction with Matter. *Phys. Rev. Lett.* **2010**, *104*, 163901. [CrossRef]
17. Mitsch, R.; Sayrin, C.; Albrecht, B.; Schneeweiss, P.; Rauschenbeutel, A. Quantum state-controlled directional spontaneous emission of photons into a nanophotonic waveguide. *Nat. Commun.* **2014**, *5*, 5713. [CrossRef]
18. Liang, Y.; Lin, H.; Koshelev, K.; Zhang, F.; Yang, Y.; Wu, J.; Kivshar, Y.; Jia, B. Full-stokes polarization perfect absorption with diatomic metasurfaces. *Nano Lett.* **2021**, *21*, 1090–1095. [CrossRef]
19. Ji, C.Y.; Chen, S.; Han, Y.; Liu, X.; Liu, J.; Li, J.; Yao, Y. Artificial Propeller Chirality and Counterintuitive Reversal of Circular Dichroism in Twisted Meta-molecules. *Nano Lett.* **2021**, *21*, 6828–6834. [CrossRef]
20. Khandekar, C.C.; Khosravi, F.; Li, Z.; Jacob, Z. New spin-resolved thermal radiation laws for nonreciprocal bianisotropic media. *New J. Phys.* **2020**, *22*, 123005. [CrossRef]
21. Khandekar, C.; Jacob, Z. Circularly Polarized Thermal Radiation From Nonequilibrium Coupled Antennas. *Phys. Rev. Appl.* **2019**, *12*, 014053. [CrossRef]
22. Khan, E.; Narimanov, E.E. Spinning Radiation from Topological Insulator. *Phys. Rev. B* **2019**, *100*, 081408. [CrossRef]

23. Dyakov, S.; Ignatov, A.; Tikhodeev, S.; Gippius, N. Circularly polarized thermal emission from chiral metasurface in the absence of magnetic field. *J. Phys. Conf. Ser.* **2018**, *1092*, 012028. [CrossRef]
24. Wu, B.; Wang, M.; Yu, P.; Wu, F.; Wu, X. Strong circular dichroism triggered by near-field perturbation. *Opt. Mater.* **2021**, *118*, 111255. [CrossRef]
25. Dyakov, S.A.; Semenenko, V.A.; Gippius, N.A.; Tikhodeev, S.G. Magnetic field free circularly polarized thermal emission from a chiral metasurface. *Phys. Rev. B* **2018**, *98*, 235416. [CrossRef]
26. Hu, L.; Chui, S.T. Characteristics of electromagnetic wave propagation in uniaxially anisotropic left-handed materials. *Phys. Rev. B* **2002**, *66*, 085108. [CrossRef]
27. Smith, D.R.; Schurig, D. Electromagnetic Wave Propagation in Media with Indefinite Permittivity and Permeability Tensors. *Phys. Rev. Lett.* **2003**, *90*, 077405. [CrossRef]
28. Smolyaninov, I.I.; Narimanov, E.E. Metric signature transitions in optical metamaterials. *Phys. Rev. Lett.* **2010**, *105*. [CrossRef]
29. Jacob, Z.; Smolyaninov, I.; Narimanov, E.E. Broadband purcell effect: Radiative decay engineering with metamaterials. *Appl. Phys. Lett.* **2012**, *100*, 181105. [CrossRef]
30. Poddubny, A.; Belov, P.A.; Kivshar, Y.S. Spontaneous radiation of a finite-size dipole emitter in hyperbolic media. *Phys. Rev. A* **2011**, *84*, 023807. [CrossRef]
31. Potemkin, A.S.; Poddubny, A.; Belov, P.A.; Kivshar, Y.S. Green function for hyperbolic media. *Phys. Rev. A* **2012**, *86*, 023848. [CrossRef]
32. Jacob, Z.; Alekseyev, L.V.; Narimanov, E. Optical Hyperlens: Far-field imaging beyond the diffraction limit. *Opt. Express* **2006**, *14*, 8247–8256. [CrossRef] [PubMed]
33. Feng, S.; Elson, J.M. Diffraction-suppressed high-resolution imaging through metallodielectric nanofilms. *Opt. Express* **2006**, *14*, 216–221. [CrossRef] [PubMed]
34. Bénédicto, J.; Centeno, E.; Moreau, A. Lens equation for flat lenses made with hyperbolic metamaterials. *Opt. Lett.* **2012**, *37*, 4786–4788. [CrossRef]
35. Smith, D.R.; Kolinko, P.; Schurig, D. Negative refraction in indefinite media. *J. Opt. Soc. Am. B* **2004**, *21*, 1032–1043. [CrossRef]
36. Hoffman, A.J.; Alekseyev, L.; Howard, S.; Franz, K.J.; Wasserman, D.; Podolskiy, V.; Narimanov, E.E.; Sivco, D.L.; Gmachl, C. Negative refraction in semiconductor metamaterials. *Nat. Mater.* **2007**, *6*, 946–950. [CrossRef]
37. Guclu, C.; Campione, S.; Capolino, F. Hyperbolic metamaterial as super absorber for scattered fields generated at its surface. *Phys. Rev. B* **2012**, *86*, 205130. [CrossRef]
38. Ramos-Ortiz, G.; Oki, Y.; Domercq, B.; Kippelen, B. Förster energy transfer from a fluorescent dye to a phosphorescent dopant: A concentration and intensity study. Phys. *Chem. Chem. Phys.* **2002**, *4*, 4109–4114. [CrossRef]
39. Masters, B.R. Paths to Förster's resonance energy transfer (FRET) theory. *Eur. Phys. J. H* **2014**, *39*, 87–139. [CrossRef]
40. Biehs, S.-A.; Menon, V.M.; Agarwal, G.S. Long-range dipole-dipole interaction and anomalous Förster energy transfer across a hyperbolic metamaterial. *Phys. Rev. B* **2016**, *93*, 245439. [CrossRef]
41. Liu, P.; Zhou, L.; Tang, J.; Wu, B.; Liu, H.; Wu, X. Spinning thermal radiation from twisted two different anisotropic materials. *Opt. Express* **2022**, *30*, 32722–32730. [CrossRef] [PubMed]
42. Wu, B.; Wang, M.; Wu, F.; Wu, X. Strong extrinsic chirality in biaxial hyperbolic material α-MoO_3 with in-plane anisotropy. *Appl. Opt.* **2021**, *60*, 4599–4605. [CrossRef] [PubMed]
43. Wu, B.; Shi, Z.; Wu, F.; Wu, X. Strong chirality in twisted bilayer β-MoO_3. *Chin. Phys. B* **2022**, *31*, 41011–41018. [CrossRef]
44. Kong, X.-T.; Khorashad, L.K.; Wang, Z.; Govorov, A.O. Photothermal circular dichroism induced by plasmon resonances in chiral metamaterial absorbers and bolometers. *Nano Lett.* **2018**, *18*, 2001–2008. [CrossRef]
45. Chen, Y.; Gao, J.; Yang, X. Chiral metamaterials of plasmonic slanted nanoapertures with symmetry breaking. *Nano Lett.* **2017**, *18*, 520–527. [CrossRef]
46. Zhang, M.; Hao, D.; Wang, S.; Li, R.; Wang, S.; Ma, Y.; Moro, R.; Ma, L. Chiral biosensing using terahertz twisted chiral metamaterial. *Opt. Express* **2022**, *30*, 14651. [CrossRef]
47. Kotov, O.V.; Lozovik, Y.E. Enhanced optical activity in hyperbolic metasurfaces. *Phys. Rev. B* **2017**, *96*, 54031–540312. [CrossRef]
48. Huo, P.; Zhang, S.; Liang, Y.; Lu, Y.; Xu, T. Hyperbolic metamaterials and metasurfaces: Fundamentals and applications. *Adv. Opti. Mater.* **2019**, *7*, 1801616. [CrossRef]
49. Huang, W.; Folland, T.G.; Sun, F.; Zheng, Z.; Xu, N.; Xing, Q.; Jiang, J.; Caldwell, J.D.; Yan, H.; Chen, H.; et al. In-plane hyperbolic polariton tuner in terahertz and long-wave infrared regimes. *arXiv* **2022**, arXiv:2206.10433.
50. Wu, X.; Fu, C.; Zhang, Z.M. Near-field radiative heat transfer between two α-MoO_3 biaxial crystals. *J. Heat Transfer* **2020**, *142*, 28021–280210. [CrossRef]
51. Zheng, Z.; Xu, N.; Oscurato, S.L.; Tamagnone, M.; Sun, F.; Jiang, Y.; Ke, Y.; Chen, J.; Huang, W.; Wilson, W.L.; et al. A mid-infrared biaxial hyperbolic van der Waals crystal. *Sci. Adv.* **2019**, *5*, eaav8690. [CrossRef] [PubMed]
52. Li, P.; Dolado, I.; Alfaro-Mozaz, F.J.; Casanova, F.; Hueso, L.E.; Liu, S.; Edgar, J.H.; Nikitin, A.Y.; Vélez, S.; Hillenbrand, R. Infrared hyperbolic metasurface based on nanostructured van der Waals materials. *Science* **2018**, *359*, 892–896. [CrossRef] [PubMed]
53. Liu, H.; Ai, Q.; Ma, M.; Wang, Z.; Xie, M. Prediction of spectral absorption of anisotropic α-MoO_3 nanostructure using deep neural networks. *Int. J. Therm. Sci.* **2022**, *177*, 107587. [CrossRef]

micromachines

Article

Design of Compact and Broadband Polarization Beam Splitters Based on Surface Plasmonic Resonance in Photonic Crystal Fibers

Chao Mei [1,*,†], Yuan Wu [1,†], Jinhui Yuan [1,2,*], Shi Qiu [2] and Xian Zhou [1]

1 Research Center for Convergence Networks and Ubiquitous Services, University of Science and Technology Beijing (USTB), Beijing 100083, China
2 State Key Laboratory of Information Photonics and Optical Communications, Beijing University of Posts and Telecommunications, Beijing 100086, China
* Correspondence: chaomei@ustb.edu.cn (C.M.); yuanjinhui81@bupt.edu.cn (J.Y.)
† These authors contributed equally to this work.

Abstract: In this work, a polarization beam splitter (PBS) based on surface plasmonic resonance is proposed and realized in a designed photonic crystal fiber (PCF). The PCF consists of two kinds of air holes with different diameters. Two solid silica cores near the center of the PCF are established by removing the cladding air holes. A gold film is plated at the external surface of the central air hole of the PCF to excite the surface plasmonic resonance. In order to minimize the length and improve the operation bandwidth of the PBS, the influences of the transversal structural parameters of the PCF are investigated in the context of both *X* and *Y* polarization beams. It was found that a 123.6-μm-long PBS with an operation bandwidth of 314 nm could be realized after the global optimization of five structural parameters. The proposed PBS may have potential applications in micro-/nano-optical systems for sensing and communications.

Keywords: photonic crystal fiber; polarization beam splitter; surface plasmonic resonance

Citation: Mei, C.; Wu, Y.; Yuan, J.; Qiu, S.; Zhou, X. Design of Compact and Broadband Polarization Beam Splitters Based on Surface Plasmonic Resonance in Photonic Crystal Fibers. *Micromachines* **2022**, *13*, 1663. https://doi.org/10.3390/mi13101663

Academic Editors: He Yang, Xinyang Su and Yizhong Huang

Received: 5 September 2022
Accepted: 29 September 2022
Published: 3 October 2022

Publisher's Note: MDPI stays neutral with regard to jurisdictional claims in published maps and institutional affiliations.

1. Introduction

Photonic crystal fiber (PCF) [1], which is made up of a central hole and surrounding periodical air holes, has been widely used in linear and nonlinear optics since it was first invented [2]. Compared to conventional step-index silica fiber, the arrangements of the air holes in PCFs enable distinct optical characteristics, such as high birefringence [3], controllable flattened dispersion [4], and endless single-mode transmission [5]. Due to the flexible engineering of the effective refractive index, n_{eff}, PCF has been commonly employed to design polarization beam splitters (PBSes) [6,7], sensors [8–10], and couplers [11]. Among them, PBSes are some of the most important optical elements in both sensing and telecommunications. This element can be utilized to divide one orthogonal polarized beam into two polarized beams, which are generally called *X*-polarized (*X*-pol) and *Y*-polarized (*Y*-pol) beams [12]. Compared to PBSes made up of conventional silica fiber, PBSes realized with a PCF structure show shorter beam-splitting lengths and higher extinction ratios [13]; thus, they have attracted much attention in recent years.

PBSes based on dual-core PCF (DCPCF) are taken as examples. In 2015, L. Jiang et al. proposed a PBS by using DCPCF, whose air holes were organized in a rhombus lattice. The resulting beam-splitting length was 4036 μm and the bandwidth was 430 nm, covering the whole telecommunication band [14]. However, its length was more than 10 times larger than that of other PBSes because of the relatively large beam-splitting distance. In 2016, J. Zi et al. proposed a 249-μm-long PBS whose beam-splitting bandwidth was 12 or 17 nm, depending on its central pump wavelengths, which were located at 1.31 or 1.55 μm, respectively [15]. While J. Zi et al. significantly reduced the length of the PBS, the bandwidth was quite narrow. The performance of a designed PBS can be defined by the ratio of operating bandwidth and the PBS length. The resulting parameter is denoted as σ, and a larger value of σ indicates better beam-splitting performance. The values of σ for Refs. [14,15] were 1×10^{-4} and 6.8×10^{-5}. In 2017, H. Wang designed a 5112-μm-long PBS in a PCF with two magnetic liquid cores. The resulting extinction ratio and bandwidth could reach −158 dB and 189 nm, respectively [16]. However, the calculated σ was only 3.7×10^{-5}, implying that the designed PBS was still too long. In 2018, Q. Xu designed a DCPCF-based PBS with a beam-splitting length of only 290 μm. While the elliptical and circular air holes alternately arranged in the Y-direction were beneficial for the length reduction, the resulting bandwidth was only 19.2 nm around the wavelength of 1.55 μm [17]. The calculated σ was 6.6×10^{-5}, which can be improved further. For the sake of larger values of σ in terms of short PBSes, an ultrashort PBS based on DCPCF filled with liquid was further proposed [18]. The resulting PBS length was only 78 μm, but was accompanied by a relatively large bandwidth of 44 nm. The corresponding σ was 5.6×10^{-4}. The PBS length can be further reduced by using PCFs with square lattices instead of hexagonal lattices. J. Lou et al. proposed a PBS based on DCPCF with square lattices [19]. The obtained beam-splitting length was only 47.26 μm, which was much shorter than that in previous work. More importantly, the bandwidth was improved to 104 nm, which was larger than that reported in Ref. [18]. The calculated σ of 2.2×10^{-3} is, to the best of our knowledge, the largest value so far. However, square-lattice PCFs face more difficulties in fabrication than hexagonal-lattice PCFs do. While the PBSes designed in Refs. [18,19] were shorter than 100 μm, further reducing the PBS length became almost impossible due to the limit of diffraction and differences in the refractive index. It is thus necessary to pay more attention to increasing the operating bandwidth of PBSes.

In this paper, we propose a compact and broadband PBS based on DCPCF consisting of seven-layer air holes. As stated above, these air holes can be divided into two categories according to the size of their diameters. In order to improve the bandwidth of the PBS, we introduce surface plasmonic resonance (SPR) in the DCPCF, which has not been fully explored in PBS design. The SPR has been widely studied in optical sensors, as it can effectively induce a sudden local variation in the refractive index. With this feature, polarized beam splitting based on SPR is possible because polarized beams are also sensitive to changes in the refractive index. Therefore, increasing the operating bandwidth of a PBS can actually be achieved by tuning the wavelength at which the SPR effect works. In PCF, the characteristic wavelength of SPR is closely related to the transversal structure of a PCF whose size is comparable to the wavelength, i.e., on the order of μm. In such a case, the n_{eff} of the PCF is very sensitive to the SPR effect. In other words, the characteristic wavelength of SPR can be effectively tuned by changing the transversal structural dimensions of PCFs. The structural influences on the PBS are discussed in detail below.

2. Design of the DCPCF

Considering the practicality of fabrication, we still adopted a traditional hexagonal-lattice PCF to design the PBS. Figure 1 shows the cross-section of the proposed DCPCF. The outermost blue ring is the perfect matched layer (PML). It can be used to reflect all rays back into the PCF and prevent energy loss in numerical simulations. The background material marked in gray is silica. Seven-layer air holes are arranged periodically. The spacing between two large air holes is denoted by Λ, while the spacing between a large and a small air hole is defined by d_2. The diameter of a large air hole is d_1. It should be noted that the air hole located at the center of the DCPCF is coated with a gold film, which is marked in yellow. Gold and silver films are the two most often used materials to excite the SPR effect due to their faster electron mobility. The diameter of the central air hole and the thickness of the gold film are denoted by d and t, respectively. The two air holes on the left- and right-hand sides of the central air hole are removed. As a result, two solid silica cores named A and B are formed to allow the propagation of two orthogonal polarized beams.

Next, we will adjust these five parameters to study the beam-splitting length for different polarized beams.

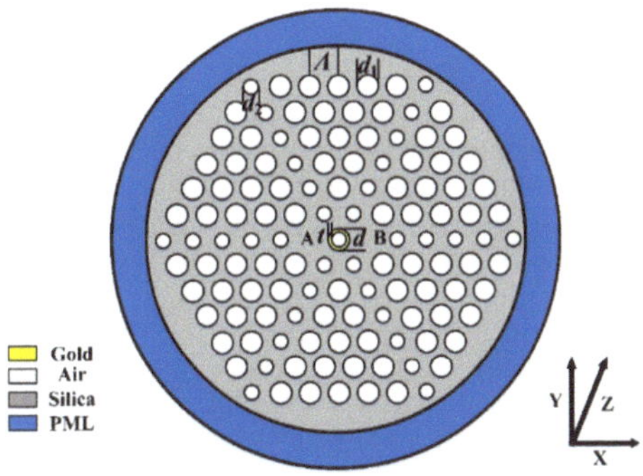

$d = 1$ μm	$d_1 = 1.2$ μm	$d_2 = 1$ μm	$\Lambda = 2$ μm	$t = 50$ nm

Figure 1. A sectional view of the proposed DCPCF with its structual parameters, d, d_1, d_2, Λ, and t.

In order to study the SPR effect n_{eff} of PCF, it is necessary to introduce the Sellmeier equation of silica, which is given as [20,21]:

$$n^2(\lambda) = 1 + \frac{A_1\lambda^2}{\lambda^2 - B_1^2} + \frac{A_2\lambda^2}{\lambda^2 - B_2^2} + \frac{A_3\lambda^2}{\lambda^2 - B_3^2}, \quad (1)$$

where λ is the incident wavelength with units of μm, and A_1, A_2, A_3 and B_1, B_2, B_3 are constants, as shown in Table 1. The dispersion effect of air is neglected in this work, as it makes almost no contributions to the n_{eff} in the wavelength range considered.

Table 1. Parameters for the Sellmeier equation of silica.

	Subscript 1	Subscript 2	Subscript 3
A	0.6961663	0.4079426	0.8974794
B	0.0684043 μm	0.1162414 μm	9.896161 μm

Unlike in semiconductor materials, the dispersion effect of metal involves the modification of the dielectric constant. In most cases, the dielectric constant of a gold film can be defined by the Drude–Lorentz model as [22,23]

$$\epsilon = \epsilon_0 - \frac{\omega_D^2}{\omega(\omega + j\gamma_D)} - \frac{\Delta\epsilon\Omega_L^2}{(\omega^2 - \Omega_L^2) + j\Gamma_L\omega}, \quad (2)$$

where $\epsilon_0 = 5.9673$ is the dielectric constant of gold, ω_{D} is the plasma frequency, $\omega_{\text{D}}/2\pi = 2113.6$ THz, γ_{D} is the damping frequency, $\gamma_{\text{D}}/2\pi = 15.92$ THz, Ω_{L} is the central frequency of the Lorentz oscillator, $\Omega_{\text{L}}/2\pi = 650.07$ THz, Γ_{L} is the spectral width of the Lorentz oscillator, $\Gamma_{\text{L}}/2\pi = 104.86$ THz, and ω is the angular frequency of the guided wave.

The coupling length *CL* is a vital parameter for describing the characteristics of a PBS, as the length of the PBS is determined by it. Generally, the CL is defined as the propagated distance when the power of X (or Y)-polarized light completely transfers from one core, i.e., A, to another core, i.e., B. The *CL* of *X* (or *Y*)-polarized light is called the *CLX* (or *CLY*), which is mathematically expressed as [24,25]

$$CLi = \frac{\lambda}{2|n_{\text{even}}^i - n_{\text{odd}}^i|}, \quad i = X, Y, \quad (3)$$

where i represents the polarization states including both the X-pol and Y-pol directions, CLi is the coupling length of the i-pol direction, and n^i_{even} and n^i_{odd} are the n_{eff} of the even and odd modes, respectively. The ratio between the CLY and CLX can be obtained from [26]

$$R = \frac{CLY}{CLX}. \tag{4}$$

When the beam propagates in a dual-core PCF, its energy will periodically transfer from core A to core B. We define that when $R = 2$ or $1/2$, the incident beam is completely separated into two polarizations. Assuming that the beam is initially injected into core A, the output power of the beam in the X-pol or Y-pol directions in core A can be calculated by [27,28]

$$P^i_{out} = P_{in}\cos^2\left(\frac{ßL_p}{2CLi}\right), \ \ i = X, Y, \tag{5}$$

where P_{in} represents the input power injected into core A. When P_{in} is assumed to be 1, the output power P_{out} is a parameter that is normalized for easy understanding. L_p is the propagation distance. The extinction ratio (ER) is used to evaluate the beam-splitting performance of the PBS. The ER of the output powers between the X-pol and Y-pol beams in core A can be defined as [29,30]

$$ER = 10\log_{10}\left(\frac{P^X_{out}}{P^Y_{out}}\right). \tag{6}$$

3. Influences of Structural Parameters on PBSes

To test whether the SPR effect can be stimulated in the designed DCPCF, we arbitrarily set the structural parameters as d = 1.1, d_1 = 1.3, d_2 = 1.0, and Λ = 2.0 μm, while the thickness of the gold film was assumed to be t = 50 nm. A numerical software, COMSOL, was employed to simulate the SPR in the designed DCPCF. Figure 2 displays the mode fields of the X-pol and Y-pol directions, respectively. Specifically, Figure 2a–c show the X-pol even mode, X-pol odd mode, and X-pol second SPR mode, respectively, when the pump wavelength was 1.239 μm. Similarly, Figure 2d–f show the Y-pol even mode, odd mode, and second SPR mode, respectively, when the pump central wavelength was 1.186 μm. As can be seen, the intensity of the even mode was much stronger than that of the odd mode and second SPR mode. More importantly, the energies of both the X-pol and Y-pol even modes were tightly confined in cores A and B, indicating that the beam splitting mostly occurred in the even mode. Figure 2 confirms that the SPR mode can be effectively excited in the designed DCPCF, but with low intensity. Nevertheless, this weak SPR effect can still modify n_{eff}.

In order to show the influence of SPR on n_{eff} for both the X-pol and Y-pol beams, the variations in n_{eff} with wavelength for the two polarized beams are shown in Figure 3a,b. It should be noted that n_{eff} values of the X-pol odd mode and second SPR mode were exchanged when the wavelength was 1.239 μm, as shown in Figure 3a. Similarly, the n_{eff} values of the Y-pol odd mode and second SPR mode were exchanged when the wavelength was 1.186 μm, as shown in Figure 3b. This kind of sudden exchange resulted from the strong couplings between the odd mode and the second SPR mode. This coupling is unfavorable for polarization splitting, as the spatial profiles will be different after beam coupling. However, considering the low intensity of the odd and second SPR modes, the performance of the PBS will not be affected. The CLX, CLY, and R were also calculated for different wavelengths by using Equations (3) and (4), as shown in Figure 3c. The inflection point for every polarized beam corresponded to the mutations in Figure 3a,b. While the X-pol and Y-pol beams had inflection points at different wavelengths (1.18 μm and 1.25 μm, respectively), the resulting R approached 2 only after 1.3 μm. More importantly, the CL for the Y-pol beam was larger than 125 μm, which can be further optimized. In the following, the influences of the structural parameters of the DCPCF are optimized for $R \to 2$ under smaller CLs.

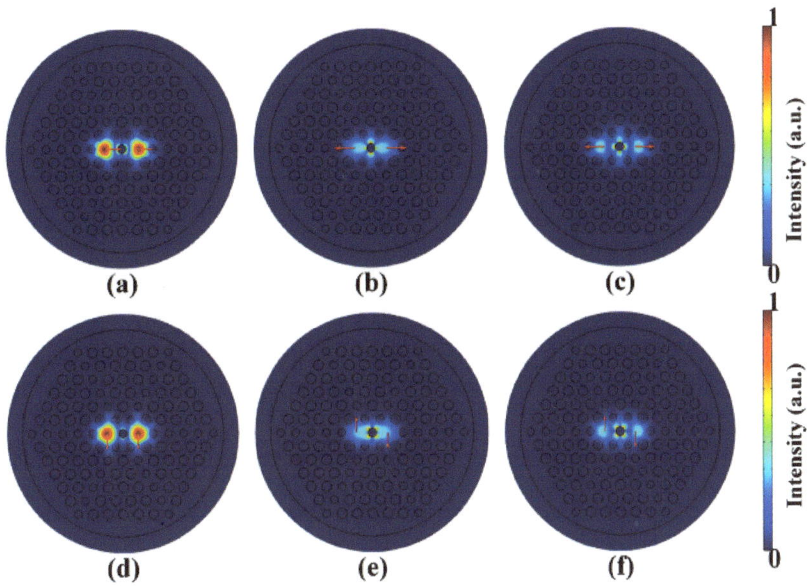

Figure 2. The mode fields of (**a**) the *X*-pol even mode, (**b**) *X*-pol odd mode, and (**c**) *X*-pol second SPR mode at the wavelength of 1.239 µm. The mode-field distribution of (**d**) the Y-pol even mode, (**e**) *Y*-pol odd mode, and (**f**) *Y*-pol second SPR mode at the wavelength of 1.186 µm.

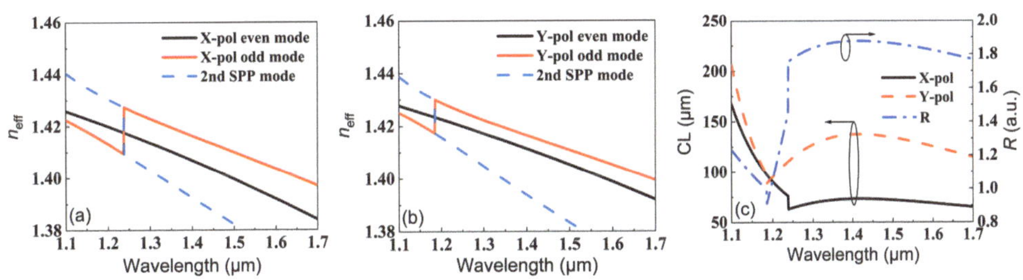

Figure 3. The effective refractive indexes of (**a**) the X-pol and (**b**) Y-pol modes. (**c**) The variations in *CLX*, *CLY*, and *R* with wavelength.

First, we considered the influence of the diameter of the central air hole, *d*. Figure 4a,b display the variations in *CLX* and *CLY* when *d* was increased from 0.9 to 1.3 µm. It can be seen that for all *d*, the *CLX* decreased with the wavelength before the inflection point and then remained almost unchanged after the inflection point. More importantly, the wavelength position of the inflection point was red-shifted when *d* was increased, indicating the reduction of the operation bandwidth. However, the reduced bandwidth was compensated by smaller *CLX*, which is beneficial when designing short PBSes. The *CLY* in Figure 4b showed a similar trend to that of *CLX*, with the small difference that the *CLY* was almost unchanged with both the wavelength and *d* between 1.6 and 1.7 µm. The corresponding *R* values were calculated according to Equation (4) and are shown in Figure 4c. Obviously, for the sake of larger values of *R*, increasing *d* was a good choice. The physical mechanism behind it is that changing the size of the central air hole essentially reshapes the distribution of the mode field. As a result, the mode couplings between the *X*-pol (*Y*-pol) odd mode and the *X*-pol (*Y*-pol) second SPR mode also vary. It should be noted that when d exceeds 1.2 μm, $R > 2$, and the operation bandwidth is decreased. Therefore, d should be chosen carefully to balance the value and operation bandwidth of R.

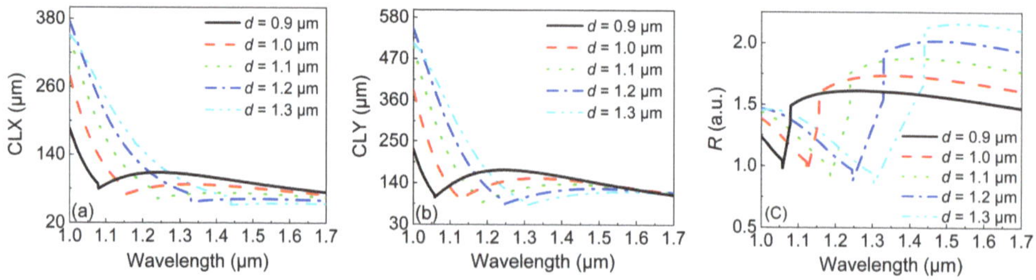

Figure 4. The variations in (**a**) the *CLX* and (**b**) *CLY* with the wavelength under different values of d. (**c**) The variations in R with the wavelength under different values of d.

We then investigated the influence of d_1 by keeping the other four parameters unchanged. Figure 5 depicts the variations in the *CLX*, *CLY*, and R with the wavelength when d_1 was increased from 1.1 to 1.5 μm. It can be seen from Figure 5a that the *CLX* remained unchanged with d_1 before the inflection points. After the inflection point, the *CLX* showed a tiny variation with the wavelength for a certain d_1. The minimum *CLX* under $d_1 = 1.5$ μm was 60 μm, which was similar to that in Figure 4a. In addition, the minimum *CLY* in Figure 5b was almost the same as that in Figure 4b. Consequently, the maximum R in Figure 5c was equal to 1.88, which was smaller than that in Figure 4c. However, the smaller value of R covered a much broader wavelength range, i.e., 1.24–1.7 μm. Therefore, d_1 has a small influence on the bandwidth and R compared with the influence of d.

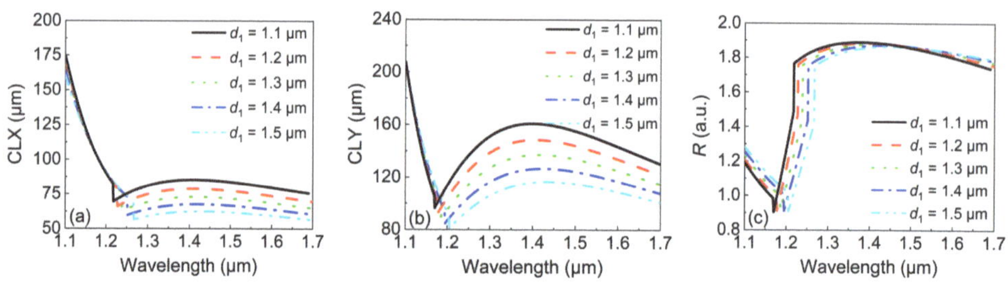

Figure 5. The variations in (**a**) the *CLX* and (**b**) *CLY* with the wavelength under different values of d_1. (**c**) The variations in R with the wavelength under different values of d_1.

Then, we studied the influence of d_2 on the *CL* and R. The relationship between the *CLX* and wavelength under different values of d_2 is shown in Figure 6a. It can be clearly seen that while similar variations in the *CLX* with wavelength could be observed, d_2 itself had almost no influence on the *CLX*. In contrast, the *CLY* in Figure 6b showed more changes with d_2. However, compared with that in Figure 5b, the variation in *CLY*was a little bit smaller. As a result, the influence of d_2 on the CL was smaller than that of d_1. However, it should be noted that the resulting R in Figure 6c showed a remarkable increment compared with that in Figure 5c when d_2 was increased. Actually, the maximum of $R = 1.95$ and the operation wavelength covered 1.25–1.7 μm, which could be attributed to the completely invariant *CLX* when d_2 was increased. Compared with that in Figure 6c, the resulting R was quite close to the ideal value of 2.

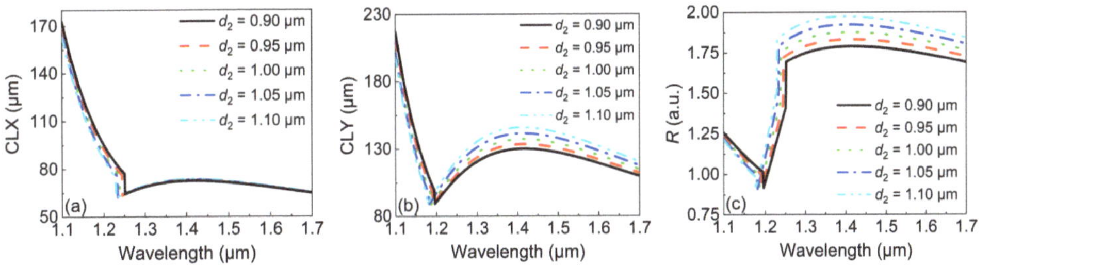

Figure 6. The variations in (**a**) the *CLX* and (**b**) *CLY* with the wavelength under different values of d_2. (**c**) The variations in *R* with the wavelength under different values of d_2.

While the optimization of d_2 was successful in obtaining a larger and broadband *R*, a short PBS is still required for a large σ. We further studied the influence of Λ on the CL and *R*. The simulation results are shown in Figure 7. Figure 7a,b depict the variations in *CLX* and *CLY* with the wavelength under different values of Λ. While both the *CLX* and *CLY* show similar trends of variation with the wavelength, the resulting minimum values of the *CLX* and *CLY* were only 40 and 65 μm, which were smaller than those in Figure 6a and Figure 6b, respectively. In general, the CL rapidly increased with Λ. This was because larger values of Λ made the mode coupling more difficult. It is worth noting that the value of *R* in Figure 7c showed an opposite trend compared with that in Figure 6c, i.e., the value of *R* decreased with Λ. Furthermore, *R* was more influenced by Λ, but at the expense of a reduced operation bandwidth. Nevertheless, changing d_2 and Λ could result in the same bandwidth and value of *R*.

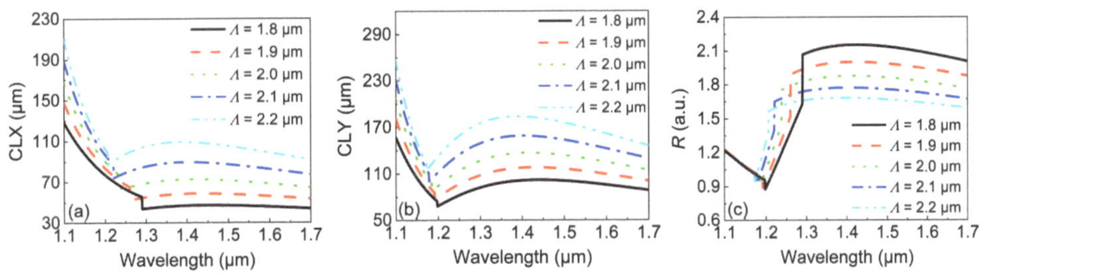

Figure 7. The variations in (**a**) the *CLX* and (**b**) *CLY* with the wavelength under different values of Λ. (**c**) The variations in *R* with the wavelength under different values of Λ.

Finally, the influence of *t* on the CL and *R* were investigated. Figure 8 shows the variations in the *CLX*, *CLY*, and *R* with the wavelength under different values of *t*. In Figure 8a,b, the *CLX* and *CLY* remained the same over the whole wavelength range under different values of *t*. This means that *t* had almost no influence on the CL. As a result, *t* also had a negligible influence on *R*, especially in the wavelength range of 1.3–1.7 μm. This is because small changes (5–20 nm) in *t* cannot influence the distribution of the optical mode. In Figure 8c, the largest values of *R* under different values of *t* showed tiny fluctuations around 1.9, which was smaller than the desired value of 2. Therefore, it is impossible to achieve $R = 2$ by only adjusting *t*.

From above simulation results, we can see that only *d*, d_2, and Λ showed a significant relationship with the CL and *R*. Actually, *d* not only affected *R*, but also determined the wavelength position of the inflection point, namely, the operation bandwidth. In addition, Λ affected not only the value of *R*, but also the CL. In contrast, d_2 mainly affected the value of *R* and showed a slight influence on the resonance wavelength. In addition, the flatness of the *R* curve after the inflection point was also crucial because it determined the bandwidth of the PBS. In the five structural parameters, only d_1 enabled a flat *R* after the inflection point. Therefore, the optimization procedure can be summarized in the following three steps: first, adjusting d and d_2 to cause R to approach the ideal value of 2; second, changing the value of Λ for small CLs; third, controlling d_1 to improve the flatness of R after the inflection point. Following the above procedure, we obtained a set of optimized structural parameters of the DCPCF: $d = 1.2$ µm, $d_1 = 1.3$ µm, $d_2 = 1.0$ µm, $\Lambda = 2.0$ µm, and $t = 50$ nm.

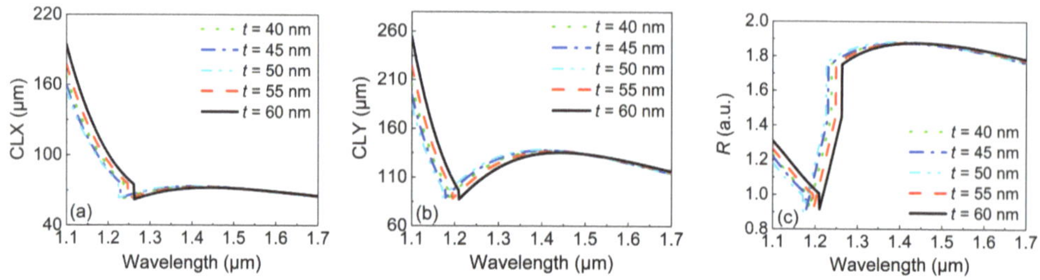

Figure 8. The variations in (**a**) the *CLX* and (**b**) *CLY* with the wavelength under different values of *t*. (**c**) The variations in *R* with the wavelength under different values of *t*.

4. Performance Analysis of the Optimized PBS

The *CLX*, *CLY*, and *R* calculated for the optimized PBS are shown in Figure 9 in the wavelength range of 1.0–1.8 µm. It can be seen from Figure 9a that both the *CLX* and *CLY* were well below 130 µm after the inflection point. More importantly, *R* slightly fluctuated around the ideal value of 2 after 1.35 µm. Figure 9b depicts the details of *R*. It is shown that *R* varied between 1.98 and 2.02 in the wavelength range of 1.39 and 1.6 µm. The maximum and minimum values of *R* were 2.015 and 1.985, which were very close to the ideal value of 2. When the pump wavelength was 1.405 or 1.568 µm, $R = 2$. For the latter wavelength, the corresponding *CLX* and *CLY* were 123.6 and 61.8 µm, respectively. Therefore, 123.6 µm could be selected as the length of the proposed PBS.

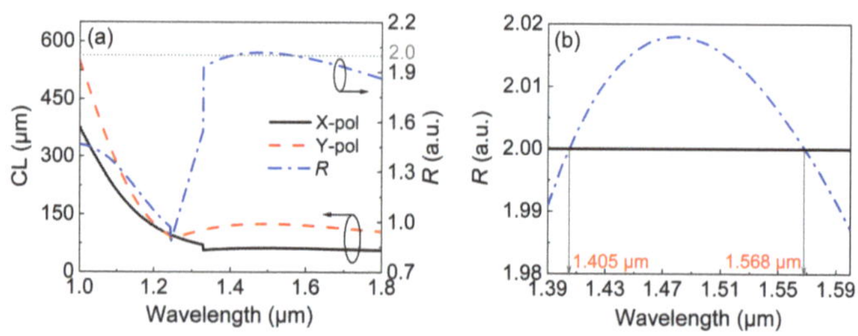

Figure 9. (**a**) The variations in *CLX*, *CLY*, and *R* with the wavelength for the optimized DCPCF. (**b**) Zooming in on *R* in the wavelength range of 1.39–1.6 µm.

By using Equations (5) and (6), Figure 10a could be obtained, presenting the relationship between the normalized output power of the *X*-pol and *Y*-pol beams along the propagation when the initial beam was injected into core A. It is shown that when the propagation distance was 123.6 µm, the normalized output power in the *X*-pol beam was 1, while that in the *Y*-pol beam was 0. At this moment, the *X*-pol and *Y*-pol beams were successfully separated into core A and core B, respectively. Figure 10b shows the relationship between the ER and the wavelength. Generally speaking, an ER larger than 20 dB can be considered an indication of successful splitting [31]. The bandwidth of a PBS is defined as the wavelength range in which the ER is larger than 20 dB. It can be seen from Figure 10b that the operation bandwidth was 314 nm and maximum ER was 78 dB at 1.57 µm. We then calculated the value of σ of our work and compared it with the values reported in other work. It can be seen from Table 2 that the value of σ in this work was the largest among all works, which strongly suggests that our PBS is better than others in terms of length and operation bandwidth.

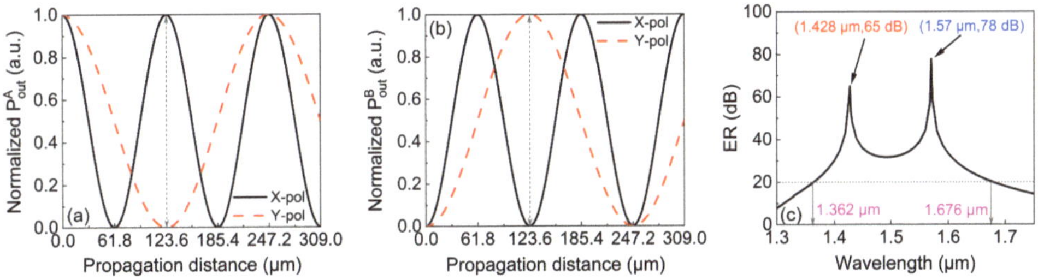

Figure 10. The relationship between the normalized output power and propagation distance in (**a**) core A and (**b**) in core B, and (**c**) the ER in core A between 1.3 and 1.75 μm with the optimized structural parameters.

Table 2. Comparison of the PBS performance between this work and other reported work.

Refs.	PBS Structure	Length (μm)	Bandwidth (nm)	Ratio σ
[14]	square PCF with gold wire	4036	430	1×10^{-4}
[15]	hexagonal PCF without gold	249	17	6.8×10^{-5}
[16]	magnetic-fluid-core PCF with gold film	5112	189	3.7×10^{-5}
[17]	PCF with elliptical holes and gold film	290	19.2	6.6×10^{-5}
[18]	PCF filled with liquid without gold	78	44	5.6×10^{-4}
[19]	octagonal PCF with gold film	47.26	104	2.2×10^{-3}
This work	hexagonal PCF with gold film	123.6	314	2.5×10^{-3}

5. Conclusions

In summary, we designed a compact and broadband PBS based on the SPR effect by using a DCPCF that we designed. By studying the influences of the transversal structural parameters, we found that, except for the thickness of the gold film, other parameters played important roles in the PBS's performance. Based on these findings, we summarized an efficient optimization procedure for quickly designing short and broadband PBSes. A set of optimized structural parameters was obtained by using this optimization procedure. The resulting PBS showed a length of 123.6 μm, an operation bandwidth of 314 nm, and a maximum ER of 78 dB. Compared with other reported work, our PBS reached the best performance concerning the length and operation bandwidth. As the operation bandwidth is located in the telecommunication band, we believe that the proposed compact and broadband PBS has potential applications in miniaturized optical systems for sensing or communication.

Author Contributions: C.M. proposed the original idea, conducted the data analysis, and wrote most of the manuscript. Y.W. designed the structure of the PCF, employed the numerical software to conduct a detailed analysis of the PCF sensors, and plotted most of the figures. J.Y. helped check the final results and conclusions obtained from the numerical simulation, revised the manuscript, and supported the funding of this work. S.Q. made a careful check of the resulting figures and revised the manuscript. X.Z. revised the manuscript and provided plenty of fair suggestions. All authors have read and agreed to the published version of the manuscript.

Funding: This work was supported by the National Key Research and Development Project of China (Grant No. 2019YFB2204001), Fundamental Research Funds for the Central Universities (Grant No. 06500228), and Guangdong Basic and Applied Basic Research Foundation (Grant No. 2021A1515110507).

Institutional Review Board Statement: Not applicable.

Informed Consent Statement: Not applicable.

Data Availability Statement: Data are available upon request from the corresponding author.

Conflicts of Interest: The authors declare no conflict of interest.

References

1. Broeng, J.; Mogilevstev, D.; Barkou, S.E. Photonic Crystal Fibers: A New Class of Optical Waveguides. *Opt. Fiber Technol.* **1999**, *5*, 305–330. [CrossRef]
2. Russell, P. Photonic Crystal Fibers. *Science* **2003**, *299*, 358–362. [CrossRef] [PubMed]
3. Lu, S.; Li, W.; Guo, H.; Lu, M. Analysis of birefringent and dispersive properties of photonic crystal fibers. *Appl. Opt.* **2011**, *50*, 5798–5802. [CrossRef]
4. Hansen, K.P. Dispersion flattened hybrid-core nonlinear photonic crystal fiber. *Opt. Express* **2003**, *11*, 1503–1509. [CrossRef]
5. Birks, T.A.; Knight, J.C.; Russell, P. Endlessly single-mode photonic crystal fiber. *Opt. Lett.* **1997**, *22*, 961–963. [CrossRef]
6. Zhao, Y.; Li, S.; Wang, X.; Wang, G.; Shi, M.; Wu, J. Design of a novel multi-channel photonic crystal fiber polarization beam splitter. *Opt. Commun.* **2017**, *400*, 79–83. [CrossRef]
7. He, F.T.; Shi, W.J.; Hui, Z.Q.; Zhan, F.; Zhang, Y.K. A dual-core PCF polarization splitter with five elliptical air holes based on tellurite glass. *Opt. Quan. Electron.* **2017**, *49*, 363. [CrossRef]
8. Lou, J.; Cheng, T.; Li, S. High sensitivity photonic crystal fiber sensor based on dual-core coupling with circular lattice. *Opt. Fiber Technol.* **2019**, *48*, 110–116. [CrossRef]
9. Liu, Q.; Xing, L.; Wu, Z. The highly sensitive magnetic field sensor based on photonic crystal fiber filled with nano-magnetic fluid. *Opt. Commun.* **2019**, *452*, 238–246. [CrossRef]
10. Chaudhary, V.S.; Kumar, D.; Mishra, R.; Sharma, S. Hybrid Dual Core Photonic Crystal Fiber as Hydrostatic Pressure Sensor. *Optik* **2020**, *210*, 164497. [CrossRef]
11. Chremmos, I.D.; Kakarantzas, G.; Uzunoglu, N.K. Modeling of a highly nonlinear chalcogenide dual-core photonic crystal fiber coupler. *Opt. Commun.* **2005**, *251*, 339–345. [CrossRef]
12. He, F.T.; Shi, W.J.; Zhang, J.L.; Hui, Z.Q.; Zhan, F . Polarization splitter based on dual-core photonic crystal fiber with tellurite glass. *Optik* **2018**, *164*, 624–631. [CrossRef]
13. Bai, Y.; Hao, R.; Yuan, B.; Cao, B. An ultrashort length and high extinction ratio polarization beam splitter based on dual-core PCF. *J. Opt.* **2021**, *50*, 257–263. [CrossRef]
14. Jiang, L.; Zheng, Y.; Hou, L.; Zheng, K.; Peng, J.; Zhao, X. An ultrabraoadband polarization splitter based on square-lattice dual-core photonic crystal fiber with a gold wire. *Opt. Commun.* **2015**, *351*, 50–56. [CrossRef]
15. Zi, J.; Li, S.; An, G.; Fan, Z. Short-length polarization splitter based on dual-core photonic crystal fiber with hexagonal lattice. *Opt. Commun.* **2016**, *363*, 80–84. [CrossRef]
16. Wang, H.; Yan, X.; Li, S.; Zhang, X. Tunable surface plasmon resonance polarization beam splitter based on dual-core photonic crystal fiber with magnetic fluid. *Opt. Quan. Electron.* **2017**, *49*, 368. [CrossRef]
17. Xu, Q.; Zhao, Y.; Xia, H.; Lin, S.B.; Zhang, Y. Ultrashort polarization splitter based on dual-core photonic crystal fibers with gold wire. *Opt. Engin.* **2018**, *57*, 046104. [CrossRef]
18. Wang, J.; Pei, L.; Weng, S.; Wu, L.; Li, J.; Ning, T. Ultrashort polarization beam splitter based on liquid-filled dual-core photonic crystal fiber. *Appl. Opt.* **2018**, *57*, 3847–3852. [CrossRef] [PubMed]
19. Lou, J.; Cheng, T.; Li, S. Ultra-short polarization beam splitter with square lattice and gold film based on dual-core photonic crystal fiber. *Optik* **2018**, *179*, 128–134. [CrossRef]
20. Wang, G.; Lu, Y.; Yang, X.; Duan, L.; Yao, J. High-sensitivity magnetic field sensor based on a dual-core photonic crystal fiber. *Appl. Opt.* **2019**, *58*, 5800–5806. [CrossRef]
21. Zhang, W.; Chen, H.; Liu, Y.; Ma, M.; Li, S. Analysis of a magnetic field sensor based on photonic crystal fiber selectively infiltrated with magnetic fluids. *Opt. Fiber Technol.* **2018**, *46*, 43–47. [CrossRef]
22. Wang, J.; Liu, C.; Wang, F.; Su, W., Yang, L.; Lv, J.; Fu, G.; Li, X.; Liu, Q.; Sun, T.; et al. Surface plasmon resonance sensor based on coupling effects of dual photonic crystal fibers for low refractive indexes detection. *Res. Phys.* **2020**, *18*, 103240. [CrossRef]
23. Wang, X.; Li, S.; Liu, Q.; Wang, G.; Zhao, Y. Design of a Single-Polarization Single-Mode Photonic Crystal Fiber Filter Based on Surface Plasmon Resonance. *Plasmonics* **2017**, *12*, 1325–1330. [CrossRef]

24. Zhao, T.; Jia, H.; Lian, Z.; Benson, T.; Lou, S . Ultra-broadband dual hollow-core anti-resonant fiber polarization splitter. *Opt. Fiber Technol.* **2019**, *53*, 102005. [CrossRef]
25. Ponnusamy, M.; Prakash, J.; Vidyasagar, S.; Boopathi, C.S.; Amiri, I.S. Design of Polarization Splitter using Elliptically Dual Core – Cladding Photonic Crystal Fiber. *Res. Phys.* **2019**, *3*, 102279.
26. Hagras, E.; Hameed, M.; Heikal, A.M.; Obayya, S.S.A. Multi-functional photonic crystal fiber splitter for the two communication bands. *Opt. Fiber Technol.* **2019**, *52*, 101986. [CrossRef]
27. Sha, X.; Li, W.; Li, H.; Li, Z. Polarization splitter based on hybrid-cladding dual-core photonic crystal fibers. *Optik* **2015**, *126*, 2331–2334. [CrossRef]
28. An, S.; Lv, J.; Yi, Z.; Liu, C.; Yang, L.; Wang, F.; Liu, Q.; Su, W.; Li, X.; Sun, T.; et al. Ultra-short and dual-core photonic crystal fiber polarization splitter composed of metal and gallium arsenide. *Optik* **2020**, *226*, 165779. [CrossRef]
29. Zhang, Y.; Liu, H.; Chen, C.; Bai, B.; Tang, S. Temperature-controlled and multi-functional splitter based on dual-core photonic crystal fiber. *Res. Phys.* **2020**, *19*, 103578. [CrossRef]
30. Shen, R.; Cheng, T.; Li, S.; Yan, X. Broadband Polarization Filter Based on Tellurite Photonic Crystal Fiber with High Extinction Ratio. *Optik* **2021**, *245*, 167721. [CrossRef]
31. Zhao, X.T.; Hua, L.; Xiong, Q.; Jiang, G.H.; Cheng, J.R. Ultra-short and broadband polarization splitter based on PCF and metal surface plasmons resonance. *Opt. Quan. Electron.* **2019**, *51*, 162. [CrossRef]

MDPI AG
Grosspeteranlage 5
4052 Basel
Switzerland
Tel.: +41 61 683 77 34

Micromachines Editorial Office
E-mail: micromachines@mdpi.com
www.mdpi.com/journal/micromachines

www.ingramcontent.com/pod-product-compliance
Lightning Source LLC
Chambersburg PA
CBHW041351210926
51850CB00013B/3048
* 9 7 8 3 7 2 5 8 3 1 3 9 5 *